Beyond the GALAXY

How humanity looked beyond our Milky Way
and discovered the entire Universe

Beyond the GALAXY

How humanity looked beyond our Milky Way and discovered the entire Universe

Ethan Siegel
Lewis & Clark College, USA

NEW JERSEY • LONDON • SINGAPORE • BEIJING • SHANGHAI • HONG KONG • TAIPEI • CHENNAI • TOKYO

Published by

World Scientific Publishing Co. Pte. Ltd.
5 Toh Tuck Link, Singapore 596224
USA office: 27 Warren Street, Suite 401-402, Hackensack, NJ 07601
UK office: 57 Shelton Street, Covent Garden, London WC2H 9HE

Library of Congress Cataloging-in-Publication Data
Siegel, Ethan, 1978– author.
Beyond the galaxy : how humanity looked beyond our milky way and discovered the entire universe / Ethan Siegel (Lewis & Clark College).
pages cm
Includes index.
ISBN 978-9814667234 (hardcover : alk. paper) -- ISBN 978-9814667166 (softcover : alk. paper)
1. Expanding universe. 2. Astronomy. 3. Astrophysics. 4. Cosmology. I. Title.
QB991.E94S54 2015
523.1--dc23

2015003689

British Library Cataloguing-in-Publication Data
A catalogue record for this book is available from the British Library.

Cover image credit: NASA, ESA and T.M. Brown (STScI).

In-house Editor: Ng Kah Fee

Typeset by Stallion Press
Email: enquiries@stallionpress.com

Printed in Singapore

Preface

When I received my class schedule in my first year as a college professor in 2009, I was overjoyed to learn I would be teaching our college's introductory astronomy class. I had remembered learning the story of the Universe in bits and pieces from a young age up through my Ph.D. studies and beyond, and being fascinated at every turn to learn the story of what the Universe was and how it came to be this way. At every level, greater details were filled in, old ideas were shown to be only approximations of a deeper truth about reality, and our knowledge of the Universe was constantly being refined with higher precision measurements and in the context of new results. The story of what we knew of our place in the Universe and how we came to know it was one I could not wait to share with my students.

Yet, when it came time to choose a textbook for the class, a book that told that story was not an option. That book, to my surprise, *did not exist*. The major astronomy textbooks that were out there were incredibly comprehensive works, teaching all about astronomical techniques, instruments, various aspects of planets, stars and galaxies, multiwavelength analyses and more. They may prove to be excellent resources for someone seeking to become a professional astronomer, laying an impressive foundation for someone who had not taken a course devoted to the endeavor before. Despite being instructive guides for students who were learning to solve a wide variety of classes of problems in astronomy, there was something that was sorely lacking from them all: that story of what we know and how we came to know it.

The reality was, on a non-textbook front, there were not even any scientifically accurate books that adequately covered this story. Your best option is to get a graduate-level text to learn about the full suite of the latest developments, curate out the heavy-lifting of the equations, and augment the more basic material with popular science books. In the end, though, what you are dealing with is an amalgam of dissatisfying options, none of which simply lay out the story of what the Universe is and how we have discovered it to be so.

Which, to my mind, really misses the point of what a first course in astronomy should be all about! Yes, a very small percentage of the people who take that course will go on to become professional astronomers, but the vast majority who take an interest in it are craving an awareness of the great cosmic story that we all share. I was not nearly as interested in what types of problems a student would be able to solve on an exam after three or four months of study as I was in what they would remember and appreciate about not only the Universe, but of the process and the enterprise of science, a year, five or even ten down the road. I wanted to begin from a place where they were totally comfortable — from the simplest, naked-eye observations here on Earth — and take them right up to the frontiers of modern scientific knowledge.

That story, the scientific story of what we know of the Universe and how we have come to know it, is one that's evolved tremendously in just the past century. As I complete the writing of this book at the end of 2014, I look back and realize that just 100 years ago, the leading physicists and astronomers of the day believed that the entire Universe consisted of the Milky Way galaxy and all the stars in it, was static and eternal, and was governed by Newton's law of gravitation. How times have changed, and how rapidly! We now know we live in a vast Universe containing hundreds of billions of galaxies, a Universe which is expanding and cooling, governed by Einstein's General Relativity, and that actually had a birthday some 13.8 billion years ago: the Big Bang. We know that the Big Bang was not the very beginning, either; there was a phase before it known as cosmic inflation. And the vast majority of the matter we know, made of protons, neutrons and electrons just like we are, makes up only 5% of the total amount of energy in the Universe. There are unseen forms of matter and energy, *dark matter* and *dark energy*, that have come to dominate our Universe today. On top of all that, we have actually learned the fate of our Universe and what our far future is going to look like.

Why is this not the story that everyone interested in astronomy learns about our Universe? And why do we not learn *how* we came to learn these things? Even those who major in physics or astronomy at many colleges miss out on this!

This book was written with the intention of correcting all of that. Whether you are learning about the Universe for the first time, taking an introductory course or looking for the latest update on modern developments in the story of what we know about the Universe and how we've come to know it, this book is designed to start at the very beginning of human exploration and take you through the most important developments that led to our present understanding of all there is. There are no problems to be solved or worked out, no formulas and no equations (except an occasional mention of $E = mc^2$), and any mathematical or physical relationships that are mentioned are described in plain English.

This is the story that is universal to us all, the story that the Universe tells us about itself. We learned it simply by looking at it and asking it the right questions, and if all of human knowledge were lost tomorrow, we could find it out again at any instant if we asked those questions once again. I hope you enjoy the journey into what we know about the Universe, and come to appreciate the great cosmic story universal to us all.

Ethan Siegel
December 23, 2014

Contents

Chapter 1

So Far, So Good: The Universe At The Start Of The 20th Century

If you were going to pick one thing in the sky to choose as the most obvious and most important, no matter where on Earth you were, you would likely choose the Sun. Imagine what it must have been like for the first humans who migrated a large distance away (say, north) from the equator. Rather than a terrain that was warm year-round, with the Sun rising in the east, passing high overhead during midday and setting in the west consistently, with only modest variations, things would appear to change dramatically. During late spring and early summer, you would have even more daylight than you had at the equator, with the Sun rising and setting much closer to the North Pole, while its path would still take it high overhead in the skies towards the south at midday. But as the year wore on, the Sun's path would shorten dramatically. It would both rise and set farther south every day, and would peak just a little lower in the sky than the day prior. As the days got shorter and darker, and the nights grew longer, the world would grow colder, as the onset of winter approached. Someone who had never experienced this before might well worry that the Sun itself would sink lower and lower as the days continued onward, perhaps disappearing below the horizon entirely.

But unless you ventured north of the Arctic Circle, that would never happen. The Sun would slow down in its descent after some time, and reach a minimum height above the horizon, which it would not drop below the next day. It would appear relatively stationary for a few days, which is

Figure 1.1 On the summer solstice from a significant latitude, the Sun appears to rise closer to the pole than on any other day, pass higher overhead, provide more hours of daylight and then set closer towards that same pole. As the year progresses, its path continues to move farther towards the opposite pole, providing fewer hours of daylight and culminating in the winter solstice, the shortest day of the year. After that, the Sun's path appears to migrate back towards the first pole again, with days lengthening and the cycle repeating. Image credit: Wikimedia Commons user Tau'olunga, under a c.c.-by-s.a.-2.5 license.

what we call the solstice: Latin for *Sun stands still*. And after that, it would begin to rise a little higher once again, signaling that a new year would indeed come, and that the Sun would eventually bring back longer days and another summer (Fig. 1.1).

This story may or may not be factually true, as it is mere anthropological conjecture, but it is illustrative of the very beginnings of astronomy, of what a person would actually see, and of science in general. By taking detailed observations and measurements of the Universe itself, we can learn about the phenomena that occur within it. By repeating these observations and measurements over time, in different locations and by independent observers, we can gain a suite of data describing what occurs. When we understand what has occurred in the past, we can use that information to make accurate predictions about what we can expect to happen in the future.

There is only one more major step required to transform this initial, primitive set of information into what we consider today to be science: The construction of a physical framework that accounts for what we see. In science, it is not enough to merely describe what happens; we want to uncover *how* it happens, and what its causes are. For that, we need an explanation of what causes our observations. We need a physical theory that underlies the reasons these phenomena occur. And we need for that theory to go out and make new predictions that *we can test*, either via observations or by experiment, to validate or falsify it. For a single object like the Sun, seen moving through the sky, the explanations are simply too numerous to be of any use. But if we turn our attention towards what we see when the Sun goes down, a whole new Universe opens up.

* * *

As the skies darken with the onset of twilight, a clear, cloudless night will bring out hundreds of stars easily visible to the naked eye, with that number rising well into the thousands if the night is moonless as well. No matter where you are on Earth, those stars will move throughout the night, as it appears the entire canopy of the sky rotates about a single point, focused either on the North or South celestial pole. Night after night, the stars appear with the same patterns, in the same relative positions, with the same brightness and always making the same motions: rotating counterclockwise about the North Pole. (Or, from the Southern hemisphere, clockwise about the South Pole.) Just like the Sun, stars appear to rise in the eastern half of the sky, move to a position high above the horizon, and set along the western side. And on nights where the Moon is visible, it, too, rises in the east, reaches a maximum position over the horizon, and sets in the west.

Why would all of this occur? The explanation that most of the early scientists defaulted to made an awful lot of sense: that all the objects in the sky were a part of a fixed sphere high above the Earth's surface, and that sphere rotated about its axis once every 24 hours, giving rise to the motion

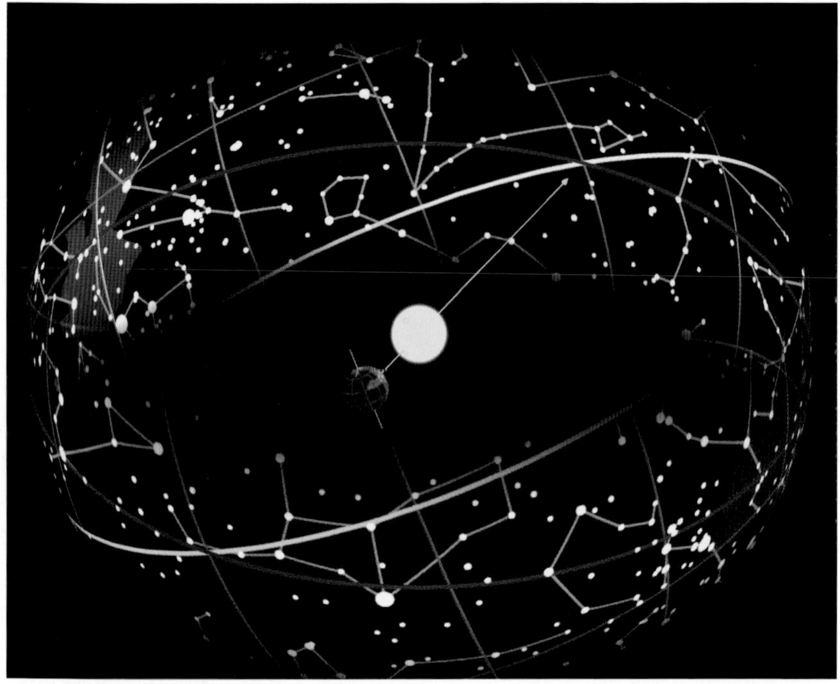

Figure 1.2 The stars appear to be located on a sphere distant from Earth, with only the constellations visible that appear opposite the Sun (i.e., on the night side). It is unclear, from this observation alone, whether the stars and Sun rotate around the Earth on a daily basis, or whether the stars and Sun are relatively stationary and the Earth rotates on its axis once per day. Image credit: E. Siegel, based on the original by Wikimedia Commons user Tau'olunga.

of the Sun, Moon and stars on a daily basis. This is a great start to a scientific theory, as it accounts for the full suite of observations available — of all the celestial objects — with one single explanation. The only things missing, preventing this from becoming a full-blown scientific theory, are the mechanism of how this happens and a new, testable prediction that should arise from this description of our Universe (Fig. 1.2).

Despite its appeal, it is not the *only* conceivable explanation for these phenomena. From our vantage point here on Earth, it certainly appears that the Sun, Moon and stars — the objects that appear in the skies — all move on an invisible sphere rotating around our world. But they *could*, just as easily, be fixed in the skies, and it could be our world that rotated instead. Certainly, from these simple observations of the objects in the sky alone, there would not be a way to tell these two scenarios apart, and there were many scientists and philosophers as early as Pythagoras and his disciples who favored this latter approach. But without a way to test one idea against the other, without differing predictions that the two ideas make, we would not yet have a scientific theory. Still, keeping these phenomena and their possible explanations in mind will help inform us about the Universe, and our place in it, as we continue to gather superior observations and information in our quest to make sense of the world.

* * *

While the fixed stars returned to the same position, night after night, relative to one another, the Sun and the Moon did no such thing. The Moon's shifts were the most notable, even more severe than the Sun's! If you were to measure the Moon's position one night at a specific time, and then look for it at the same exact time on the next day, you would find that its position would have shifted by *twelve degrees* (12°), or roughly the amount of space between your index and little finger if you throw heavy metal horns with your arm extended (Fig. 1.3). Similarly, the Sun's position shifts a little bit each day, which is why the positions of the visible stars shift ever so slightly from night-to-night. It appears, on average, that the sky is off by just *one degree* from one night to the next if you mark

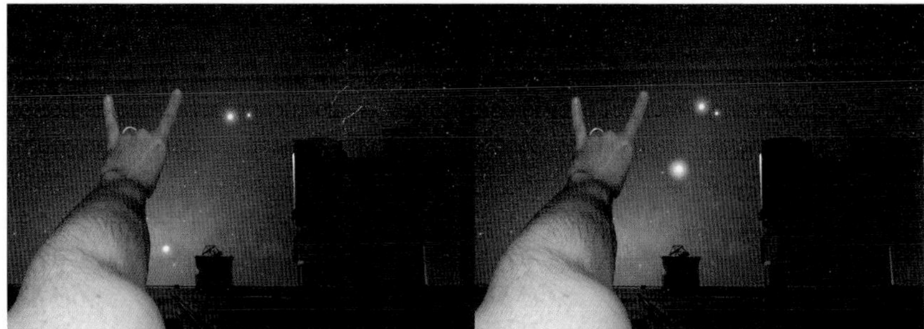

Figure 1.3 The Moon is the brightest object in the picture that appears to shift by the greatest amount, with two planets — Venus (brighter) and Jupiter (slightly fainter) — that shift by much smaller amounts relative to the background of static stars. Image credit: ESO/Y. Beletsky, with the author's hand superimposed to show 12° on the sky, the amount the Moon appears to shift its location in a 24 hour timespan.

positions at the same time on successive nights. This is why different constellations are visible during different parts of the year, as the Sun's position relative to the background of stars appears to shift through the sky.

In addition to this, the Sun and Moon appear to be related in a particularly interesting way: through the Moon's phases. When the Moon appears very close to the Sun in the sky, it appears as a very thin crescent. With each passing day, it falls farther and farther behind the Sun, with progressively more of it becoming illuminated over time. It transitions from a crescent phase to halfway filled, then becomes a gibbous and, eventually, completely full, as it appears to be a complete circle. This process, as the Moon appears to fill in over a roughly two-week time period, is known as *waxing*. But once the Moon becomes full, it does not remain that way for long. Practically immediately, it begins *waning*, or emptying out, from its west side to its east side, the same way it waxed earlier. Over the same timeframe that it took to reach its full phase, it becomes a gibbous, then half-full, then a crescent once again, all the while getting closer and closer to the Sun in the sky. Finally, the Moon becomes new once again, a crescent so thin and close to the Sun it is usually invisible, and the lunar cycle repeats, beginning once again.

Rather than moving along with the rest of the stars, the Sun and the Moon's motions must be independent of them, as their positions change relative to all the other bodies, fixed in the sky. We can learn that the Moon must be closer to Earth than the stars are, since when a crescent moon passes into the same location as a star, the Moon passes in front of the star, blocking its light, a phenomenon known as *occultation*. We can also conclude, quite remarkably, that the Moon's phases are caused by reflected light from the Sun! When the Moon's phase appears full, it appears on the opposite side of the Earth from where the Sun appears, indicating that the lunar hemisphere illuminated by the Sun is completely visible to our eyes. On the other hand, when the Moon appears in a new phase, it is on the same side of Earth as the Sun, indicating that not only is the hemisphere facing away from our planet the one that is illuminated, but also suggesting that the Moon is closer to our world than the Sun is (Fig. 1.4).

You can imagine a very strong light source coming from a single bulb down at one end of a hallway as the Sun. You can imagine that your head

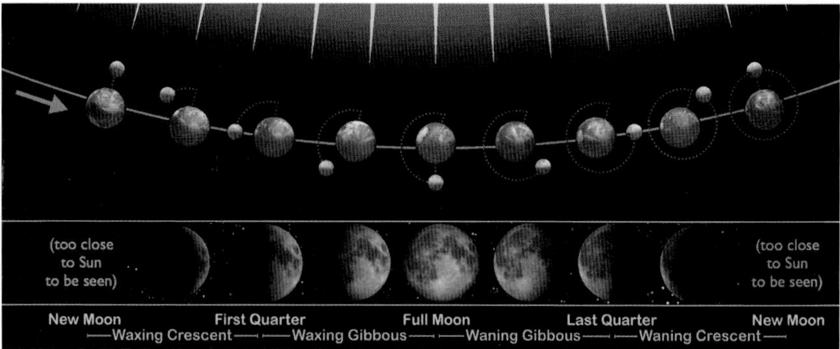

Figure 1.4 Relative to the Earth–Sun distance, the Moon orbits the Earth at a vastly smaller scale, with the Moon being the closest celestial object to our world. The reflected sunlight off of the Moon is what causes its phases as seen from Earth, as the portion of the Moon that's illuminated is what changes over the course of a month from our vantage point. Note how the full disc of the Moon always obscures the objects behind it, even when that disc is not visible itself. Image credit: E. Siegel, based on an original by Wikimedia Commons user Orion 8, under c.c.-by-s.a.-3.0.

is the Earth, and you can imagine that a ball — held at arm's length — is the Moon. If you face towards the light source, just a tiny bit to your left, and hold out the ball, what do you see when you look at it? You know that half of your "Moon" is lit up by the Sun, but you can only see a tiny sliver illuminated; the rest of the side of the Moon that faces you is in shadow. Now, rotate your arm farther to the left, and watch how more and more of the ball appears illuminated to you. Eventually, the crescent fills in to become halfway full (known as a "quarter" phase), which occurs when you make a 90° angle with respect to both the ball and the light source. As you move the ball to the opposite side of your head, the entire illuminated side is visible to you, provided you keep your head's shadow out of the way. Finally, you can complete the revolution, coming back so that the ball is once again between you and the light source. An entire trip of the Moon around the Earth, just like that, is what causes not only the phases that we see, but is where the idea (and the name) of a "month" comes from, with a full lunar revolution about Earth taking 29.5 days on average (Fig. 1.5).

You will notice, if you do this demonstration yourself, that you will have to take care at two different times to avoid blocking out the lights!

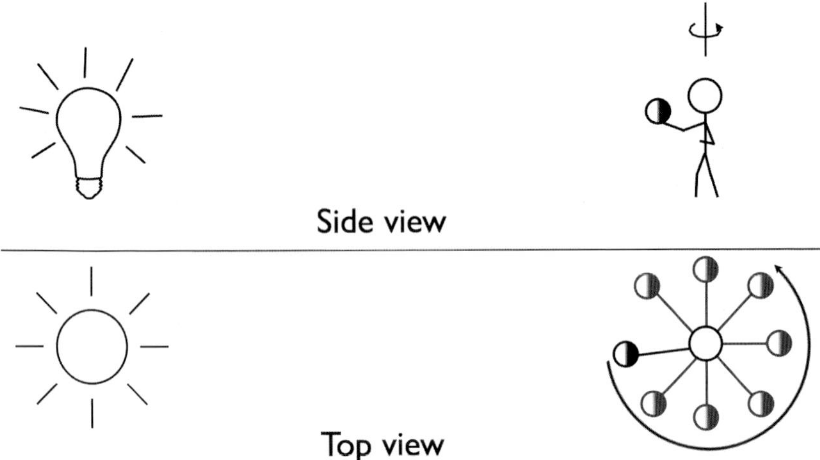

Figure 1.5 If you take a light source and place it far away in a room while holding out a ball with your left hand and rotating, you can simulate the phases of the Moon that you would see from Earth, including when it is in the new, crescent, quarter, gibbous and full phases. Image credit: E. Siegel.

Once when the Moon passes in between the Earth and the Sun, otherwise the sunlight will be blocked from reaching Earth (represented by your eyes), and a second time when the Moon lines up so that it is in the Earth's shadow, as the Sun's light can be blocked (by your head) from reaching it. In reality, these alignments do happen occasionally — about twice a year each, on average — and are known as eclipses. Most months see no eclipses happen, as the Moon's orbit around Earth is inclined at about 5° (about the width of your three middle fingers at arm's length) to the Earth–Sun plane, while the Moon and Sun take up only about 0.5° (or half the width of your littlest finger at arm's length) in angular diameter apiece. During the majority of months, the Moon passes either above or below the line connecting the Sun and Earth during its new and full phases, so no eclipses occur. But twice a year, the Moon comes close enough to crossing that imaginary line that, when it passes between the Sun and the Earth, some or all of the sunlight that should fall on our planet is blocked by the Moon instead. This is known as a *solar eclipse* (or an eclipse of the Sun), and it comes in three varieties, depending on both how good the alignment between the Sun and Moon is and also on their relative angular sizes. The three varieties are as follows:

1) **Partial Solar Eclipse**. When the alignment between the Moon, Earth and Sun is such that an observer on Earth will see the Moon's disk cover only a portion of the Sun's surface, we get a partial solar eclipse. The brightness of the Sun is so incredible that even blocking 90% of the light results in imperceptible optical changes; the Sun's brightness remains undiminished to the naked eye. However, you may notice that the heat from the Sun is much less than normal. Both of the other varieties of solar eclipse are preceded and superseded by partial solar eclipses as well.
2) **Total Solar Eclipse**. When the Moon is seen to block the entirety of the Sun's disk from Earth, we call that a total solar eclipse. If we were to view this from space, we would see a dark shadow — the Moon's shadow — pass across the surface of the Earth. As seen from Earth, the sky darkens to the equivalent of night, with not only the (rarely-seen) solar corona visible, but also the normally-invisible-during-the-day stars! This can only occur when both the alignment between the Moon, Earth and Sun is perfect, and additionally when the angular size of the Moon appears larger than the angular size of the Sun. In reality, the angular size of both the Moon and Sun vary significantly throughout each month (for the Moon) and over the course of the year (for the Sun), so that either body will appear to grow or shrink in size if viewed over substantial periods of time.
3) **Annular Solar Eclipse**. This is the counterpart to a total eclipse, and occurs when the alignment between the Moon, Earth and Sun is perfect, but the Moon does not appear to be large enough to block out the Sun's disk. Instead, as the Moon passes completely inside the Sun's disk, it creates an annulus — Latin for ring — at the moment of maximum eclipse. As seen from space, this corresponds to the Moon's shadow being lined up with Earth, but falling short of the planet itself. In perhaps an odd twist, an astronaut in orbit around Earth could experience a total eclipse while those on the ground experienced only an annular one! (Fig. 1.6.)

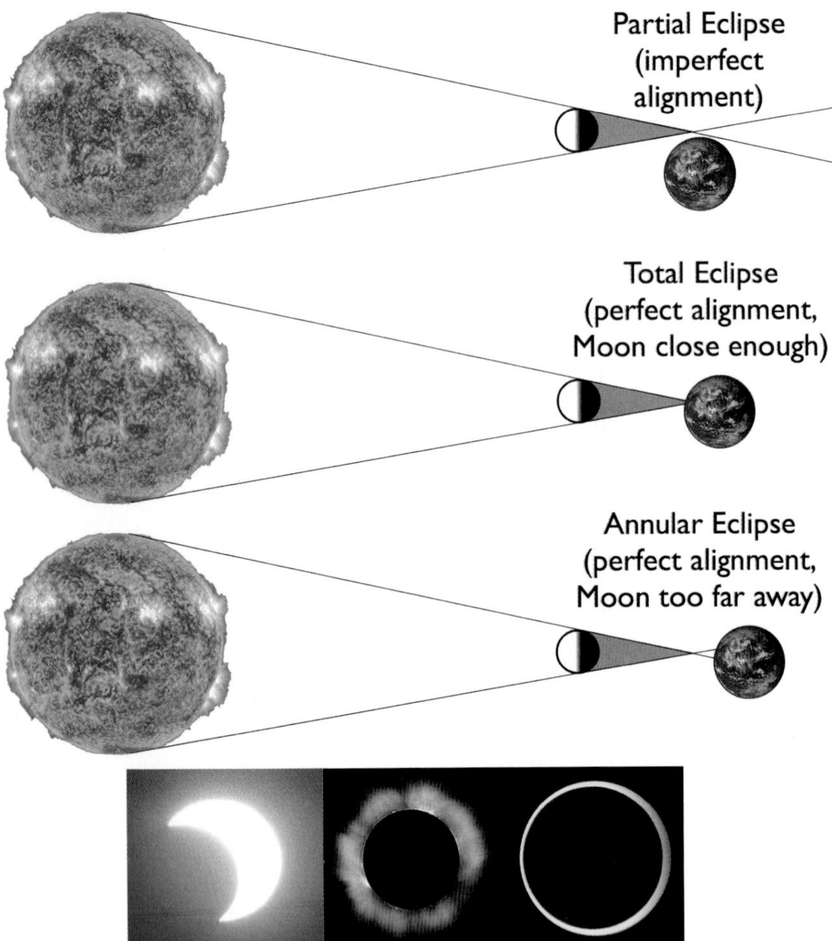

Figure 1.6 When the Moon appears to block a portion of the Sun's disc as seen from Earth, we get a partial solar eclipse. When the Moon's shadow falls directly on the Earth's surface, observers who happen to be in the path of the shadow see a total solar eclipse, which occurs when the Moon is close enough to appear larger (in angular size) than the Sun. But when the Moon is too far away and hence appears smaller than the Sun, even a perfect alignment will only result in an annular solar eclipse. For observers on Earth near but outside the path of totality or annularity, a partial solar eclipse will still be visible. Image credit: E. Siegel.

There are also *lunar eclipses*, or eclipses of the Moon, which occur when the full Moon passes behind the Earth's shadow. Unlike a solar eclipse, which can only be viewed at certain places on Earth, *anyone* on the Earth's night side can view a lunar eclipse. When the alignment between the Moon and Earth's shadow is imperfect, and the Moon passes through only a portion of Earth's shadow, we get a partial lunar eclipse, where a portion of the Moon becomes dark, thanks to the Earth blocking the sunlight from reaching it. If the alignment *is* perfect, however, and the Moon passes entirely into Earth's shadow, we get a total lunar eclipse.

There are three tremendously interesting things about lunar eclipses that surprise most people the first time they see them. First, total lunar eclipses can last for a very long period of time. While solar eclipses last mere minutes in the total or annular phases (up to about 7.5 and 12 minutes, respectively), a total lunar eclipse can last up to an hour and 46 minutes! You might not think it would be interesting to see the Moon in the Earth's shadow, but it turns out the Moon is not invisible during this time; it actually appears faint and red, which is our second remarkable fact. While there is no direct sunlight falling on the Moon at this time, there is the sunlight that passes through the Earth's atmosphere in places that are experiencing either sunrise or sunset. From the Moon's vantage point, that occurs in a "ring" around the Earth. Just as the sky turns a reddish color during sunrises and sunsets — thanks to blue light getting scattered away — some of that red light gets bent by the Earth's atmosphere, and falls on the Moon while it's in the Earth's shadow! So not only does the Moon appear red during a total lunar eclipse, but it actually appears a deeper shade of red the closer it gets to the very center of Earth's shadow. And finally, there is the third fact that comes from a lunar eclipse, perhaps the most surprising one of all: we can use it to measure the shape of the Earth! During either a partial lunar eclipse or the partial phase of a total lunar eclipse, the outline of the Earth's shadow can be seen to fall on the Moon. If we could accurately depict or record what this outline looked like, we could reconstruct the silhouette of the Earth, just from viewing its shadow on the Moon alone. To no one's surprise, just like the Sun and the Moon, the Earth's shadow appears as a nearly perfect circle (Fig. 1.7).

* * *

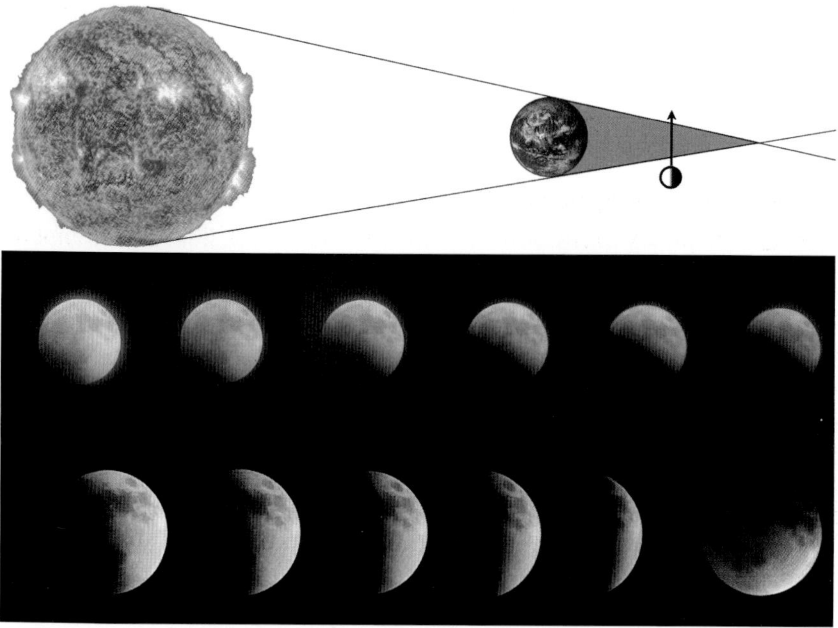

Figure 1.7 The Moon is always close enough to Earth's shadow that lunar eclipses can only be partial or total, with the Moon turning red and appearing quite dim during the period of totality. During the partial phases, the shadow of Earth can be seen on the surface of the Moon, indicating quite clearly that it casts a roughly circular shape. Image credit: E. Siegel, with eclipse sequences by Wikimedia Commons users Zaereth and Javier Sánchez.

But what does having a circular shadow mean for the shape of our world? In principle, it could mean a *lot* of things, but there are two extremely simple explanations:

1) Earth could be a circular disk, where the Sun (on one side) casts a circular shadow on the Moon (on the other side), with the Earth's presence blocking sunlight from reaching the Moon.
2) Earth could be a sphere, where the Sun (on one side) casts a circular shadow on the Moon (on the other side), with the Earth's presence blocking sunlight from reaching the Moon.

From solely looking at the shadow we cast on the Moon, there is no good way to tell the difference between these two scenarios, since both circles and spheres cast circular shadows if the geometry is right. Just as most people assumed (incorrectly) that it was the sky whose rotation caused the apparent motion of the stars around Earth, most people also assumed that the Earth was a circular disk. The earliest written records, including in ancient Greek and Hebrew texts, make reference to the "circle" of the Earth, surrounded by water all around.

There are plenty of hints that the world is shaped as a sphere and not as a disk, as there are a number of observations that support an Earth whose surface is curved. For one, if you watch a ship sail out into the sea or ocean, you can observe its mast progressively disappear below the horizon, something you would not expect if the Earth's surface were flat. For another, you can climb up to the top of the highest mountain, and see only a limited distance across a surface (like an ocean) that you know to be level, yet the higher you climb, the farther you can see. And finally, if you travel farther South, you can see stars, constellations and even deep-sky objects — such as the Magellanic Clouds — in the night sky that are invisible from more northern latitudes (Fig. 1.8). All of this makes for some very strong evidence that the surface of the Earth must be curved. But the fact remains that even from the highest terrestrial altitudes, the curvature of Earth's surface cannot be seen, and that was the piece of evidence that people most strongly connected with. For centuries, a flat Earth was the prevailing dogma.

But it was in the third century B.C.E. that someone not only figured out how to prove that the Earth was spherical (or close to it), but to actually *measure* how big it was! At that period in time, the greatest scholars in the world lived in Alexandria, Egypt, where the fabled Library of Alexandria was located. One of those scholars was Eratosthenes of Cyrene, which was a Greek city located in modern-day Libya. Alexandria and Cyrene were located at roughly the same latitudes, so Eratosthenes likely enjoyed a consistent experience of the night skies throughout his life. Which is why it must have been shocking for him to receive a piece of correspondence from Elephantine Island in the city of Syene, located in southern Egypt (modern-day Aswan). The letter declared that, on the summer solstice:

> "[T]he shadow of someone looking down a deep well would block the reflection of the Sun at noon."

Figure 1.8 The Large Magellanic Cloud and the Small Magellanic Cloud, now known to be small, satellite galaxies of the Milky Way, were all but unknown to most Europeans, as they are invisible, due to the curvature of the Earth, from most northern latitudes. Image credit: Wikimedia Commons user Markrosenrosen.

Who could believe such a thing? Sure, the path of the Sun through the sky varied throughout the year, reaching its minimum height above the horizon on the winter solstice and its peak height on the summer solstice, but even at its highest, it was *never* overhead! At least, not in Alexandria.

Yet, according to this letter-writer on Elephantine Island, the Sun was directly overhead, and hence your own head would block the Sun's light when you were looking down a well. If you hammered a perfectly vertical stick into the ground, it would cast no shadow at all at noon on the summer solstice. Not even a single degree to the north, south, east or west would you see in the form of a shadow; that is what the implications of having the Sun directly at the zenith are. So Eratosthenes made plans that, the very next year when the solstice came around, he would hammer a vertical stick directly into the ground, and measure — to the greatest accuracy possible — the angle that the Sun made when it was directly overhead.

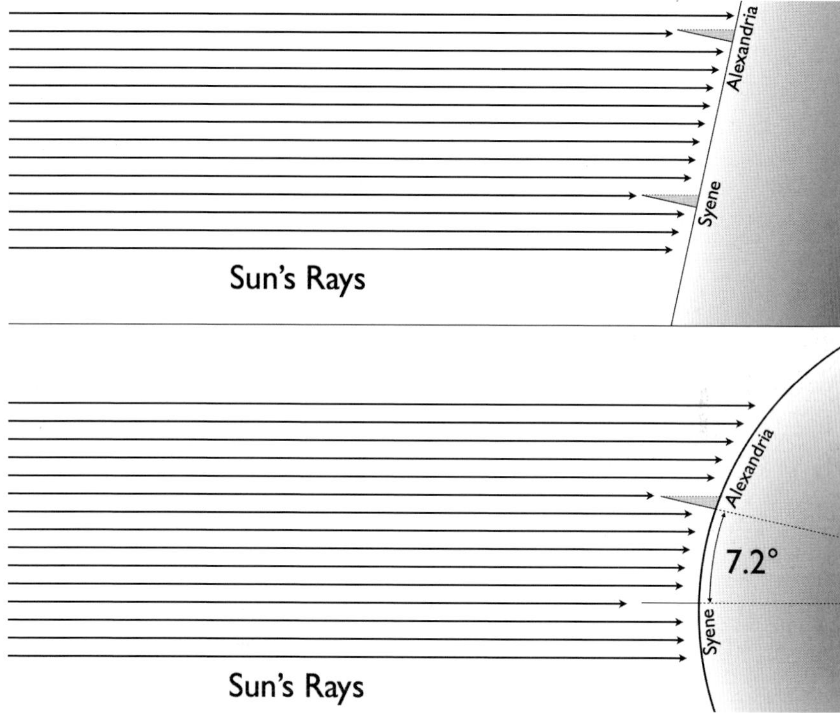

Figure 1.9 If the Earth were perfectly flat, then the Sun's rays would cast identical shadows at noon on the solstice everywhere on Earth (top), no matter where you were located. But if the Earth's surface were curved (bottom), shadows at different locations would cast different shadows on the same day, depending on the angle that the Sun's rays struck the object in question. By measuring the difference in shadow angle between two points on Earth's surface, it became possible to measure the size of the Earth for the first time. Image credit: E. Siegel.

He painstakingly made this measurement as accurately as possible, and concluded that the minimum angle that a stick's shadow cast did indeed occur at noon on the summer solstice, which at Alexandria corresponded to *one-fiftieth* of a circle. Since a circle was 360°, a little math leads to the conclusion that the shadow's angle cast by the stick was 7.2°. Yet if the correspondence were true, the shadow cast at Syene would be exactly 0°, since the Sun appeared directly overhead.

How could this be? (Fig. 1.9)

Eratosthenes realized that this could not reasonably be the case if the Earth were flat, like a circular disk. But if the Sun's rays were all parallel to one another, perhaps because it was a great distance away, this difference in shadows cast by the Sun could be caused if the surface of the Earth itself were curved like a sphere! In perhaps a stroke of brilliance, Eratosthenes had an epiphany: the geometry that he had measured showed that Syene and Alexandria differed by 7.2°, or one-fiftieth of a complete circle, in their latitudes. If he could figure out the distance between these two cities, all he would have to do was multiply by 50 (to form a complete circle, or 360°) and he would have a measurement for the Earth's circumference.

If only Eratosthenes had a graduate student; this would have been the perfect project for the world's first Ph.D. dissertation! Instead, he was forced to rely on the best measurement of the distance available between the two cities at the time: estimates of the distance based on travel time by camel. The most precise estimates for that distance at the time were that Alexandria and Syene were separated by 5,000 stadia, which means the entire Earth should have a circumference of 250,000 stadia. We can convert that into a modern measurement, simply by converting a stadium into conventional units, like kilometers or miles. As it turns out, this is a conversion which historians still argue over today. Eratosthenes, as we have covered, was a Greek living in Egypt. If he was using Greek units of measurement — what is known as an Attic stadium — a stadium corresponds to 185 meters, for a planetary circumference of **46,620 kilometers**, a figure just 16% larger than the accepted value today. However, an alternate possibility is that he was making his measurements in Egyptian stadia. Egyptian stadia were shorter, at only 157.5 meters apiece, which would give a global circumference of **39,375 kilometers**, a figure that differs from today's accepted value of 40,041 km by less than 2%!

This measurement had certainly been carried out by 240 B.C.E., and Eratosthenes went on to become the first accurate geographer, mapping out the world known to him at the time and becoming the first to introduce the modern concepts of latitude and longitude. He wound up accurately depicting the locations of more than 400 cities in the world, linking them together by a unified, objective concept of position on our world. He even divided Earth into five climate zones: two frozen ones around each pole, two temperate ones at intermediate latitudes, and a tropical one at the pole. The scientific fields of Earth science and geography had just been born.

* * *

But as monumental as his contributions to our understanding of Earth were, there was an even greater leap that he helped enable. Eratosthenes' measurement of the circumference of the Earth and his discovery of its sphericity relied on only one assumption: that the Sun was far enough away that the light rays emanating from it could be assumed to be parallel. Since it was known from eclipses that the Moon must be much closer than the Sun, this allowed the very first calculation of an *astronomical* distance: the distance to the Moon.

A contemporary of Eratosthenes, Aristarchus of Samos, wrote a book (that still survives!) called "On the Sizes and Distances," which is all about the sizes of and distances to the Sun and Moon. Using the radius of the Earth, something that could now easily be derived from Eratosthenes' measurement of our planet's circumference (by dividing by 2π), Aristarchus devised a simple method to calculate the size of the Moon and our distance to it simply by making measurements of a lunar eclipse. Not only does the shadow of the Earth appear on the Moon during an eclipse, but simply by measuring how much larger Earth's shadow is than the Moon itself, we can come up with a ratio for the size of the Moon to the size of Earth. Aristarchus determined, based on these shadow measurements, that the Moon's size was 35% of Earth's, in terms of its radius (Fig. 1.10).

As it turns out, the Moon's diameter is about 3,470 km, which places it at approximately 27% the size of Earth. And once you have an estimate for the physical size of the Moon, getting an estimate for the distance *to* the Moon is actually straightforward. You see, Aristarchus was also able to measure that both the Sun and the Moon took up approximately 0.5° each, in diameter, on the sky. (Archimedes credits him with this as well.) If you know the Moon's physical diameter, and you know that a circle takes up a full 360°, then you can figure out how physically large the Moon's orbit is around Earth. Then it is as simple as turning a

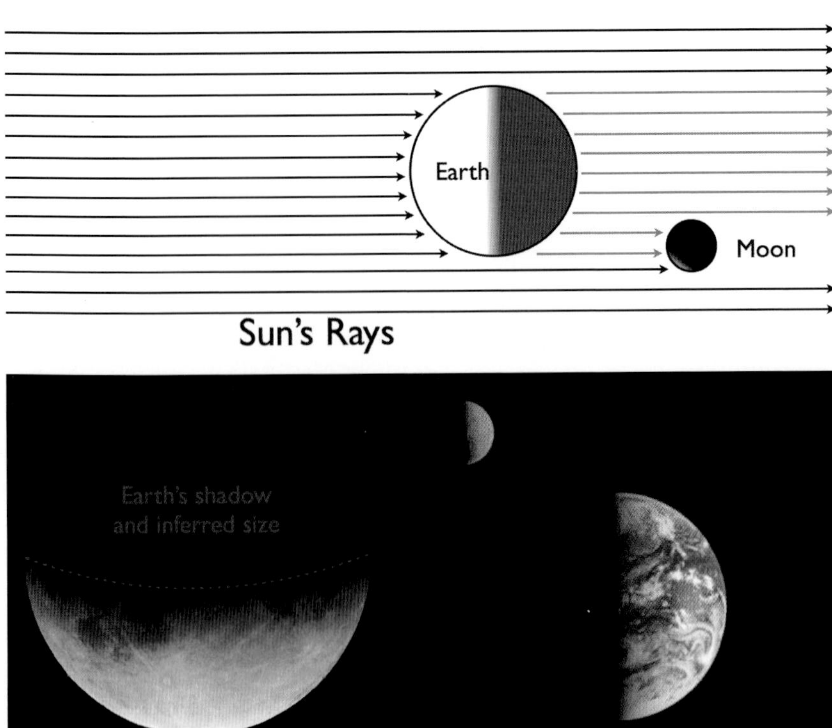

Figure 1.10 If you know the physical size of the Earth, you can take a look at the size of the shadow it casts upon the Moon during a lunar eclipse to understand, proportionally, how big the size of the Moon is relative to Earth. The inferred size of the Earth relative to the Moon is shown superimposed over an actual lunar eclipse (lower left) along with a photo of the Earth and Moon (lower right) taken by NASA's Galileo spacecraft. Realistically, this method allows us to determine the average Earth–Moon distance to an accuracy of a few percent (less than 10,000 km). Modern measurements, using laser ranging (reflecting a laser off of mirrors that we placed on the Moon) can measure the Earth–Moon distance to an accuracy of better than 1 cm. Image credit: E. Siegel (top), Wikimedia Commons user Mdekool (lower left); NASA/Galileo spacecraft (lower right).

circumference into a radius to get the Moon's distance from Earth. Using this method with the modern value of the Moon's diameter, we get 397,600 km for the Moon's distance from Earth. This falls well within the actual range of values that the Moon's distance actually takes on, which varies from 363,104–406,696 km.

Aristarchus' initial measurements were not the greatest, but his calculational methods were nothing short of brilliant. These measurements were improved upon in the ancient world by Hipparchus and Ptolemy, who were able to use Aristarchus' methods to calculate the Earth–Moon distance to an accuracy of 98%, with only a 2% error.

While his methods provided great insight for calculating the distance to and size of the Moon, Aristarchus also attempted to calculate the distance to the Sun, which proved to be a disaster. His method of attempting to do so was quite clever: he thought one should wait until the Sun and Moon appeared to make a 90° angle with one another with respect to an observer on Earth. You could then measure the Moon's phase as precisely as possible, and determine how much of the Moon appeared "filled" by the Sun's light. Whatever number you got that was less than 50%, you could convert that into an angle, and then — since you knew the distance to the Moon — use the properties of a right triangle to determine the distance to the Sun. Unfortunately, this is a task that is beyond the limit of human vision, as the Moon would appear 49.7% full at that moment, a number indistinguishable from 50%. Aristarchus, on the other hand, records the equivalent of 48.3%, and so estimated a Sun–Earth distance that was much too small: combined with his too-close measurement of the size of (and distance to) the Moon, his estimate was off by a factor of *sixty*. This type of error, where an observer *thinks* they are measuring a real signal that is, in fact, right at the limit of what can be measured with the equipment available, is one that still plagues scientists thousands of years after Aristarchus.

* * *

Nevertheless, thanks to the work of Eratosthenes and Aristarchus, it became firmly established that the Earth was a round planet, of a measurable and finite size, orbited by the Moon whose size and distance could be well-measured, with the Sun and stars all much more distant than that. In addition to the two most prominent objects in the sky, though, there are five points of light that appear different from all the fixed stars: the planets. They are so named from the Greek word for wanderer (πλανητης), because unlike the other fixed stars in the sky, they do not remain in the same position from night-to-night. Instead, they appear to move, with Mercury's motion occurring the most quickly, followed by Venus, Mars, Jupiter and finally the slowest planet, Saturn, nicknamed "the old man of the skies."

The planets share a few things in common with the Moon that separate them from the stars. For one, just like the Moon, planets do not appear to twinkle as stars do. If you see a point of light shining constantly in the sky, not twinkling at all like the points surrounding it, it is almost a certainty you are staring at one of the five naked-eye planets in our Solar System. Furthermore, just as the Moon moves around the Earth in a counterclockwise fashion as viewed from the North Pole, and hence migrates an extra 12° eastward relative to its position the prior day, the planets generally appear to wander towards the east as well relative to their positions the prior night. The motion of the planets is far smaller in magnitude so long as they migrate towards the eastern direction, which is known as *prograde motion*. However, every once in a while, the planets will slow down in their motion across the sky, come to a relative standstill, and then temporarily reverse their motion for a time, migrating back towards the west during a period of *retrograde motion*. After a few nights, that western migration will slow and pause, with prograde motion continuing once again (Fig. 1.11).

For the planets Mercury and Venus, they appear to reach a maximum angular separation (known as elongation) from the Sun, then enter a period where they rapidly migrate back towards the west, passing very close to (or even transiting across) the Sun in the process. For Mercury, retrograde occurs for a three week period every 116 days; for Venus, a six week period every 584 days. For the other worlds, Mars, Jupiter and Saturn, retrograde lasts much longer, occurs when the planets are opposed to the Sun, and their apparent night-to-night position shifts are much slower than those of Mercury and Venus.

There were two competing explanations as to why this would occur. The first was the *Geocentric Model*, first put forth by Anaximander in the 6th century B.C.E., which reckoned that the Earth was stationary, and that the planets, Sun and Moon all orbited it separately. Modifications were made to this model over time to account for the retrograde motions of planets, with the concept of epicycles and deferents introduced by

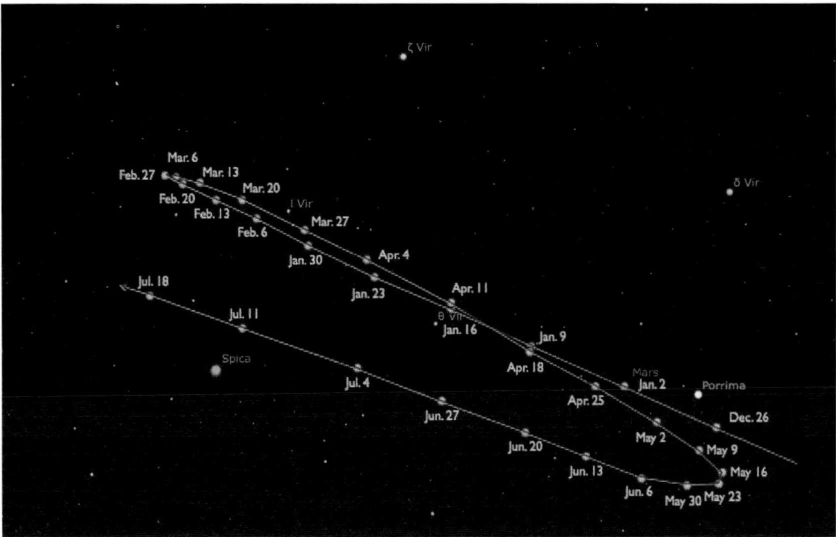

Figure 1.11 Planets like Mars (shown) normally migrate eastwards, ever so slightly, from night-to-night relative to their previous position. This motion is known as prograde or direct motion. Periodically, however, that motion will slow, reverse, slow, and continue in its original direction again, with the period of reverse motion known as retrograde motion. Shown in the figure here is the retrograde motion of Mars relative to the background stars from March to May, 2014. Image credit: E. Siegel, using Stellarium.

Apollonius of Perga in the 3rd century B.C.E. Instead of orbiting on circles with the Earth at the center, planets could orbit on *epicycles* or small circles that moved along larger circles (known as deferents) that were off-center from the Earth. Hipparchus and Ptolemy each improved this later by better calculating the orbital parameters that predicted their positions over time, with Ptolemy's predictions remaining so accurate that the positions of the planets differed by no more than 2° *at most* by the time of Copernicus more than 1,000 years later!

Partially because of the incredible success of this model, but also because alternative lines of thought were not only unexplored but outright suppressed for many centuries during the middle ages, the Geocentric Model was virtually unopposed in western thought. As a result, when corrections were discovered, or new observations were made that defied Ptolemy's predictions, the standard approach became to simply add extra, smaller epicycles to "fix" the model. Scientifically, this is incredibly dissatisfying, as there is no predictive power to this. It is merely a descriptive model, not a theory that explains how or why this motion would occur. The idea that Earth was fixed, unmoving and the center of the Universe became the accepted dogma of its time.

But scientifically, there was a second explanation that could account for this apparent retrograde motion as well: a *Heliocentric Model.* It was first put forth by Aristarchus as an alternative to the Geocentric Model, although that particular work has been lost with the destruction of the Library at Alexandria; we will never know how satisfying of an explanation Aristarchus' heliocentric model provided. It was not until the 16th century C.E. that Nicolaus Copernicus published his great work, *De revolutionibus orbium coelestium* (On the revolutions of the celestial spheres), that retrograde motion could successfully be explained without epicycles and deferents. Instead, the Sun would be at the center of the Solar System, and the planets — ordered Mercury, Venus, Earth, Mars, Jupiter and Saturn, moving outwards — would all orbit it, with the inner worlds moving more quickly and the outer ones moving more slowly. When an inner world, moving faster, overtook an outer world in its orbit, it would be seen to move backwards from its normal motion, causing the apparent retrograde motion (Fig. 1.12).

Copernicus' original model used circular orbits only, and so wound up being less accurate than Ptolemy's model for most predictions. The only way Copernicus saw to salvage it was to reintroduce the use of epicycles. Despite eliminating their need for the most obvious of applications, a huge step forward, he had not solved the problem of planetary motion entirely.

* * *

Copernicus' work — published upon his death in 1543 — got people thinking that perhaps there was a better model than this geocentric theory for describing the motions of the planets. Perhaps the heavens were not fixed, static and eternal, with the Earth unmoving at the center of it all. The generation of astronomers who came after Copernicus grew up considering that there might be better ways of conceiving of the Universe than Ptolemy and Aristotle did more than a millennium before. One of

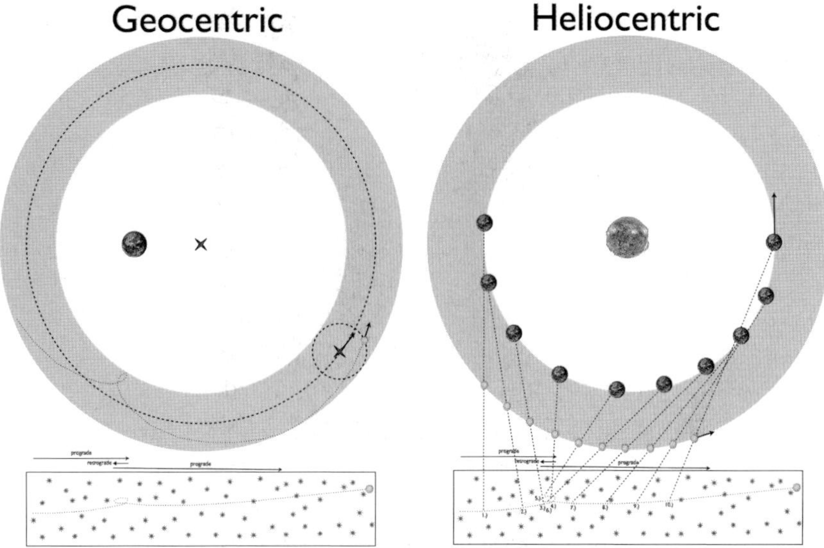

Figure 1.12 In the Geocentric Model, planets experience an apparent retrograde motion because their orbits reverse course as they follow a path laid out by epicycles and deferents, implying that there was an actual, physical reversal of motion. In geocentrism, the planets move along the small circle, which in turn moves along the larger circle, with the Earth offset from the center of the larger circle. In the Heliocentric Model, the apparent retrograde motion is entirely due to the relative motion of planetary bodies through space, as inner worlds overtake outer ones in their orbits, causing the temporary appearance of a backwards motion. The planets all orbit the Sun, and it is only due to the faster speeds of the inner worlds that this retrograde phenomenon appears to occur. Image credit: E. Siegel.

those astronomers was Tycho Brahe, the greatest naked-eye astronomer of all. Known for his incredible visual acuity, Tycho catalogued the positions and brightnesses of thousands of stars as well as the five planets over time with incredible precision. This led to him rejecting geocentrism outright, developing his own Tychonian system of the Universe, where the planets all orbited the Sun, which in turn itself orbited the Earth. He also became the first to note that comets had two tails: one that always pointed directly away from the Sun (which we now know to be made of ions), while another curved away from the comet's path at an angle (now known to be made of dust). But perhaps the greatest observation of all was his

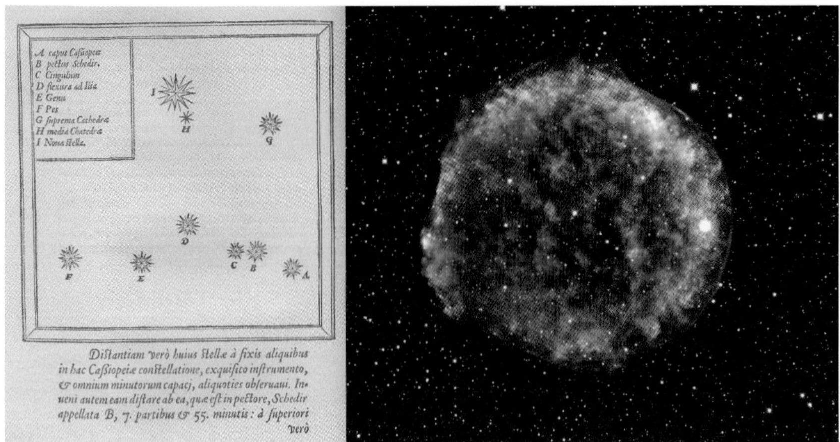

Figure 1.13 The "new star," labeled "I" in the figure at left, outshone all the others in the sky at its peak brightness, eventually fading from view and becoming invisible once again. The supernova remnant was rediscovered centuries later, and is shown against the background star field where it is located with its presence highlighted in X-ray and infrared light, as taken by NASA's Chandra X-ray and Spitzer infrared space telescopes. Image credit: Tycho Brahe, *De nova stella*, 1573 (L); X-ray: NASA/CXC/SAO, Infrared: NASA/JPL-Caltech, Optical: MPIA, Calar Alto, O. Krause *et al.* (R).

cataloguing and popularizing of the 1572 appearance of what he called a *stella nova*, or new star:

> "When, according to habit, I was contemplating the stars in a clear sky, I noticed a new and unusual star, surpassing the other stars in brilliancy. There had never before been any star in that place in the sky."

On November 6th of that year, a star appeared in the constellation of Cassiopeia, and over the coming days, continued to brighten, eventually outshining not only all the other stars and planets in the sky, but becoming so bright that it was even visible in the daytime! (Fig. 1.13)

While many geocentrists of the time claimed it must be an atmospheric phenomenon, Tycho's painstaking observations showed that the position of the "new star" remained unchanged during its entire reign of visibility. There was a zero measurement of a *parallax*, or a change in the apparent position of an object when viewed from varying locations. (For example, when you hold your thumb at arm's length and close your left eye, then open your left eye and close your right, the position of your thumb appears to shift against the background. *That* is a parallax.) Over a time period of months, the star dimmed and eventually became invisible again, providing the first well-known, concrete evidence that the stars *were not* eternal and unchanging, but rather could appear, brighten, and disappear all at once. What Tycho saw turned out to be a supernova, or the catastrophic death of a star at the end of its life cycle. Based on Tycho's accurate position measurements of 1572, we can locate and measure the supernova remnant that exists today. Additionally, in modern times, we have been able to locate supernova remnants whose appearances were catalogued in other sources (e.g., Chinese, Egyptian, Japanese, and Middle Eastern texts) from the years 185, 393, 1006, 1054 and 1181. The stars — once thought to be fixed and eternal — were suddenly shown to have finite lives.

After Brahe's sudden death in 1601, his contemporary, Johannes Kepler, took up his studies as his successor. Kepler had his own beautiful theory of planetary motion. Like Copernicus, Kepler imagined that the planets orbited in circles about the Sun, but Kepler had his own unique take on it. Noting that there were five planets (other than Earth) and five perfect mathematical solids — that is, five three-dimensional shapes that could be constructed solely out of regular polygons (triangles, squares and pentagons) — he developed a model whereby each of the planets were "held up" by spheres either inscribed or circumscribed around these perfect solids. Mercury, the most interior planet, would rest on a sphere inscribed within an octahedron, while Venus would orbit on a sphere circumscribed around it. Simultaneously, Venus would be inscribed within a dodecahedron, with Earth circumscribed around it while inscribed within an icosahedron beyond that, followed by Mars, a tetrahedron, Jupiter, a cube, and finally Saturn. He called the model the *Mysterium Cosmographicum*, published the work in 1596, and set out to test it using the outstanding suite of observations left to him by Tycho Brahe (Fig. 1.14).

The annals of history are filled with scientists who had great, elegant and beautiful ideas that did not quite match the observations. How does one react when that happens, when you have the "idea of a lifetime" only

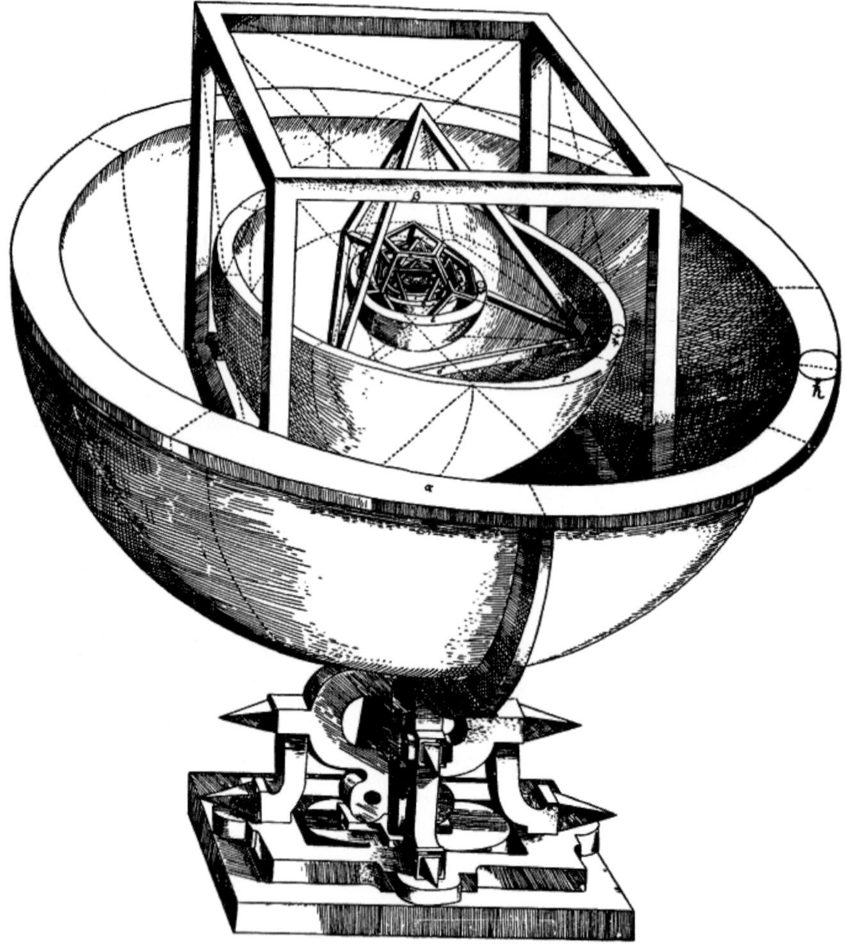

Figure 1.14 In Kepler's original model, the planets moved in circles about the Sun, where the radius of each planet's orbit was prescribed by a sphere either inscribed within or circumscribed around one of the five perfect, Platonic solids. Image credit: Johannes Kepler, (1596). *Mysterium Cosmographicum.*

to find it does not quite give you what you thought it would. Most frequently, the scientists who had that idea spend their lives trying to salvage their theories by modifying their hypotheses, adding proverbial epicycles (or *ad hoc* fixes), or taking continued observations, hoping that they will get different results in the future. What makes Kepler such an admirable figure is that when the observations failed to align with his model, as beautiful and beloved as it was, he sought out alternative explanations that might better fit the actual data. Much like Copernicus, he noticed that circular orbits were not satisfactory for describing the motions of the planets — particularly for Mercury and Mars — and spent years experimenting with alternative orbits, such as ovals. Finally, in 1605, he hit upon the successful solution: that there was absolutely no need of any deferents or epicycles to account for the motion of the Moon, the planets or of the Earth around the Sun if it was assumed that planets orbited in ellipses around the Sun (and the Moon in an ellipse around the Earth), with the Sun not at the center, but rather at one focus of the ellipse.

This new work was finally published in 1609 (after a legal battle with Brahe's heirs over the use of his observations), and it was immediately transformative. By the next year, Galileo's telescopic observations of four moons orbiting Jupiter (and not Earth) lent further credence to the idea that the Earth could not be the center of all motion in the Solar System, and his additional telescopic discovery of the phases of Venus — and the fact that it ran through the full gamut, from a crescent when nearest Earth to a full phase when it was opposite the Sun — was all the evidence the scientific world needed to establish that the heliocentric view was a far superior description of the Solar System than the geocentric one. With just a few years of science, more than a millennium of dogma had been overthrown (Fig. 1.15).

* * *

History may not have been kind to Galileo and Kepler, with Galileo found guilty of heresy and placed under house arrest for the final decades of his life and Kepler's mother charged with witchcraft as a result of his publications, but the success of their scientific advances was incontrovertible. In addition to his observation that planets moved about the Sun in ellipses, Kepler subsequently published two more laws of planetary motion: that if you calculated the area swept out by a planet's motion around the Sun, the area swept out in any given amount of time along that ellipse was always a constant, and that the period of a planet's orbit squared was proportional to the semi-major axis (or half the "long axis" of the ellipse) of the orbit cubed. These laws were then used to calculate — for the first time — the predicted transits of the inner worlds, Mercury and Venus, across the face of the

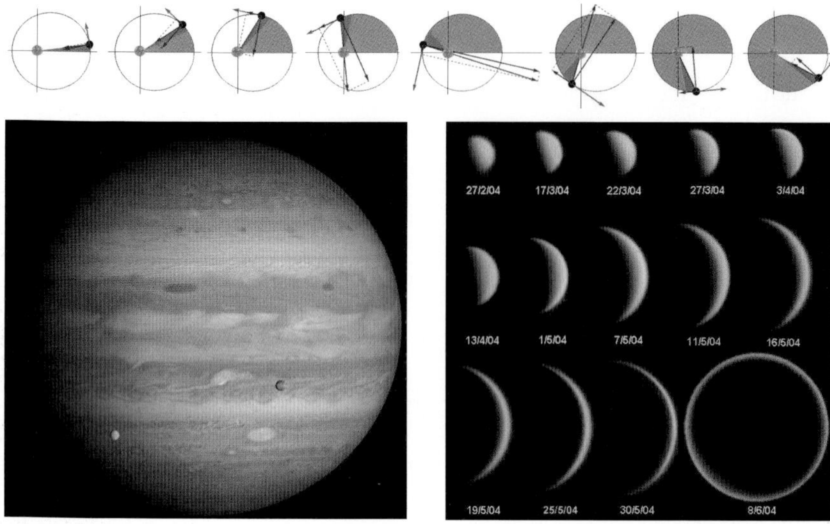

Figure 1.15 With the discovery that planets moved in ellipses around the Sun (top), that Jupiter had its own moons that orbited neither Earth nor the Sun (lower left, with Io and Europa shown), and that Venus ran the full gamut of phases (lower right), appearing with a small, full phase (when it was opposite the Sun from Earth) and a large, crescent phase (when it was near the Earth), the predictions of the Heliocentric Model were validated. What we learned from this is that the Earth is not so different from the other planets with respect to the Sun. Note also, atop, that the "blue" section in every one of the ellipse sections represent equal amounts of area, which happen to be swept out in equal amounts of time in each individual planet's orbit around the Sun. This is Kepler's second law of planetary motion, where the first was that planets move around the Sun in ellipses. Image credit: Wikimedia Commons user Gonfer (top); NASA/JPL/Voyager 1/Björn Jónsson (lower left); Statis Kalyvas — VT-2004 program (lower right).

Sun. These predictions were then verified in the 1630s, with the 1631 transit of Mercury becoming the first transit of Mercury ever observed (Fig. 1.16).

At last, astronomy was becoming a bona fide science, with theories that were not only capable of *describing* reality, but making observable predictions that were different from alternative descriptions. But the most important step was yet to come. By the late 1600s, heliocentrism governed by Kepler's laws were all the rage in astronomy. The telescope had opened up a whole new suite of discoveries, including

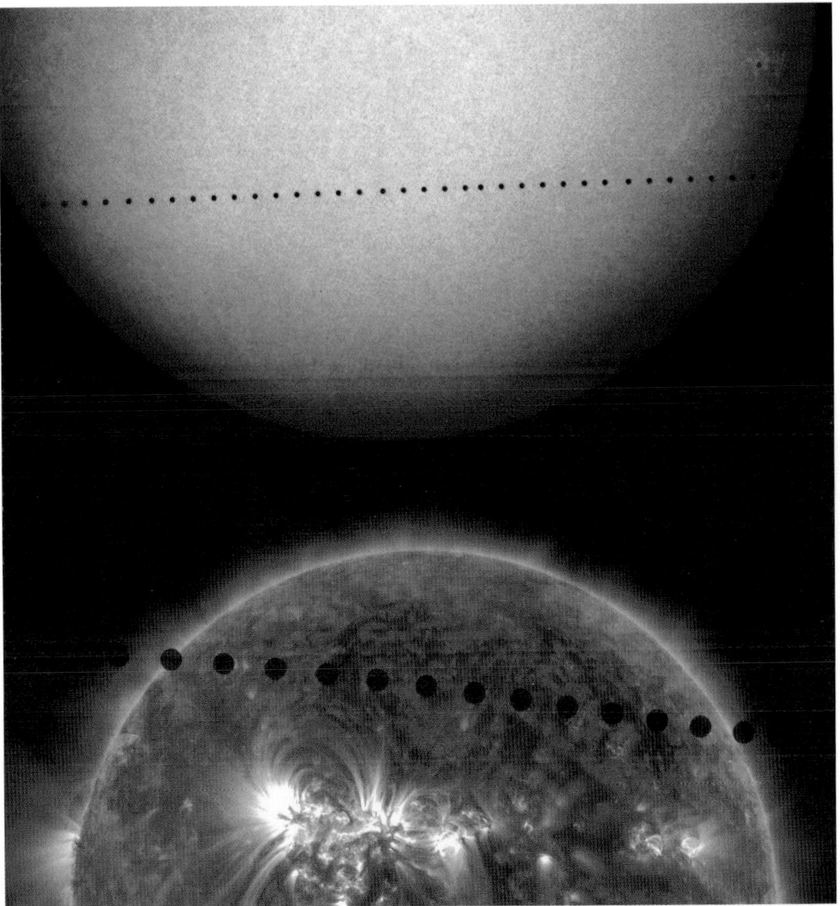

Figure 1.16 Transits of the inner planets across the disk of the Sun — Mercury atop (in 2006) and Venus at bottom (in 2012) — had never been successfully predicted before the discovery of Kepler's laws of planetary motion. With them, these events can be predicted accurately, down to the minute, millennia in advance. Image credit: ESA/NASA/SOHO (top); NASA/Goddard Space Flight Center/SDO (bottom).

moons around Saturn, the discovery of the first fixed, extended deep-sky objects and the birth of comet hunting as a scientific endeavor. Edmond Halley — of the famed Halley's comet — was studying the motion of comets (notably, Comet Kirch) in the late 17th century, attempting to match up their historical sightings with orbital motions, and using Kepler's laws to help guide him. There was a difficult problem he was working on related to these orbits: attempting to figure out what type of physical force law would account for these types of motions.

While attempting to work out a proof of Kepler's laws of planetary motion in 1684, he sought out the advice of Isaac Newton, inquiring as to what sort of force law would cause elliptical orbits. Newton immediately provided him with the answer, and although he claimed to have lost the work, said he would work it out again and send it to Halley. Awestruck by the derivation sent to him, Halley arranged for its publication and served as Newton's patron, while the latter composed the most world-changing treatise in the history of science: the *Philosophiæ Naturalis Principia Mathematica*, in which Newton put forth his famed universal law of gravitation. This law claimed that gravitation followed an inverse-square force law, where the gravitational force between any two objects was directly proportional to both the objects' respective masses and inversely proportional to the square of the distance between them. Using this single rule, not only could Kepler's laws all be derived, but many other phenomena as well, including the varying period of a pendulum based on its location on Earth, the force that kept our atmosphere from flying away, and the motions of not only Earth's moon but the moons of all the planets. Following the *Principia*'s publication in 1687, Halley was able to work out the gravitational effects of Jupiter and Saturn on the long-period comets in our Solar System, and hence was able to account for not only their elliptical orbits, but the slight *corrections* that the gravitational forces of the planets provided. The identification of a 1682 comet with a prior sighting in 1607 (by Kepler) and one previously in 1531 led to Halley's epiphany that these were all the same comet, which would once again return in 1758. Its re-discovery on December 25, 1758 led to not only its naming in Halley's honor, but another vindication for Newton's theory of gravity (Fig. 1.17). As it turned out, the Heliocentric Model was merely an inevitable consequence of a more fundamental, underlying theory of gravity, one that would go on to make a great many more predictions that could be tested experimentally and observationally.

Figure 1.17 Halley's comet has continued to return approximately every 75 to 76 years, most recently in 1986, as shown in the figure. It is scheduled to next return to the inner Solar System in 2062, possibly first becoming visible in December of 2061. Image credit: NASA/JPL.

* * *

Gravitation, telescopic discoveries and the ability to describe the motions within our Solar System were not the only astronomical advances that were made during the 17th century. Christiaan Huygens, the discoverer of Saturn's first moon (Titan) and the scientist who showed that Earth's gravitational field varied by building a pendulum clock that ran slower closer to the equator, considered — as Aristarchus did before him — that the stars in the sky were not fixed points of light orbiting on a celestial sphere, but rather were distant suns not so different from our own.

Unlike Aristarchus, however, Huygens set out to use this assumption to figure out how far away the stars might actually be.

In astronomy, we define a star's brightness by its visual magnitude, which works on a logarithmic scale. There are a handful of stars that are among the brightest: Sirius, Canopus, Alpha Centauri, Arcturus, Vega and Capella are the top six (in order), where Vega was used as the definition of the magnitude scale (at magnitude 0) for centuries. A difference of one magnitude corresponds to a difference in brightness of a factor of 2.5, so a star of magnitude −1 would be 2.5 times as bright as Vega, while one of magnitude +5 would be only 1% as bright. The brightest star in the night sky, Sirius, comes in at a visual magnitude of −1.4, exceeded only by a few of the planets, the Moon and the Sun. The faintest stars visible to the naked eye under typical dark skies are of approximately magnitude +6.5, which means there are approximately 9,000 stars visible from Earth without any sort of astronomical aid.

The Sun, however, is *much* brighter than anything else in the sky, a full 400,000 times brighter than even the full Moon. Thankfully, there is a simple relationship between distance and brightness that Huygens was able to figure out: if you double the distance that a light source is from you, its brightness drops down to *one-fourth* the original brightness; if you triple that distance, the brightness drops to *one-ninth*; if you quadruple the original distance, the brightness becomes *one-sixteenth* the original. The reason for this is that when a star emits its light, that light spreads out in a sphere. As that sphere moves farther and farther away from the original source, the amount of light contained in any given area decreases as the inverse of the distance squared. Therefore, you need an ever greater surface to pick up the same amount of light, overall, the farther away you are (Fig. 1.18).

Huygens sought to take advantage of this relationship, reasoning that if he could figure out how many times brighter than Sirius the Sun appeared to be, he could then use the distance-brightness relationship, and the (now known) Earth–Sun distance to calculate the distance to the night sky's brightest star. His first attempt involved drilling a series of holes in a brass plate, each hole successively smaller and smaller, with the idea that eventually, he would create a hole small enough that the amount of sunlight that came through would show the same brightness during the

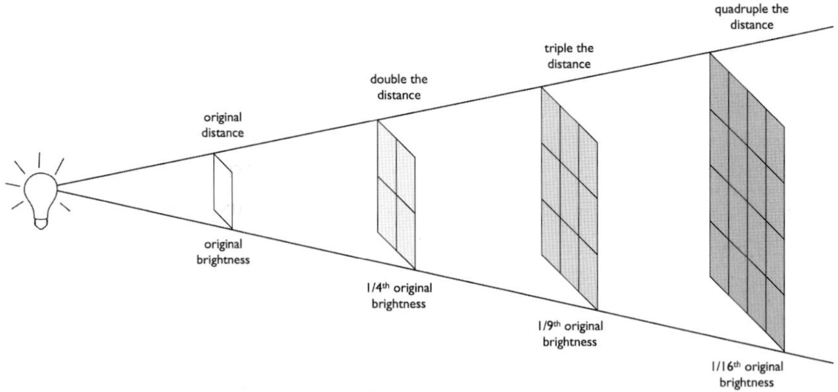

Figure 1.18 The farther you are away from a light source, the dimmer it appears. The relationship between distance and brightness is very straightforward: it follows an inverse-square relationship, similarly to gravity. If you could figure out how many times dimmer a star was than the Sun, then — assuming the star had the same intrinsic properties as our Sun — you could figure out how far away it must be. Image credit: E. Siegel.

day as Sirius showed at night. Unfortunately (and unsurprisingly), even the smallest hole that could be manufactured still led to an amount of light that far exceeded that of Sirius. So to augment even his smallest hole, Huygens obtained a large number of semi-opaque glass beads that he fit to the hole. By reducing the amount of light that came through even further, he was finally able to create an artificial "star" from sunlight that rivaled the brightness of Sirius. All it took was a reduction of the Sun's original brightness by a factor of 800 million!

But with the relative brightnesses of the Sun and Sirius known, as well as the distance to the Sun, Huygens could use the physical relationship between brightness and distance to infer the distance to a star for the first time. Sirius, assuming it was identical to the Sun, was 28,000 times more distant. With the Sun located a mean distance of 150 million kilometers away, that would put Sirius at a distance of 4.2×10^{12} km, or about 0.4 light years. Admittedly, this was a very crude way of measuring brightnesses, as Huygens' first estimate was off by quite a bit: Sirius is actually 8.6 light years from us. However, what Huygens did not know is that Sirius is intrinsically 25 times brighter than our Sun is, a fact that, if we take it into account, gives Huygens' distance measurement an error of less than 20% from the true value, a remarkable achievement for "eyeballing" brightness!

* * *

As telescope technology continued to improve, a number of new phenomena began to reveal themselves to astronomers. In addition to the discovery of numerous new moons around planets and hundreds of thousands of stars that were beyond the limit of human vision, astronomers began to experience frustration in their searches for comets. Comet-hunting had become a sort of hobby among wealthy men of leisure, with the great comet of 1744 having become the first to be spotted by a number of astronomers *before* becoming visible to the naked eye. While still far from the Sun, comets would only appear as faint, extended and fuzzy smudges on the sky. Although they would not develop noticeable tails until they were interior to the orbit of Jupiter, distant comets would still be easily identifiable by the fact that they would move, night-to-night, against the backdrop of fixed stars. However, there are a number of faint, extended and fuzzy objects that were *stationary* in the night sky: the nebulae. If you were hunting for a new comet and saw one of these nebulae, you might incorrectly identify it for a number of nights, until its stationary nature convinced you it wasn't a comet after all.

Charles Messier was 13 when the great comet of 1744 came close to Earth, and he was both awed and inspired by it. Having spent the next 14 years studying to be an astronomer, he was looking for the soon-to-be-returning Halley's comet in 1758. Sure enough, he became frustrated at his misidentification of what we now know to be the Crab Nebula — the remnant of the supernova of 1054 — with the comet in question. There was no clue available to him that the faint, fuzzy smudge he saw through his eyepiece was not the comet he was seeking, but was a nebula instead. It was only the fact that the nebula he discovered remained in the same location with the same visual features night-after-night that indicated to him he was seeing a fixed object in the night sky (Fig. 1.19).

Although Messier went on to discover *thirteen* new comets over his lifetime, his greatest contribution to astronomy was the development of the Messier Catalogue, the first comprehensive collection of deep-sky

Figure 1.19 Long-exposure astrophotography allows us to gather far more light than our simple eyes can through a telescope, bringing out the structure in deep-sky objects like the Crab Nebula. As viewed through a telescope's eyepiece, this object would have been identifiable only as a faint, extended smudge, with little detail visible. It would have been indistinguishable from a newly brightening comet. Image credit: Wikimedia Commons user Rawastrodata, under c.c.-by-s.a.-3.0.

objects with accurate positions and descriptions. Although he composed his catalogue to assist comet hunters by giving them the set of bright, fixed objects to avoid confusing with transient comets, his catalogue wound up becoming something much more important. The objects he recorded were a combination of open star clusters, globular clusters, stellar remnants, star-forming regions, and spiral and elliptical galaxies. Although Messier did not know what he was observing at the time, what he was seeing were pieces of evidence of the Universe outside of our Solar System that were beyond mere stars. Instead, each of these represented something remarkable in our Universe:

- **Open star clusters**: when stars form, they do not do so in isolation, but rather in groups of hundreds, thousands or more, usually in the plane of spiral galaxies. The vast majority of open clusters are found

along the plane of our Milky Way, and range from just a few million to many billions of years in age.
- **Globular clusters**: these are collections of typically a few hundred thousand stars, found in-and-out of the galactic plane, orbiting the centers of galaxies. These contain some of the oldest stars in the known Universe, and the only hot, young, massive blue stars that they contain formed recently, from the merger of smaller stars.
- **Stellar remnants**: these are planetary nebulae and supernova remnants, stars that have recently ended their lives, with their centers becoming white dwarfs, neutron stars or black holes, while their outer layers are expelled back into interstellar space.
- **Star-forming regions**: these are nebulae in the true sense of the word, clouds of molecular and ionized gas that are actively forming stars, right now, in their inner regions. Over tens of millions of years, their gas will be completely blown off and evaporated by the hot, young stars present inside, at which point they will resemble an open star cluster.
- **Spiral and elliptical galaxies**: these are the two main classes of galaxy found in the Universe, although no one knew they were objects outside of our own Milky Way until the 20th century. These are typically collections of hundreds of billions (or more) stars, organized into either a disk-like structure with spiral arms or an elliptical structure whose density decreases as you move away from the center (Fig. 1.20).

While Messier's original catalogue consisted of just over 100 objects, continued telescope improvements led to a vast wealth of discoveries of what was out there in the Universe. In 1781, William Herschel discovered a bright, blue disk in the skies *that did not twinkle* like the other stars. What he had discovered turned out to be the first planet in our Solar System beyond Saturn: Uranus. His reward was the first telescope that was over a meter in aperture size, the first of its kind. With the use of it, the number of deep-sky objects skyrocketed into the thousands in just a few years.

As technology continued to improve — larger telescopes, improved optics, and the development of astrophotography — new astronomical

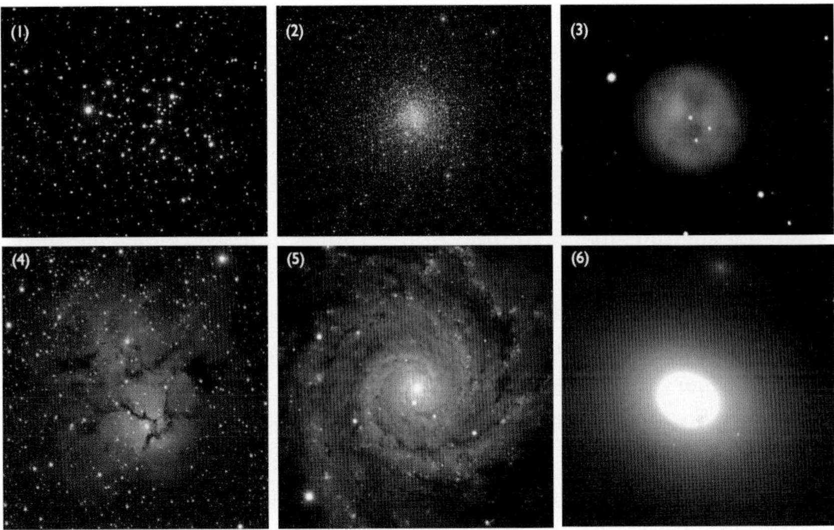

Figure 1.20 These six objects — Messiers 6, 4, 97, 20, 74 and 86, respectively — represent the major classes of deep-sky objects, organized into their now-known categories. The 110 objects classified by Messier are still avidly used today as objects to be avoided by comet hunters as well as common targets for astronomers, both amateur and professional. Image credit: Ole Nielsen (1); ESO/ESO Imaging Survey (2); Wikimedia Commons user Fryns (3); Hunter Wilson/Wikimedia Commons user Hewholooks (4); ESO/PESSTO/S. Smartt (5); NASA/STScI/Wikisky (6).

discoveries continued to pile up. Many star-forming regions were discovered to contain clusters of stars and it was uncovered that they extended for great lengths into space, well-beyond what could be seen with human vision alone. Many globular clusters were resolved into individual stars, and their individual properties could now be studied on a star-by-star basis. In the early 1800s, the first asteroids were discovered, and by mid-century it was recognized that there was an entire belt of them present. In 1846, a planet beyond Uranus was discovered: Neptune. Telescopes became large enough that a spiral structure to galaxies could be observed for the first time, and in 1887, a nebula whose structure could not be seen with the naked eye had its spiral shape revealed through photography. The nature of these spiral nebulae was hotly debated, with the consensus view that they were stars forming within our own galaxy countered by a minority viewpoint that they were "island Universes" far beyond our own Milky Way.

Modern astronomy was well underway.

* * *

The last predictions of a heliocentric Universe that had not yet been observed finally came to fruition in the 19th century as well. One of the arguments used against heliocentrism was that the stars truly appeared fixed in their positions, which should not be the case if the Earth orbited the Sun and the stars were actually all at varying distances. For one, if Newton's law of gravitation was truly universal, all the stars must be in motion relative to one another, as the gravitational forces between them would cause them to accelerate, even if they started off stationary. The closest stars to us should, therefore, be seen to move over time. For another, if the Earth orbited the Sun, the closest stars to us should experience a parallax at six month intervals, since the Earth should now be on the opposite side of the Sun, providing a long baseline for observing a potential parallax (Fig. 1.21).

In the early 1800s, Giuseppe Piazzi — already famous for having discovered Ceres, the first asteroid — noticed that a star, 61 Cygni, appeared to shift gradually but noticeably in its position over a 10-year time period. Follow-up observations by Friedrich Bessel confirmed this: the fixed stars were not so fixed after all. Perhaps inspired by how nearby this star must be, even if it were moving terribly rapidly, Bessel decided to monitor it for evidence of a parallax. The parallax would be incredibly tiny, and rather than measuring it in degrees, we would need to break each degree up into smaller units: arcminutes (symbolized by $'$), so that there are $60'$ in each 1°, and then arcseconds (symbolized by $''$), so there are $60''$ in each $1'$. Although there were others racing to measure parallax with other prominent stars — Friedrich Struve choosing Vega and Thomas Henderson choosing Alpha Centauri — Bessel got there first, measuring a parallax of just $0.314''$ for 61 Cygni. (No wonder it took so long, and so many telescopic advances, to discover a stellar parallax!) This corresponded to a distance for the star of 10.3 light years, off from the modern value of 11.4 light years by less than 10%. In a historical curiosity, Bessel published his results in 1838, but Henderson had made

Figure 1.21 As the Earth orbits the Sun, the incredible distance differences during six month intervals of some 300 million km cause slight apparent motions among the nearest stars. A parsec — which happens to be about 3.26 light years — is defined by a theoretical object that exhibits a parallax of 1″ as measured from Earth from perihelion to aphelion. Image credit: Wikimedia Commons user Abeshenkov/public domain.

his observations of Alpha Centauri back in 1832–1833, deriving a parallax but declining to publish his results for fear that his instruments were inaccurate, and he would be ridiculed for announcing a specious discovery. He wound up publishing his work only in 1839, after Bessel's work had been accepted. Too bad, as the Alpha Centauri system (consisting of Alpha Centauri A, Alpha Centauri B and the red dwarf

Figure 1.22 Alpha Centauri A & B (main) are shown alongside the nearly-as-bright Beta Centauri (inset), which is deceptively many times farther away at 350 light years. Proxima Centauri, gravitationally bound to Alpha Centauri A & B, is located in the small red circle. A low mass star that burns faint and red, it contains just 12% the mass and emits just 0.17% the light of the Sun. At 4.24 light years distant, it is the closest star to Earth today. Image credit: Wikimedia Commons user Skatebiker (main); ESA/Hubble & NASA (inset).

Proxima Centauri, which is slightly closer than the other two) turned out to be the nearest star system to Earth beyond our Sun, and hence the one with the greatest parallax: more than twice as large as the one experienced by 61 Cygni! (Fig. 1.22)

The final nail in the coffin for geocentrism came in 1851, when the French physicist Léon Foucault devised a brilliant demonstration to show the rotation of the Earth. By attaching a very heavy pendulum bob to a thin wire suspended many stories up, he created a physical system that would respond in a macroscopically visible way to our 24 hour rotation. If the Earth were stationary, the pendulum would simply swing back-and-forth in the same plane, eventually coming to rest without having shifted at all. But if the Earth were rotating, it would shift in proportion to the location of the experiment's latitude, with a maximum shift of 360° per 24 hours predicted at the Earth's poles. Demonstrating his pendulum for all the world to see — a 28 kilogram brass-coated lead weight suspended from a wire 67 meters long from the Panthéon in Paris — the plane of the pendulum's swing rotated very slowly, at 11° per hour, in full agreement with the prediction of a planet that rotated on its axis once per day at the latitude of Paris. A version of the experiment has subsequently been constructed and performed at the South Pole itself, where the rotation has been observed to occur at a rate of 15° per hour: a full 360° per day. At last, all the predictions of a heliocentric Universe had been verified.

* * *

The years leading up to the early 1900s brought with them not only a revolution in our understanding of the Universe and tremendous developments in astronomy, but also in our fundamental understanding of the physical laws governing the world itself. Matter was determined to be made of atoms, with those in turn made of heavy, positively charged atomic nuclei and light, negatively charged electrons orbiting them. The amount of positive charge present in an atom's nucleus determined its identity and its physical and chemical properties, defining what type of element it was. These elements could be organized into a table, periodically, where elements within a given column displayed similarities in their bonding properties and chemical reactivity.

Additionally, some of these elements — predominantly the heavier ones — were found to be *radioactive*, in that they would spontaneously decay into other elements. By emitting smaller, lighter particles, usually either helium nuclei (alpha particles) or single electrons (beta particles), these elements would transform into other elements, often in a chain reaction, to become more stable particles. As it turns out, *all* elements heavier than lead (element 82) are unstable on long enough timescales, and will decay into lead (or lighter) given enough time. Perhaps the most bizarre thing about these decays, however, is that they violated what was previously thought to be one of the most sacrosanct laws of the Universe: the law of conservation of mass. In all other reactions that had been observed previously — chemical, mechanical and electrical — whether you gained or lost energy from that reaction, the amount of mass that you began with and the amount of mass that you wound up with, if you accounted for all

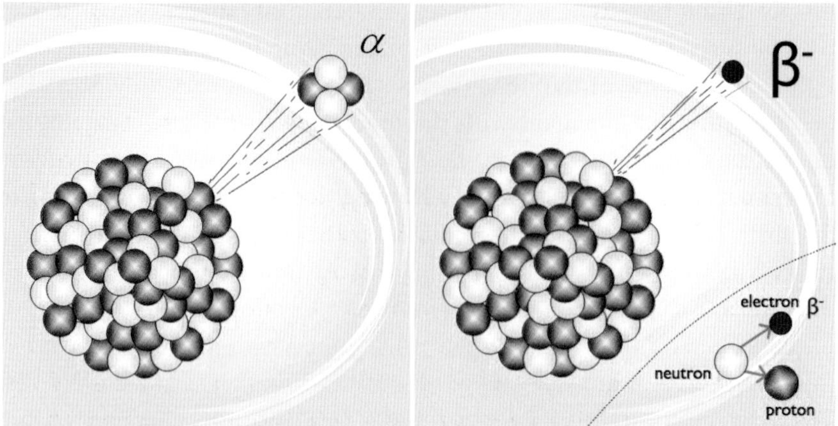

Figure 1.23 In the two most common types of radioactive decay, α-decay and β-decay, the mass of the particles after the reaction was completed was *less* than the mass of the particle that existed before the radioactive decay took place. This discovery led to a refinement of the conservation of mass: that it was only approximate, and only applicable in the larger context of conservation of energy. In many ways, the discovery of radioactivity paved the way for the discovery of mass-energy equivalence, and $E = mc^2$. Image credit: E. Siegel/Wikimedia Commons user Burkhard Heuel–Fabianek under c.c.-generic-2.5.

the eactants and products, were identical. But in the case of radioactive decays, the products always weighed slightly less than the reactants, a sign that something was different from our best current understanding (Fig. 1.23).

In addition, although it had been experimentally established that the speed of light in a vacuum was finite, it had always been assumed that light, like all other waves, needed a medium to travel through. The argument over whether light was a wave or not goes all the way back to the time of Newton (who argued it was not) and Huygens (who argued it was), with phenomena like diffraction and interference pointing to the fact that it was. In the 1860s, Maxwell's electromagnetic theory showed that light, in fact, was nothing more than an electromagnetic wave. Like all waves, all that was needed to complete our understanding of light was a medium for it to travel through. While sound needed matter to compress and rarify, and water waves needed the water itself to travel through, light could travel even through the vacuum of empty space. Since light does exactly this to reach our eyes after being emitted from distant stars, it was assumed that there was a medium inherent to space itself for light to travel through: the luminiferous æther.

But if there was a medium that light traveled through, it should be possible to detect at what speed we move with respect to that medium. If you dropped a rock into a river flowing downstream, the downstream ripples would move faster, while the upstream ripples would move more slowly, as the overall motion of the river would affect them in all directions. As we now knew, the Earth is not stationary, but both rotates on its axis and orbits the Sun, the latter providing a motion through our Solar System at a mean speed of 30 km/s. While that is a small number compared to the speed of light at around 300,000 km/s, it is large enough that a clever experiment could, in principle, detect a motion that small.

Even though the luminiferous æther had never been directly observed or detected, the scientist Albert A. Michelson devised an ingenious experiment to detect its effects. It was based on a simple principle of waves: that waves interfere, and that if light was truly an electromagnetic wave, it must interfere both constructively and destructively with other electromagnetic waves. He developed a special device — the Michelson interferometer — that took a beam of light and split it into two identical beams that would go off in completely perpendicular directions, reflect after traveling identical distances, and return to the same location, where they would then interfere with one another. If the entire apparatus were stationary, as in, if there were *no* motion through the æther, the interference would be perfectly constructive, and you would observe no shift at all. But if you *were* moving through the æther, and based on the Earth's motion through the Solar System, you would expect to be moving at a speed of *at least* 30 km/s, it would take the light just a little bit longer to travel through the direction you were moving relative to it. By 1881, Michelson was ready to perform his experiment for the first time. His original design was unable to detect any shift, but with an arm length of just 1.2 meters, his expected shift of 0.04 fringes was just above the limit of what his apparatus could detect, which was about 0.02 fringes. Michelson performed the experiment at multiple times throughout the day, as the rotating Earth would have to be oriented at different angles with respect to the æther, but still was unable to detect the expected effect.

The null result was interesting, but not completely convincing. Over the subsequent six years, he designed an interferometer 10 times as large (and hence, ten times as precise) with Edward Morley, and the two of them in 1887 performed what is now known as the Michelson–Morley experiment. They expected a fringe-shift throughout the day of up to 0.40 fringes, with an accuracy down to 0.01 fringes. This time, the results were compelling: there was no detectable motion through the æther. Whatever light was doing in its propagation, it was not traveling through a medium that the Earth was moving through (Fig. 1.24).

The solution to both of these puzzles, the loss of mass in radioactive decays and the lack of any detectable motion through the æther, were provided by Albert Einstein in 1905. The reason the mass of the products in radioactive decays was less than the mass of the reactants is because that mass was getting converted into energy, and the amount of energy that was liberated was given by his most famous equation, $E = mc^2$. The mass by itself was not conserved because it was only part of a more fundamental conservation law: the law of conservation of energy. If you were to measure the mass of the products, convert it into an equivalent amount of energy, and combine it with the energy liberated in the reaction, you would find that it exactly matched the equivalent energy in the mass of the reactants. As for the lack of fringe shifts in the Michelson–Morley experiment, that was due to the fact that there was no such thing as the æther, but rather that light always moved at a constant speed relative to any observer: the speed of light in a vacuum. That big idea was one cornerstone of Einstein's special theory of relativity, which changed our way of thinking about three major concepts:

1) The speed of light in a vacuum was a constant to all observers in any and all reference frames.
2) There was no such thing as absolute space or absolute time; anyone in any location moving at any speed would have equal claim to the laws of physics being the same and the speed of light being its exact value.
3) And finally, there was no special medium that light needed to travel through. Simply existing in space and time was enough.

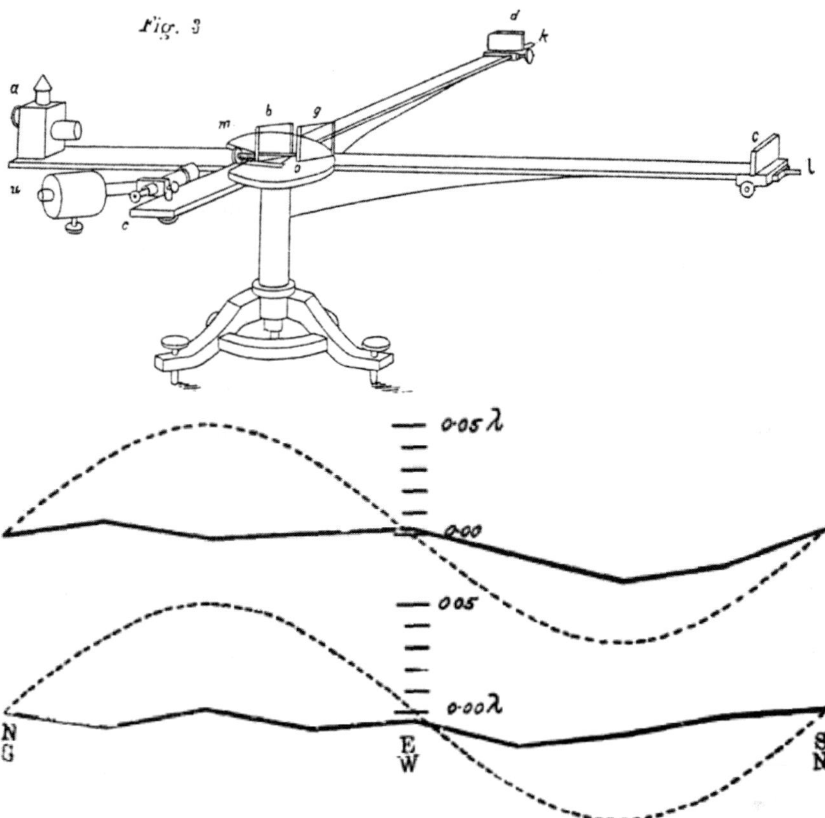

Figure 1.24 The top image shows the design of the Michelson interferometer used to measure the shift in the fringes of light due to the motion through the æther. By splitting a beam of light and sending it down two perpendicular pathways, then reflecting them and recombining them, a shift in the reconstructed interference pattern was predicted. The magnitude of the minimum prediction, shown in dotted lines at bottom, are compared with the null results obtained by their 1887 experiment. The experiment, despite the painstaking predictions and detailed experimental setup, failed to detect the æther. Image credit: Albert A. Michelson (1881); A. A. Michelson and E. Morley (1887). On the Relative Motion of the Earth and the Luminiferous Ether. *American Journal of Science*, 34 (203): 333.

Figure 1.25 Coudé Auxiliary Telescope, ESO. With all the known objects — planets, moons, comets and asteroids — of our Solar System, along with the stars and nebulae beyond, the laws of electromagnetism and Newtonian gravity seemed to account for everything we knew of. Many top scientists of the day predicted that soon, fundamental discoveries in science would cease, as there would be no new laws, particles or properties left to discover. Image credit: Y. Beletsky (LCO)/ESO.

* * *

With these developments in mind, we come to our understanding of the Universe as it was approximately a century ago. It consisted of our Solar System, complete with the Sun, the four inner, rocky worlds, an asteroid belt, the four gas giants and their moons beyond that, and comets originating from the outer Solar System. The Sun was at the center, governed by Newtonian gravity and the laws of motion that could all be derived from that. In addition, there were millions (and probably billions) of stars in the night sky: distant suns that varied in color, mass and lifetime. Some of these stars were close by — just a few light years distant — while others were so far away they appeared to be utterly fixed in their positions, even over timespans of thousands of years. Stars were born in nebulae and died in planetary nebulae and supernovae, not fixed and eternal as we once thought. And, as far as we knew, everything that was in our Milky Way comprised the full extent of our Universe, which was neither contracting nor expanding, but rather existed in a relatively static state (Fig. 1.25).

It was amazing how far we had come, from ancient times to the rapid growth of knowledge that the prior few centuries had brought. But the subsequent 100 years, from 1915 to 2015, would make the prior advances of human history look like chump change compared to the riches that were coming. Over the course of the 20th and into the 21st century, our understanding of the entire Universe changed forever, including what its structure is on the largest scales, what it's made of on the smallest scales, where it all came from and, finally, what its ultimate fate will someday be.

Chapter 2

A Relatively Different Story: How Einstein's Relativity Revolutionized Space, Time, And The Universe

With the serendipitous discovery of Uranus in 1781, we not only realized that our Solar System was larger than we had previously suspected, extending out twice the distance from the Sun to Saturn, but we also gained a new opportunity to test our most cherished physical law: Newton's law of universal gravitation. With centuries of observations of the other planets behind us, we had extraordinary confirmation that they all matched Kepler's three laws of motion:

- They all moved in ellipses with the Sun as one focus.
- They all swept out equal amounts of area in equivalent intervals of time.
- They all orbited with periods related to their semi-major axes (in proportion to the $\frac{3}{2}$ power).

All of Kepler's laws can be derived from Newton's single law of universal gravitation, so if a planet was found to move differently from the predictions of Kepler's/Newton's laws, that would present a problem for our understanding of how gravity behaves. The discovery of a new planet much farther out than all the others gave us exactly that opportunity: to test our most sacrosanct laws in a new, unexplored regime (Fig. 2.1).

Based on Uranus' motion, we quickly figured out how far away from the Sun it was, what its orbital shape looked like, and what its orbital

Figure 2.1 The disk of the planet Uranus against a background of fixed stars. Image credit: Leo Taylor, 2010. More of Leo's work can be found at http://astrophotoleo.com/.

period would have to be. All the observations agreed very well with the predictions of Kepler's laws, except for one: for the first few decades of our observations of Uranus, it appeared to be moving slightly *too quickly* in its orbit, meaning it swept out slightly more area than it was expected to in a given amount of time. Over the next 20 years or so, however, Uranus returned to the predicted speed, sweeping out exactly the right amount of area predicted by Kepler's laws. Many were willing to discount the earlier data if the current data consistently matched the predictions, and all seemed well for a time. But by the 1830s and 1840s, Uranus appeared to slow down further, sweeping out too little area for the time periods observed. The conclusion was inescapable: there was a problem with Uranus' motion after all.

In particular, Uranus did not appear to be following Kepler's second law: it moved more swiftly than the law of gravity predicted, then at the predicted speed, and then more slowly than was predicted. This meant the area swept out in a given amount of time was at first too great, then decreased to the predicted value, and then became too little, again defying predictions. These observations left many scientists wondering if there were not some flaw in Newton's law of gravitation at large distances in the Solar System.

But there was another possibility advanced by a few theorists: if there were one large, massive planet out beyond Saturn, could not there be still *another* one? Uranus takes 84 years to complete an orbit around the Sun, but a planet even more remote than Uranus would orbit our Sun even more slowly than that. Just as Mars is overtaken by Earth in its orbit every so often, a hypothesized *eighth* planet would be overtaken periodically by Uranus in its orbit around the Sun. And if that planet were a massive, gas giant as well, would it be possible for its gravitational force to cause Uranus to accelerate ever-so-slightly from our point of view?

Indeed it could. Picture Uranus, a blue planet some 1.8 billion miles distant, orbiting the Sun in a nearly perfect circle. The planets of the inner Solar System race around their orbits, completing revolution after revolution, while Uranus slowly lumbers through space. Even Saturn, known since antiquity as the "old man of the skies" because of its slow apparent motion relative to the fixed stars, completes *three* orbits in the time it takes Uranus to complete just one.

But now imagine that there is an outer world orbiting the Sun even more slowly, and Uranus begins to approach it. Instead of Newton's law of gravitation being a problem, that very law could instead *be the solution*, as approaching another rather large mass would cause a small gravitational attraction between the two. If the more remote planet were ahead of Uranus in its orbit, Uranus would be accelerated in its current direction of motion, causing it to temporarily move faster than the predictions of Kepler's 2nd law!

As Uranus began to overtake the outer world, Newton's law of gravity would still give it an extra bit of acceleration. However, it would be slightly accelerated *not* along its direction of motion, but rather away from the Sun, a motion practically imperceptible from the perspective of the inner Solar System. It is far easier to mark a planet's transverse speed, as its position relative to the fixed stars changes night-to-night, than it is to measure slight changes in its radial speed. As Uranus closely approaches and overtakes an outer world in its orbit, its transverse speed will be unaffected by the gravity of that extra planet, and so Kepler's 2nd law should appear to work exactly as predicted during this time.

And finally, once Uranus has overtaken the outer world, it ought to be pulled backwards, towards the planet's position behind it. This time, the extra acceleration should work to *slow* Uranus, and so it ought to move with a velocity slightly less than Kepler's 2nd law predicted (Fig. 2.2).

Remember that Kepler's laws are a special case of Newton's, where only two masses — a very large, stationary mass and a much smaller, orbiting mass — are present. Once additional objects (such as planets in the Solar System) start showing up, Kepler's laws are no longer exact, but are approximations to the actual motions of planets. Corrections were used previously by Halley to explain the orbits of long-period comets; perhaps an additional correction was necessary to explain the orbit of Uranus. The known planets and asteroids could not have explained Uranus' orbit on their own, but adding one extra large, outer world could potentially fix the entire problem.

That was the thought process that went through the mind of Urbain Le Verrier, who in 1846, after months of painstaking calculations, determined what orbital properties such a planet would need to have. On August 31, 1846, Le Verrier publicly announced to the French Academy the results of his calculations, determining the mass, semi-major axis and present position of the never-before-seen planet. This marked the first time that the laws of gravity were applied to infer the presence of a mass that had never before been seen.

Le Verrier's prediction was transmitted via letter to the Berlin Observatory, and arrived on the 23rd of September. That evening, Johann Galle and Heinrich d'Arrest searched for Le Verrier's predicted object, and a faint point of light that matched no known star was discovered less than 1° away from Le Verrier's prediction. This was the story of the discovery of Neptune; for the first time, an object's very existence in space was physically predicted *before* it was observed! (Fig. 2.3)

* * *

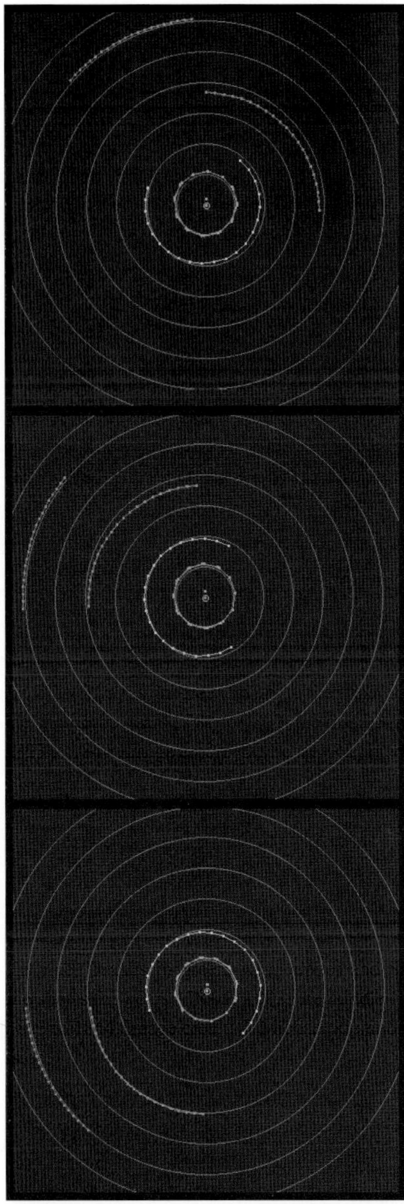

Figure 2.2 Neptune is in blue, Uranus in green, with Jupiter and Saturn in cyan and orange, respectively. The upper frame shows the years 1781 to 1802; the middle frame shows the years 1802 to 1823; the bottom shows 1823 to 1844. Image credit: © Michael Richmond of R.I.T., generated with XEphem, used with permission.

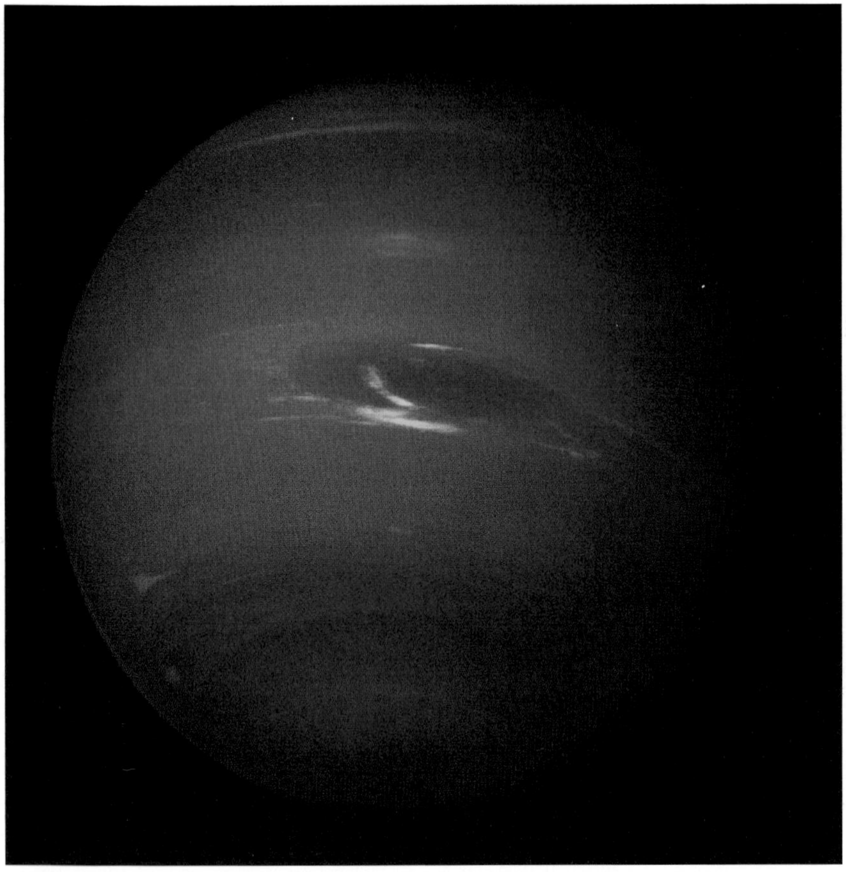

Figure 2.3 Neptune, the eighth planet in our Solar System. Image credit: NASA/Voyager 2 spacecraft, August 20, 1989.

But the story of Uranus' orbit and Neptune's discovery were far from the last unresolved puzzles of our Solar System. The innermost planet, Mercury, also happens to be the most eccentric planet of the eight, which means that the "long" direction of the ellipse it traces out is significantly longer than the "short" (or minor) axis. According to Kepler's laws of motion, that ellipse should be *closed*, meaning that each time Mercury makes a complete orbit around the Sun, it ought to return to its initial position in space: exactly where it was 87.9691 Earth-days ago. The way astronomers would measure this is by tracking the position of Mercury's perihelion over time, or its orbital point-of-closest-approach to the Sun.

With centuries of high-precision observations of Mercury to go off of (back to the 1500s and Tycho Brahe), it was clear that Mercury's perihelion was precessing, or returning to a slightly different position than the one it started at. Of course, Kepler's laws assume three things:

- That Newtonian gravitation is the exact law that governs bodies in our Solar System,
- That Mercury and the Sun are the only two masses that matter in this system,
- And that the position in space that you are observing from is not changing over time.

The second and third of these were very clearly not true: there were seven other planets (and an asteroid belt) to consider, and our observations are from Earth, which moves with its own orbital intricacies. The question would be whether incorporating the presence of the other masses in the Solar System and the Earth's motion through space would account for Mercury's observed precession or not (Fig. 2.4).

Earth's motion through space is actually the largest of these effects that cause the apparent precession of Mercury's orbit. We normally think of a year as the amount of time it takes for the Earth to pass through all four seasons: winter, spring, summer and fall. But we also think of a year as being the amount of time it takes the Earth to make a complete revolution about the Sun. In reality these two definitions of a year — the former being a tropical year and the latter being a sidereal year — are *slightly* different from one another. A sidereal year, it turns out, is 20 minutes and 24 seconds longer than a tropical year. Although this might not seem like much, it means that when you go from midnight on January 1st of one year to midnight on January 1st of the next, the Earth has not moved 360° in its orbit around the Sun, but has come up ever-so-slightly short, at 359.98604°.

This effect adds up over time, so that after 72 years, the difference between a calendar based on tropical years and sidereal years would be a full day, and after 26,000 years, your sidereal calendar would be a full year behind. Because the positions of all astronomical objects are based on their positions relative to the distant stars, using our conventional (tropical) year rather than a sidereal year for timekeeping

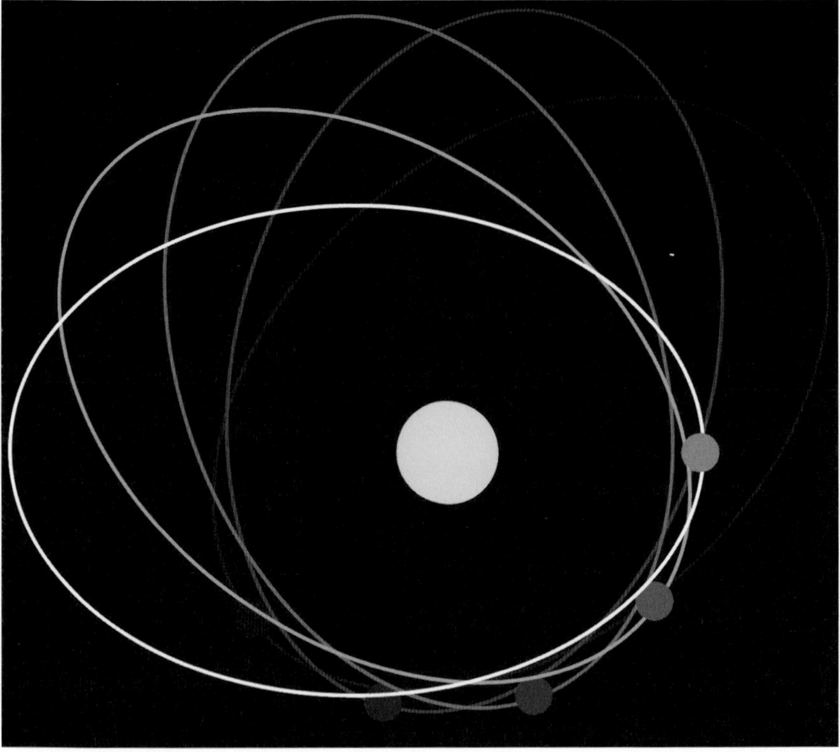

Figure 2.4 Planetary precession refers to the fact that bodies do not return to their exact same location in space with each orbit, but rather that the orbital path slowly rotates around its parent body. The effect shown here is greatly exaggerated. Image credit: Rainer Zenz, from the public domain and retrieved via Wikimedia Commons.

means that the position of Mercury's perihelion will appear to precess by 1.396°-per-century. To deal with small differences in apparent position, astronomers divide up a degree into 60 arcminutes (1° = 60′) and each arcminute into 60 arcseconds (1′ = 60″), so over the course of a century, simply observing Mercury from the Earth will cause an apparent precession of 5025″-per-century.

In addition to this effect, there is also the gravitational impact of the other planets on Mercury's orbit. Just as Uranus had its orbit deviate from Kepler's laws thanks to Neptune's relatively large mass and close proximity to Uranus, the other planets, *particularly* the closest and most massive ones, have a substantial impact on the precession of Mercury's orbit. It was actually Le Verrier — the same Le Verrier who predicted Neptune's mass and position — who first worked out the contributions of the other planets and asteroids to the orbit of Mercury. Although his values were slightly off (primarily due to uncertainties in the masses of some of the planets), his painstaking calculations were brilliant and robust. Venus was determined to be the most impactful planet, followed by Jupiter, Earth, Saturn and Mars (in decreasing order). Even Uranus and Neptune contributed slightly! With the proper, present-day values for the masses and positions of the planets put in, the other bodies in the Solar System account for an additional precession of 532″-per-century, on top of the 5025″-per-century caused by the precession of the equinoxes.

This predicted amount of precession for Mercury's orbit — 5557″-per-century — was *very* close to the observed value of 5600″-per-century. But that very small difference of just 43″-per-century, just 0.77% of the total amount, was a real problem. It could not have been an error in the observations, as there were simply too many high-precision (and high-accuracy) measurements of Mercury's position for too long a time. It could not be that another world out near Neptune (or beyond) was causing this effect, as the contribution of those distant worlds would be insufficient to cause such a large effect; such a distant world would have to be massive enough to ruin the orbit of Neptune. And so scientists started searching for something that could contribute that extra 43″-per-century.

The planet Venus could have been slightly heavier than expected, for one. If the mass of Venus were just 14% higher than we would have expected, that could explain Mercury's orbital precession just fine. But if that were the case, it would be ruinous for *Earth's* orbital precession, something that was known far too precisely for this to be a feasible option. It was then thought that if there were a world *interior* to the planet Mercury in the Solar System (or a series of small worlds), that could explain the additional 43″-per-century. The idea was so popular that the hypothetical planet even acquired a name: Vulcan, after the Roman god of fire (Fig. 2.5). But exhaustive searches — both amateur and professional — failed to turn up any bodies interior to Mercury, much less one substantial enough to cause the desired precession. A final alternative (that could not be tested observationally at the time) was put forth by Hugo von Seelinger, that the Sun's corona was quite

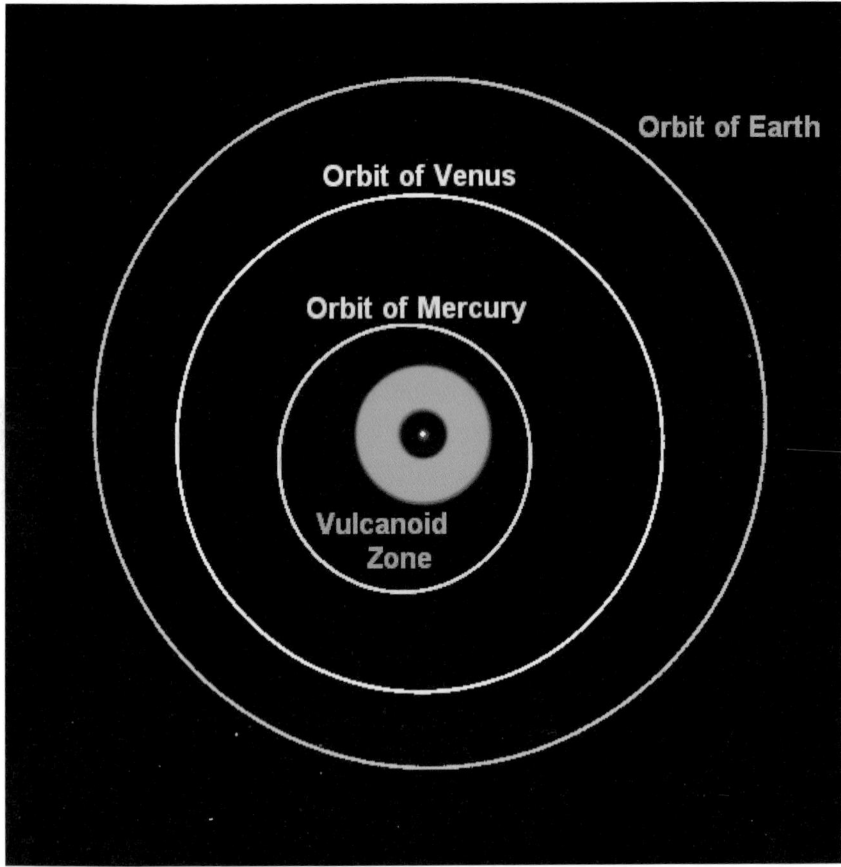

Figure 2.5 This image shows the possible locations of the hypothetical planet Vulcan, whose existence was ruled out in the late 1800s. Image credit: user Reyk of Wikimedia Commons, released into the public domain.

massive, and that the additional mass caused the anomalous precession of Mercury's orbit.

Without any viable candidates for additional masses that could cause this precession, it became imperative to question whether Newton's law of universal gravitation was, in fact, the correct theory of gravity. Newton's law insisted that objects were attracted to one another in proportion to their masses and inversely proportional to the distance between them squared, or to the power of 2. Simon Newcomb and Asaph Hall noted that if objects were instead attracted to one another inversely proportional to the distance between them to the 2.000000157 power, that would explain the precession of Mercury's perihelion. But this solution was very *ad hoc*, and offered no predictive power beyond solving the one problem it was designed to solve. (It is now known that the Newcomb–Hall gravitational law disagrees with observations of Venus' and Earth's precessions, but this could not be tested at the time.)

But there was a hint that going beyond Newton's laws would, in fact, solve this problem. You see, according to Newton, there was no special behavior for matter as it traveled close to the speed of light; speeds could be arbitrarily fast and the adherence to Newton's laws of motion would not be altered. But Einstein's special theory of relativity dictated that not only was an object's motion limited by the speed of light, but that lengths would contract and time would dilate for an object the closer it moved to that limiting speed. While all the objects in our Solar System move relatively slow compared to that speed, the planet Mercury is the swiftest, orbiting the Sun at an average speed of 47.87 km/s, or 0.01597% the speed of light. While this is negligible for most measurable effects on shorter timescales, over the course of 100 years these effects would cause an additional precession of 7″-per-century, as determined by Henri Poincaré in 1908. While Special Relativity could not be ultimately responsible for the full discrepancy of 43″-per-century, it was very much a contribution in the right direction, and one that went beyond Newtonian physics. There was a suspicion among many that the solution to this anomaly would take us beyond Newtonian gravitation as well.

* * *

Newtonian gravitation was simple and intuitive: you take any two masses in the Universe, put them in any two locations anywhere you like, and they will attract one another with equal and opposite forces. If you give Newton's law of gravitation the positions and masses of every object in the entire Universe, it can tell you, *exactly*, what the gravitational force on every single mass is. It is a simple, straightforward and incredibly powerful law of nature, describing everything from the motions on Earth's surface to that of the moons, planets, comets and stars (Fig. 2.6).

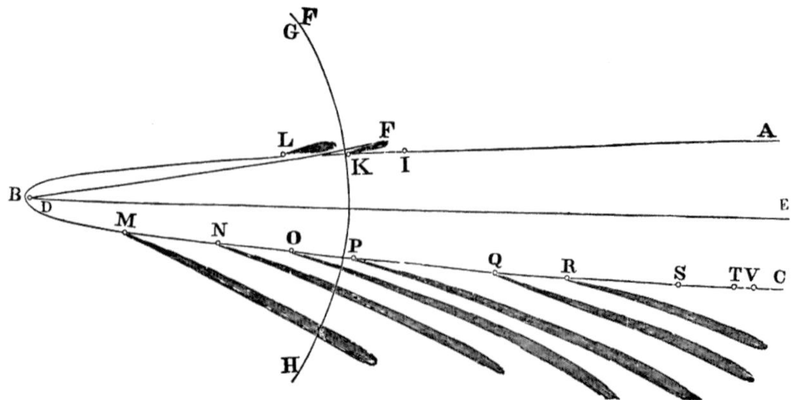

Figure 2.6 This shows the orbit of the great comet of 1680, fitted to a parabola according to Newton's laws. Image credit: Isaac Newton's *Philosophiæ Naturalis Principia Mathematica*, original publication 1687; this work published 1846 by Daniel Adee.

Yet, despite how frequently Newton's laws gave answers that agreed with observations, when it came to the particulars of Mercury's orbit, it was also clearly insufficient, if not outright wrong.

Ever since Newton discovered this law in the 1600s, the motions of all objects governed by gravitation, ranging from pendula to projectiles to ocean tides to the planets and moons, were successfully predicted by Newtonian mechanics. But if Mercury's orbit failed to match what was predicted, perhaps this failure was only the tip of the iceberg. Perhaps there was a fundamental problem with Newton's whole conception of how the Universe worked.

And there were, indeed, some troubling aspects to Newton's gravity, if you thought about them. For example, what would happen if the Sun simply ceased to exist? If somehow, it were just plucked completely out of the Solar System and removed from the Universe entirely? Because the light generated and released by the Sun has to traverse some 150 million kilometers (93 million miles), here on Earth we would not know of its disappearance for slightly more than eight minutes. But what about its *gravity*? Would the Earth (and all the planets) just fly off in a straight line, like a twirled poi ball that was suddenly released? Or would it continue in its elliptical orbit for some time, until that change in the gravitational message reached it? Newton's gravity suggested the former,

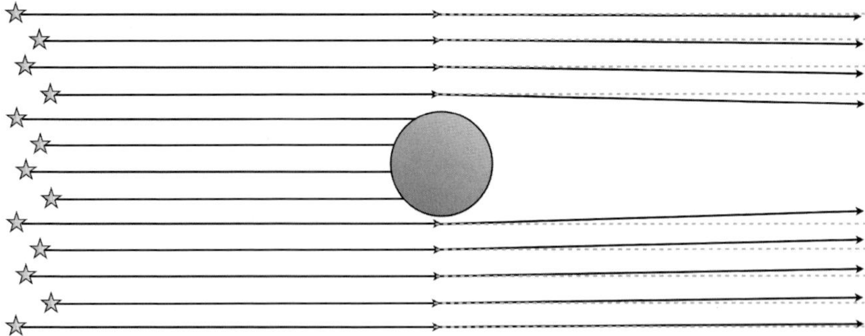

Figure 2.7 If light, being massless, were unaffected by gravitation, light rays that passed near a star would remain unbent (dotted lines), while if it were to behave as other forms of energy do, it should experience gravitation according to Newton's laws (solid lines), with the photon's mass equivalent (E/c^2) substituted for its mass, and follow the arrows. Newton's gravitation is ambiguous. Image credit: E. Siegel.

but made no definitive predictions one way or the other. This was a problem unaddressed by Newtonian gravity.

With the discovery of Special Relativity, there were a few aspects that were even more troubling than the presumed instantaneous action of gravity. If, as Einstein's famous $E = mc^2$ suggested, mass was just one form of energy, does that mean other forms of energy experienced gravitation as well? Could a beam of light attract matter, and would it be attracted by matter as well? Newton's gravity again had nothing to say about this (Fig. 2.7).

Furthermore, if one mass was in motion relative to another one, would the gravitational force between them be the same as if both masses were stationary? We know that lengths contract when an object moves, so if one is in motion while the other remains stationary, what is the gravitational force between these objects? Would the gravitational force between them depend on the contracted distance (for the moving observer) or the uncontracted distance (for the stationary one) and would the force of attraction depend only on the mass of the object, or the mass plus the equivalent amount of kinetic energy?

And finally, as a mass changes position relative to another one, the gravitational force acting on it will change over time. When two electrically charged particles do this, they emit radiation. If mass is just another word for "gravitational charge," is there any type of gravitational radiation that is emitted when two masses move relative to one another?

These are all good questions, and they are important questions to ask of the theory of gravity set forth by Newton. Why? Because it is not just that Newtonian gravity gives the *wrong* answer to these questions, it is that in most cases, it does not give *any* answer to them at all! Newtonian gravity is rooted in the assumption that space is a fixed quantity, and that if you have an object in that space, it will exert a gravitational force infinitely and instantaneously on every other location in space. Although it was never stated by Newton, his theory of gravitation implicitly assumed that just like space, time is also a fixed, absolute quantity.

Yet, we know that both distances in space and the passage of time are dependent on an object's relative motion: *that* is what is meant by the phrase "relativity." Observationally, Newton's gravity ran into a problem with the orbit of Mercury. But theoretically, the big problem of Newtonian gravity is that its very conception is incompatible with the observationally and experimentally confirmed theory of Special Relativity!

Einstein's great hope was that these two important phenomena — gravitation and the principle of relativity — could be generalized to be made compatible with one another. But to reconcile these two concepts would require a revolution in physics. It would require a general theory of relativity.

* * *

Perhaps the biggest conceptual problem with Newtonian gravity was the idea of *action-at-a-distance*. How can two objects — regardless of how far away they are from one another — affect one another and exert forces on each other if they never even come into contact with one another? Newton simply named this effect *action-at-a-distance*, and left it at that. But Einstein was wholly dissatisfied with that explanation, and sought to find a way to explain exactly how this gravitational interaction took place.

The big idea of Special Relativity was that all objects — including you and I — do not just move through space and move forward through time; our movement through space and our movement through time are inextricably linked together. When one object is in motion relative to one another, clocks (and hence, all measures of time) appear to run differently for the moving object when viewed from the stationary one, and vice versa. This led to the new concept of *spacetime*.

General Relativity went a step further, and sought to address the question of just what the relationship was between spacetime and all the mass, matter and energy in the Universe! Once we realized that all objects and particles in the Universe do not just move through space and time but through spacetime, there were two big questions that needed to be answered:

1) How do mass, matter and energy affect this spacetime that they exist in?
2) How does this spacetime — full of mass, matter and energy — affect the motion of everything in it?

It took Einstein eight years of full-time work (from 1907–1915, and even he needed help from a number of top mathematicians) to figure out a self-consistent physical and mathematical framework for General Relativity. When it was complete, a revolutionary picture of the Universe emerged.

Every particle, mass and quantum of radiation in the Universe plays a role. Without them, spacetime would be perfectly flat and there would be no gravitation; Special Relativity would suffice. But in a Universe with even *one* particle — one point-like instance of energy — spacetime would no longer be so simple. Instead, the presence of any form of energy, whether matter, radiation, or something more exotic, actually *alters* the structure of spacetime itself (Fig. 2.8).

In other words, unlike a Newtonian Universe, space and time are no longer static, fixed, unchanging quantities. Spacetime can expand, contract or remain static, dependent on a number of variables. An object at rest will not necessarily remain at rest and an object in motion will no longer remain in constant motion. Instead, all objects will follow a path dictated by the curvature of spacetime, where the curvature of spacetime is determined from the presence of every component of mass and energy present in your Universe.

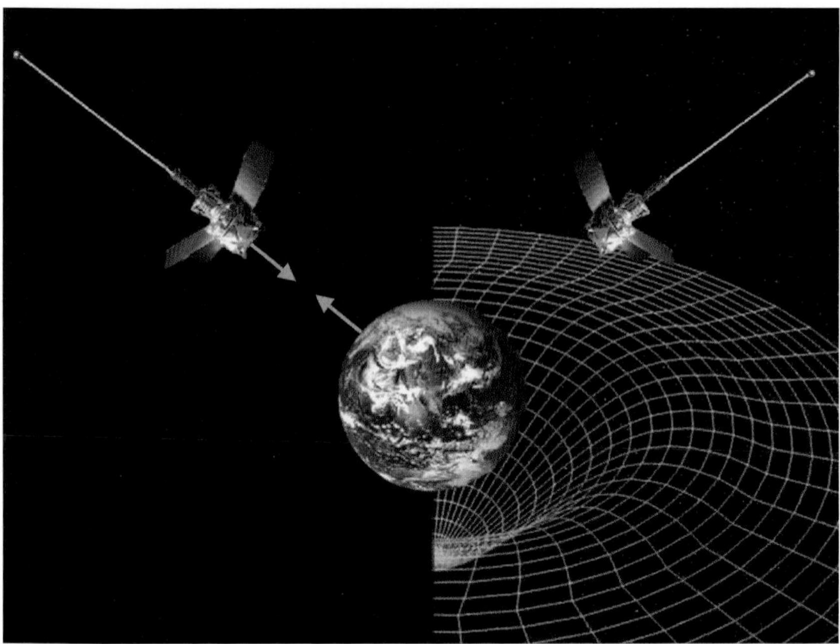

Figure 2.8 Instead of a Newtonian conception of gravitation (L), where forces are instantaneous and occur at a distance, Einstein's relativity (R) states that mass and all forms of energy warp the fabric of spacetime itself and then all particles follow the shape of that curved spacetime. Image credit: NASA/Gravity Probe B; modifications by E. Siegel.

Planets would no longer orbit the Sun because of some invisible, infinitely-reaching action-at-a-distance, but rather because the Sun dominates the curvature of spacetime within our Solar System. And the gravitational force that we feel would no longer be instantaneous, but would be limited by the speed of light. If you visualize our Solar System as a two-dimensional sheet, stretched flat, every major source of energy (i.e., the Sun, planets, moons, asteroids, etc.) deforms that sheet, which we experience as gravitational attraction. Reality is a little more complex in that it has three spatial dimensions, but the same principle applies: everything in the Universe that contains mass and/or energy will alter the fabric of your spacetime, and affect how every other object experiences gravitation.

* * *

Before it was ever presented to the public, there were three challenges that this new theory — Einstein's General Relativity — needed to rise to:

1) It needed to demonstrate that it encompassed Newtonian gravity, and that under certain special cases or approximations, Newton's law of gravity could emerge from General Relativity.
2) It needed to account for the anomalous precession of Mercury's perihelion, *without* causing any other conflicts with known observations.
3) And finally, and perhaps most importantly, it needed to predict something new and novel: something that could be tested experimentally.

If ever you want to replace an established scientific theory with a new one, these are the three basic burdens-of-proof for your new theory: you must show that it reproduces what the previously established theory predicts over a certain range of validity, you must show that it successfully solves the problem it was intended to solve, and you must make a new prediction that can be either verified or contradicted.

The first part was the easiest: if you placed a single point mass in empty spacetime, you would wind up with a certain amount of curvature to your spacetime's fabric. If you then asked about the behavior of any test particle, whether at rest or in motion, you would get the same behavior as predicted by Newton's laws so long as you were *sufficiently far away* from the mass that caused the curvature. In other words, where distances were large relative to the masses involved and the force of gravity was sufficiently weak, Newton's laws of gravity emerged naturally from General Relativity. Both General Relativity and Newtonian gravity work extraordinarily well for predicting, for example, the motion of the Earth around the Sun and how objects accelerate at the Earth's surface. For these well-understood cases, the two theories — Newton's old laws and Einstein's new proposal — were in near-perfect agreement, and certainly both were consistent with the best observations we had at the time (Fig. 2.9).

But there were also small *differences* between General Relativity and Newtonian gravity, and these differences were most important when objects were close to a large, dense mass and gravitational fields were

Figure 2.9 For weak gravitational fields, the predictions of Newton's gravity (L) and Einstein's gravity (R) are indistinguishable for most applications. Image credit: Norbert Bartel (2004). From his film, *Testing Einstein's Universe: The Gravity Probe B Mission.*

stronger. This meant that General Relativity *could* be a reasonable explanation for Mercury's anomalous precession; the regime where the theories differed aligned with the location in our Solar System where they needed to differ. The critical calculation would need to be done explicitly to find out whether its predictions agreed with the observations.

This was no easy calculation to do, mind you. General Relativity may be an incredibly powerful framework *in theory*, capable of describing how the entire Universe is shaped and curved simply by knowing the properties of all the matter and energy in it. But in practice, it is incredibly difficult to solve for even the simplest, most idealized cases. When Einstein first published his theory in 1915, it took a little over a month for the first exact, non-trivial solution to emerge: the solution for a single, stationary point-mass in an otherwise empty spacetime, discovered by Karl Schwarzschild. If that same mass is also allowed to rotate, a solution for that went undiscovered for nearly half-a-century, until Roy Kerr solved it in 1963. In fact, only a dozen-or-so exact solutions are known even today, a full century after Einstein. If you want to describe a spacetime with two orbiting point masses in it, for example, that solution has not yet been found, and the best one can do is calculate it by a series of numerical approximations.

That is exactly the case for a planet orbiting the Sun. For that system, approximate, numerical calculations are still a necessity today, just as they were in Einstein's time. By treating the Sun as a stationary, non-rotating mass responsible for the curvature of spacetime and a planet as a massless object traveling through that spacetime, a first-order approximation will give identical predictions to Newtonian gravity. However, the *second*-order approximation gives an additional precession term, which can only be explicitly determined with knowledge of the specific orbital properties of the system in question. For the planet Mercury, that came out to an additional precession of 43.0″-per-century, a value not only calculated by Einstein himself, but one that is all the more remarkable considering that the value of the anomaly had fluctuated from 38″ to 45″-per-century from Le Verrier's time to Einstein's! Yet Einstein's calculated value still stands today, and explains not only the anomalous precession of Mercury, but also the now-known post-Newtonian contributions to the orbits of Venus (at 8.6″-per-century), Earth (at 3.8″-per-century) and Mars (1.4″-per-century), all in perfect agreement with observations.

But in order to distinguish itself from other alternatives and establish itself as a superior scientific theory, it was not enough to merely include Newtonian gravity and account for one new observation. It needed to meet the third (and most difficult) challenge for a new scientific theory: to predict a new, novel effect that could be detected experimentally! Although Einstein was unskilled in observational astronomy, he suggested two possibilities for where to look to verify his theory of General Relativity: at the predicted shift in the wavelength of light as it moved into either a deeper or more shallow gravitational field, or at the predicted deflection of starlight as it passed by a nearby, massive object (Fig. 2.10).

While the first test was also, arguably, predicted by Special Relativity (by applying the equivalence principle), the latter was a pure prediction of General Relativity, due to its unique gravitational properties. An observation of whether light was deflected according to Einstein's predictions could easily distinguish General Relativity from all other competitors. The key would lie in finding the right conditions to test this effect.

* * *

When a massive object passes near another mass in space, Newton's law of gravitation tells you what the force of attraction will be at every single

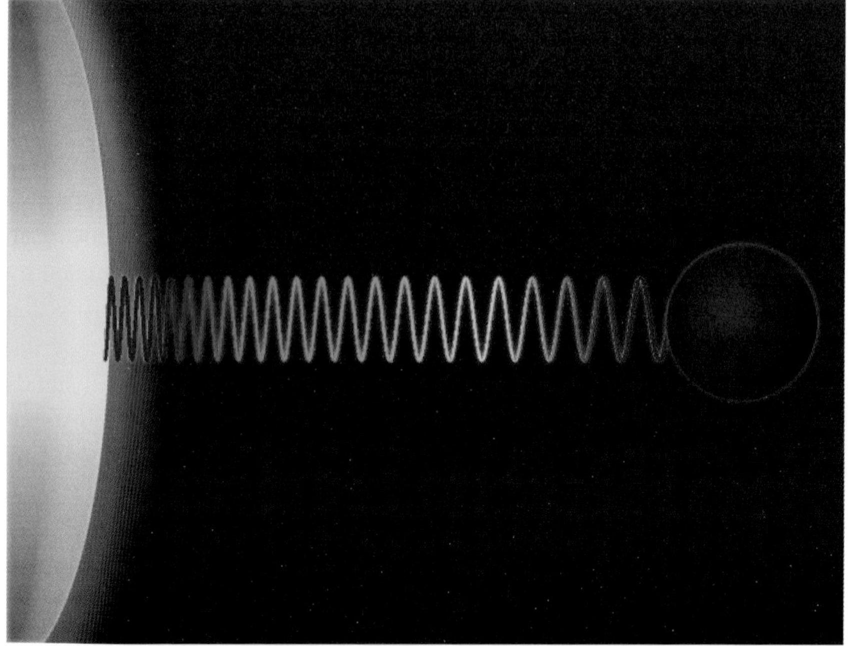

Figure 2.10 As photons leave a region where space is strongly curved for one that is less curved, they redshift, losing energy; as they enter a region where the curvature is stronger, they blueshift, gaining energy. Image credit: Wikimedia Commons user Vlad2i, c.c.-by-s.a.-3.0, under the GFDL.

point along the particle's trajectory. But what about *massless* objects that move at the speed of light?

As far as we are able to tell — both in Einstein's time and even today — every quantum and ray of light is completely massless, and always moves precisely at the speed of light. Strictly speaking, in Newtonian gravity, there is no force on any object without mass. However, since the discovery of mass-energy equivalence (via $E = mc^2$), we could assign light an *effective* mass based on this relation. Once we do that, it would be a straightforward exercise to calculate how much light ought to be deflected by when it passes by another mass.

We can also do the equivalent calculation for light deflected by a mass in General Relativity. Unlike in the Newtonian case, there is no need for any additional assumptions, since massless objects moving at the speed of light fit naturally into Einstein's theory of gravity. But what is really fantastic — and why Einstein suggested it in the first place — is that this is one case where the two different theories give *dramatically* different predictions. Take, for example, how starlight would be affected by the most massive object in our Solar System: the Sun.

When starlight travels across interstellar space, leaving the star that emitted it, traveling at light speed across the galaxy, and arriving years later at your eye, you normally think of that light as traveling in a straight line. And that is mostly true, so long as there are not any extra significant sources of gravity near the photon's light-travel path. But imagine a star that, tonight, would appear exactly 180° opposed to the Sun's position. The star is visible because the Sun is on the opposite side of the Earth from where the star is. But six months from now, the Sun and the star will still be in roughly the same place, but the Earth will have migrated to the other side of the Sun! Now, you might think, you would not be able to see that star at all, since the Sun will be in the way of the line-of-sight from the star to your eye. Furthermore, stars are not visible during the day, anyway.

But two effects conspire to make the observation of this star possible. First, dependent on which theory of gravity is correct, light from the distant star that would otherwise miss the Earth can instead pass nearby the limb of the Sun *and be bent* by the Sun's gravity, where it will then arrive at Earth after all. The star will *appear* to be in a different location than normal, relative to the other background stars, due to this gravitational deflection of light. While Newton's theory, based on the effective mass one can give to light, predicted a small but non-negligible deflection of 0.87″ at maximum, Einstein's theory predicted *twice* that number: 1.75″. And this could be tested under the one condition where stars are visible during the day: during a total solar eclipse (Fig. 2.11).

Total solar eclipses happen every year or two here on Earth, and this would be the ideal testbed for Einstein's General Relativity. The eclipse of February 3, 1916, came too quickly on the heels of Einstein's publication and would pass almost entirely over the ocean, making it a poor candidate to test the new theory. The total eclipse of June 8, 1918, was the next great candidate, as the Moon's shadow would pass directly over many parts of the United States, and teams of physicists and astronomers made extensive

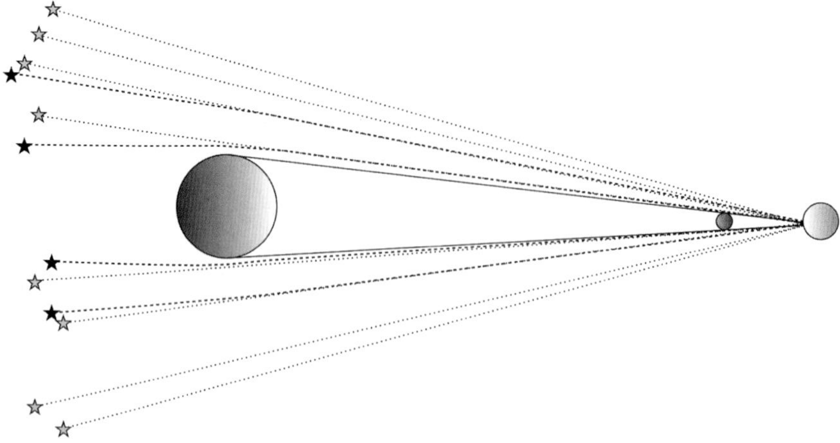

Figure 2.11 During a solar eclipse, the Moon blocks the Sun's light, rendering the stars visible during the day. However, the Sun is also a great source of mass and hence, is capable of significantly curving space. The stars that appear very close to the limb of the Sun (in black) will have their light-paths significantly bent, and will appear shifted to the location shown by the dotted lined. Stars farther away will have their positions unaffected. Image credit: E. Siegel.

plans to observe it. Unfortunately, the clouds would not cooperate, and the opportunity to discern whether Newton or Einstein had it correct was lost. The world would have to wait one more year, until May 29, 1919, which happened to be one of the *longest* total solar eclipses — at 6 minutes and 51 seconds — of the 20th Century.

This eclipse was also a fortuitous candidate for testing these two theories, as there would be many bright stars visible around the location of the Sun during totality. Additionally, Arthur Eddington, one of the few astronomers with the mathematical skills necessary to understand Einstein's theory, had the academic and political clout to make the critical observations. Along with Frank Watson Dyson, he organized two expeditions — one to Brazil and one to Principe — to precisely measure the deflection of starlight by the Sun.

Although observing conditions were less than ideal in both locations, the teams persevered. Andrew Crommelin, leading the expedition in Brazil, had to resort to his backup telescope due to temperature issues, while Eddington faced a thunderstorm earlier in the day and had to make his observations through the clearing clouds. Nevertheless, usable photographic plates were gathered at the critical times. By analyzing the positions of the stars relative to their locations in the absence of the gravitationally deflective Sun, Eddington and his collaborators concluded that the deflection of starlight due to the Sun was 1.61″ ± 0.30″, in excellent agreement with Einstein's predictions and well outside the allowed predictions of Newton's theory (Fig. 2.12).

All over the world, newspaper headlines hailed the success of Einstein's new theory of gravity. Physicists were more measured, as they were keenly aware of the difficulties associated with Eddington's observations, where some of the data was left out for technical reasons. Nevertheless, each subsequent eclipse confirmed Eddington's conclusions, and modern analyses have found that Eddington's results and conclusions were neither cherry-picked nor serve as an example of confirmation bias; they were the best any scientist could have done under those conditions. Newtonian physics was still a highly valid approximation under a vast array of conditions, but could no longer be considered the best physical description of gravitation known to the world. That honor now belonged to General Relativity, thanks to the new — and newly confirmed — prediction concerning the deflection of light by massive bodies (Fig. 2.13).

As Eddington himself wrote years later

Oh leave the Wise our measures to collate
One thing at least is certain, LIGHT has WEIGHT
One thing is certain, and the rest debate —
Light-rays, when near the Sun, DO NOT GO STRAIGHT.

* * *

Perhaps because it so drastically changed our picture of the Universe, or perhaps because it runs so counterintuitive to our everyday notions of fixed space and a constant passage of time, General Relativity has, over the past century, been one of the most frequently challenged and most thoroughly scrutinized theories in all of modern science. Yes, it achieved all of the successes of Newtonian gravity, plus explained the orbital anomaly of Mercury *and* made new — and verified — predictions about the bending of starlight by the Sun. But there are many other new

Figure 2.12 The positions of the stars are marked with horizontal lines, and were measured to have shifted by the amount predicted by Einstein's theory, in disagreement with either no deflection or a Newtonian deflection. Image credit: F. W. Dyson, A. S. Eddington, and C. Davidson, (1919). A Determination of the Deflection of Light by the Sun's Gravitational Field, from Observations Made at the Total Eclipse of May 29, 1919, *Philosophical Transactions of the Royal Society of London*. Series A, Containing Papers of a Mathematical or Physical Character (1920): 291–333.

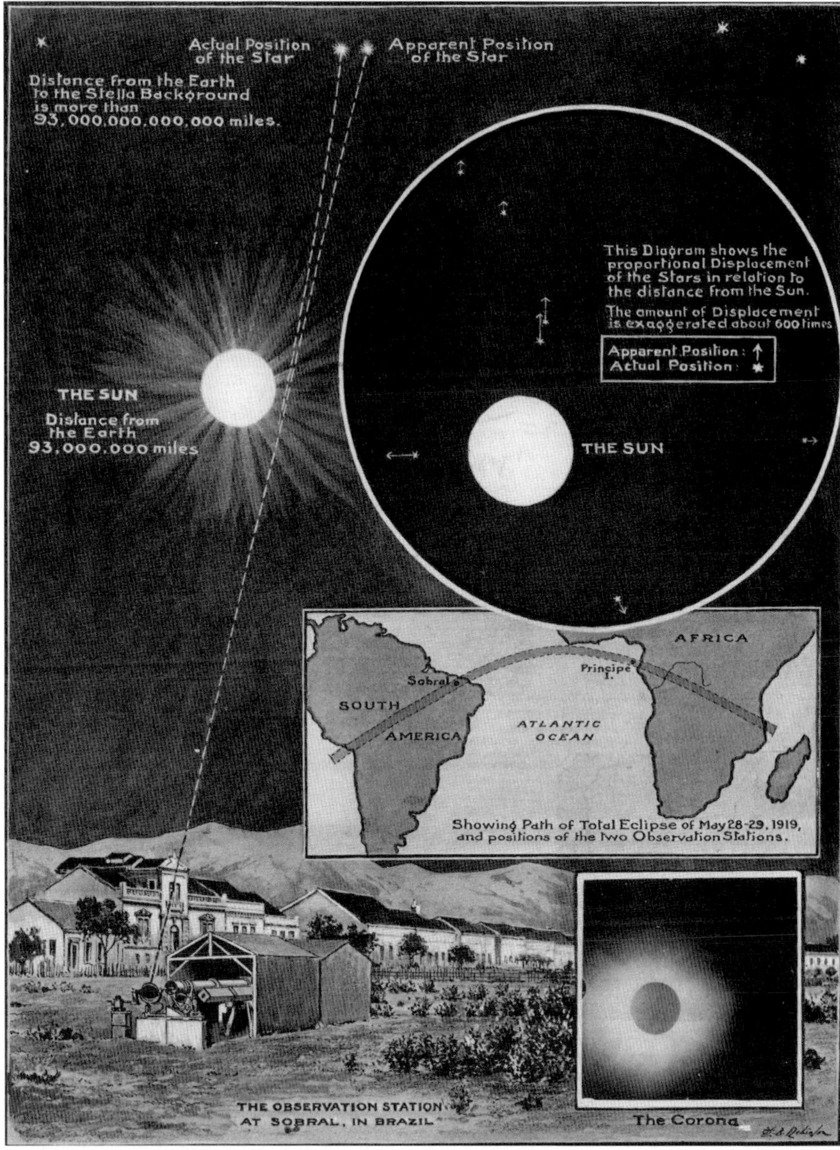

Figure 2.13 The Illustrated London News, November 22, 1919, p. 815. Image credit: W.B. Robinson, from material supplied by Andrew Crommelin.

predictions that come along with this new theory as well, and those need to stand up to observations and experiments, too. If we are going to try and understand our Universe in the context of this new framework, it is up to us to make *sure* it is correct!

The last prediction that Einstein himself made about General Relativity was that light itself, as it moved deeper into a gravitational field, would gain energy. And conversely, if it tried to escape from a gravitational field, it would lose energy while doing so. This gain-or-loss of energy would show up as a shift in the wavelength of the light itself, where a gain in energy corresponded to a shift towards bluer (or higher frequency) light, while a loss would cause the light to shift towards a redder wavelength (or lower frequency). While this gravitational redshift or blueshift was never successfully tested during Einstein's lifetime, it was finally confirmed in brilliant fashion in 1959 by the Pound–Rebka experiment.

Because of the way atomic nuclei work, electron transitions within atoms, from excited states to ground states, can only occur at very specific energies, which mean they can emit and absorb photons of very precise wavelengths. If one atomic system is made to emit a photon of a particular wavelength, then an identical system should be able to absorb that photon, so long as the wavelength of that photon has not changed. But if the wavelength *has* changed, such as by falling deeper into (or climbing out of) a gravitational field, the light will fail to be absorbed by the new atom.

This prediction was tested by holding the absorbing atomic system at a higher altitude — and hence in a less-curved region of spacetime — than the emitting one. As was expected, the atomic system failed to absorb the photon. However, if the emitting atom was in motion at a very particular speed when it emitted the photon, the extra shift in wavelength due to that motion (i.e., a Doppler shift) could cancel out the gravitational wavelength-shift predicted by Einstein's theory. Lo and behold, using a technique discovered in 1958 known as the Mössbauer effect, the atom on the receiving end was once again able to successfully absorb the photon. When the experiment was concluded, it showed that the motion of the emitting atom was exactly what General Relativity predicted was necessary to achieve the desired outcome (Fig. 2.14).

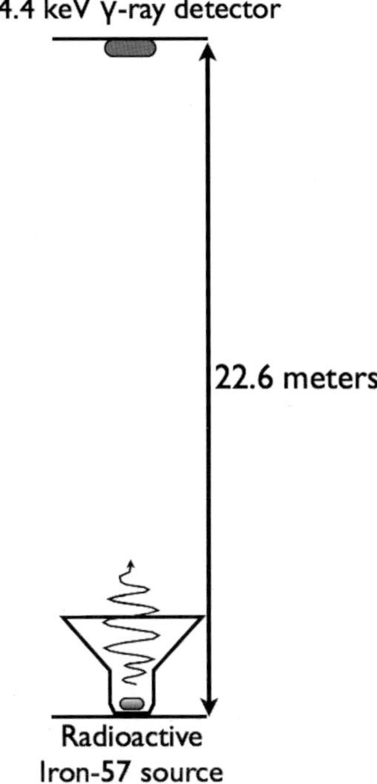

Figure 2.14 If the source — radioactive Iron-57 — emits a photon of a given energy and a detector is placed at a height of 22.6 meters, the gravitational redshift will be substantial enough that the detector would not absorb that photon. But if an additional velocity "kick" is given to the photon by hooking the source up to a speaker cone, the absorption will occur. This was the first verification of the gravitational redshift being of the exact magnitude that Einstein's General Relativity predicted. Image credit: E. Siegel.

Since Einstein's time, more subtle effects of General Relativity were predicted, including a relativistic time-delay of photons when they passed by a massive body, known as the Shapiro time delay. Radar reflections done between various planets in the Solar System have verified this, with the best results coming from the Cassini mission orbiting Saturn, whose results agree with General Relativity to an astounding 99.998% accuracy! Other predictions include an explicit prediction for the delay in GPS satellites, a slight distortion in the apparent positions of stars in the night sky, relativistic corrections for both spinning and moving objects in a gravitational field, the gravitational decay of very massive objects closely orbiting one another, and — perhaps most spectacularly — the bending of light from ultra-distant

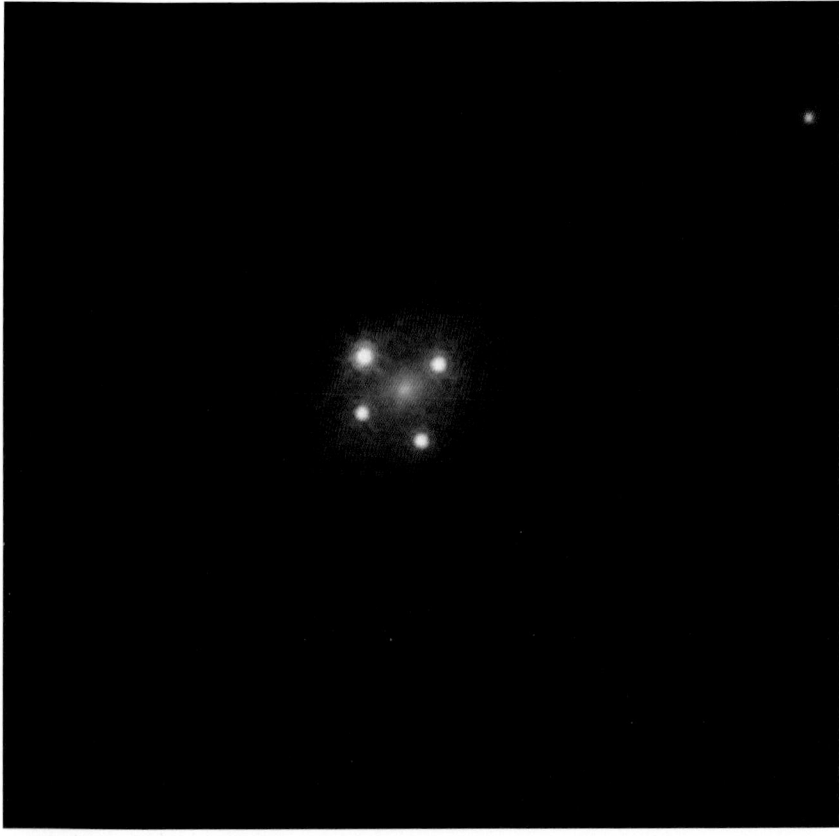

Figure 2.15 This incredible example of gravitational lensing shows *four separate images* of the same quasar, QSO 2237+0305, as it is gravitationally lensed by a foreground galaxy. The phenomenon is known as Einstein's Cross, and is located roughly 8 billion light years away. Image credit: ESA/Hubble and NASA.

objects in space by an intervening mass known as gravitational lensing. All of these predictions have been scrutinized either experimentally or observationally, and General Relativity has been the only gravitational theory to stand up to each and every test. As far as we have been able to tell, General Relativity — with spacetime containing (and curved by) matter-and-energy — is the theory of gravity that describes our physical Universe (Fig. 2.15).

* * *

So, then, what did this *mean* for our Universe? At the time, remember, we didn't know about galaxies, the Big Bang, or the expanding Universe. The only objects we knew about were stars and a variety of extended, deep-sky objects, all of which appeared to be roughly uniformly distributed in our sky. Yes, there were slightly more star clusters aligned with the Milky Way than in other parts of the sky, but for the most part, all the objects beyond our Solar System — particularly the stars — appeared in roughly equal numbers in all directions. This means the observable Universe appeared to be **isotropic**, or the same in all directions, as well as **homogeneous**, or the same at all locations in space. It is possible to be one without the other — a hollow sphere is isotropic but not homogeneous, while a uniformly westward wind is homogeneous but not isotropic — but our Universe appears to be both, simultaneously, everywhere and at all times.

Under the old picture of Newtonian gravity, there were already some puzzles related to this. For example, if stars were almost perfectly evenly distributed throughout space, that would be an *unstable* configuration according to Newton's laws. In a timespan of just a few thousand years, tiny gravitational imperfections would grow, causing the stars to attract the ones closest to them, forming large clumps of star clusters separated by great voids of interstellar space. And yet, our night sky shows us that stars clumped and clustered together in large groups were far and away the rarity, with isolated systems — like our own — being the norm. In other words, an isotropic, homogeneous Universe is *unstable* under Newtonian gravity.

You might think that General Relativity — just as it was able to meet many challenges that Newtonian gravity could not — might once again come to the rescue. But in Einstein's theory of gravity, this problem actually gets *worse*! In General Relativity, even if you start with an *arbitrary* distribution of mass, you will find that it is unstable against gravitational collapse. That means the masses could have begun in a configuration like points in a lattice, in a sphere, pyramid or cube, or they could be clumped together or uniformly smooth; it is truly arbitrary. So long as those masses are moving slowly (or not at all) compared to the speed of light, you can calculate what's going to happen to them in the future. And what General Relativity tells you is, surprisingly, that not only will these masses wind up

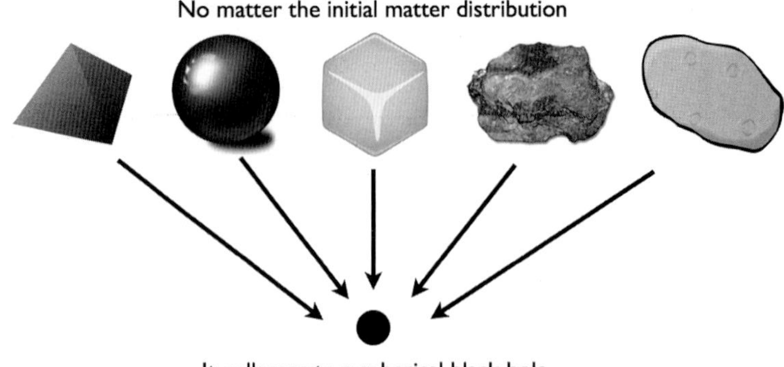

Figure 2.16 So long as the only force at play is gravitation, any initial configuration of mass, originally at rest, will collapse to form a spherical black hole. Image credit: E. Siegel.

in a massive clump, but that the masses will collapse all the way into a *black hole*! (Fig. 2.16)

Clearly, that *has not* happened to our Universe. And Einstein knew that this claim of the theory was unreasonable, so he needed some way to stabilize our Universe against gravitationally collapsing. The "fix" he proposed was to add something called a cosmological constant, or an intrinsic amount of energy *inherent to spacetime itself*. If you did not have quite enough of this energy, the collapse problem would still be inevitable, while if you had even slightly too much of it, it would push all the other matter away, leaving our Solar System as the only thing left in the Universe. But just the right amount would balance gravity's attraction out, and leave us with a Universe — filled with matter — that neither collapsed nor pushed everything away.

This type of solution, where either too much or too little of something would cause catastrophe, with only one particular value that's a viable solution, is known as a fine-tuning problem. In this case, it is dissatisfying for a number of reasons, including that a slight change in the distance to the nearest stars would render this solution unstable. For a number of years, this was the only conceivable solution to the problem of how matter could appear to be uniformly distributed throughout the Universe, and yet did not collapse into a gigantic black hole.

However, it was only in the 1920s that a preferable solution was brought up as a possibility. Through the independent work of four theorists — Alexander Friedmann, Georges Lemaître, Howard Percy Robertson and Arthur Geoffrey Walker — it was shown that perhaps there was no cosmological constant, but rather that the Universe was *evolving in scale* over time. For what they found was rather than having a delicate balance between gravitational attraction and the intrinsic, outward-pushing energy of spacetime, the spacetime of the Universe itself could either be expanding or contracting. Either one of these dynamic, changing solutions could give rise to an isotropic and homogeneous Universe.

As always, the ultimate arbiter of scientific truth would have to come from asking questions, experimentally and observationally, of the Universe itself.

Chapter 3

Beyond The Milky Way: A Giant Leap Into An Expanding Universe

As our understanding of the Universe progressed forward in great leaps from Copernicus to Newton to Einstein, so did the instruments with which we viewed the curiosities of the night sky. We went from using our naked eye to small refracting telescopes to larger and larger reflecting telescopes. The maximum diameter of your eye's pupil — when fully dilated — is up to 9 millimeters, which sets the limit to what an unaided human (with otherwise perfect vision) can see. Early telescopes, such as Galileo's, improved this only slightly; Galileo's first telescope was just 15 millimeters in diameter. But in time, this improved significantly, and the increase in aperture brought along with it an increase in the amount of light that could be gathered. By the late 1600s, both refracting and reflecting telescopes had reached the size of an outstretched human hand, and became widespread among astronomers. And as the light-gathering power of these telescopes increased, they became capable of seeing progressively fainter, dimmer and more distant objects. This led to the identification of what appeared to be many permanent, fixed, deep-sky objects that appeared as extended smudges on the sky. They were classified, broadly, as *nebulae*.

But it was really in the late 1700s and into the 1800s that telescopes became large enough for us to begin to discover the structures of many of these objects in exquisite detail. The Great Forty-Foot Telescope, constructed by William Herschel in the 1780s (because when you discover the first new planet *ever* — Uranus in 1781 — people are more than happy

Figure 3.1 Leviathan of Parsonstown. At six feet (1.8 meters) in aperture, it was the largest telescope ever constructed at its time, and became the first telescope large enough that the human eye could discern structural patterns in galaxies by looking through it. This would hold the world's record until 1917, when finally broken by the 100″ (2.54 meter) Hooker Telescope. Image credit: copperplate engraving, *circa* 1860.

to give you unprecedented amounts of funding), had a diameter of 48 inches (1.2 meters), and was not surpassed until the 1845 completion of the 72-inch (1.8 meter) Leviathan of Parsonstown. By this time, the number of known nebulae had swelled from around 100 to many thousands, and had been classified into many different types (Fig. 3.1).

Rather than just appearing as faint, fuzzy but fixed clouds in the sky, these large-aperture telescopes revealed these nebulae to come in numerous, distinct forms. There are open star clusters, or collections of anywhere between hundreds-to-thousands of stars in a single region of space. Often populated by bright, blue stars, these nebulae are sometimes (but not always) accompanied by large amounts of nebulosity, or dust, and are primarily located in the plane of the Milky Way. There are also globular clusters, which are spherical in appearance, diffuse around the edges and denser towards the center. For a long time, we were unable to resolve individual stars in nearly all of these globulars, but the advent of large telescopes with superior light-gathering power allowed us to determine that these objects are much larger than typical open clusters, containing anywhere from hundreds of thousands to many millions of stars. Unlike the open star clusters, globular clusters appear randomly distributed in all directions, with no preference for or against the plane of the Milky Way. Based on what we knew about stars at the time, we were able to conclude that these clusters — both the open clusters and the globulars — were all located within at most 100,000 light years, placing them in or around our own Milky Way.

But there were other types of nebulae besides these clusters. Planetary nebulae and supernova remnants were some of the most colorful and spectacular extended objects, a few of which were eventually identifiable as remnants of specific historical events! (Messier 1, the Crab Nebula, is the remnant of a supernova that was visible across the world back in 1054.) There were nebulae that varied in color from red-to-blue, with the red regions indicative of ionized gas, underwritten by the process of active star formation, while the blue regions indicate places where neutral gas is reflecting light from the young, bright blue stars. The great nebula in Orion, also known as Messier 42, is an archetypal example of both types (Fig. 3.2).

There were giant, elliptical monstrosities that *were not* resolvable into individual stars, even when the most powerful telescopes in the world were pointed at them. And finally, there were nebulae observed to possess a unique spiral structure, whose origins were unknown. The first one discovered was the Whirlpool Galaxy — Messier 51 — immortalized in this sketch by Lord Rosse in 1845, using the 72-inch Leviathan. A total of 14 others (such as the Pinwheel Galaxy, Messier 101) were soon identified, although just what this spiral structure indicated was still a mystery, which is to say that no one *knew* they were galaxies at the time (Fig. 3.3).

By far, the largest of these spirals in our night sky is the great nebula Messier 31, also known as Andromeda. Unlike the other spirals, whose structure is clearly visible through a large enough telescope due to their face-on orientation relative to us, Andromeda's spiral structure was only

Figure 3.2 The Orion Nebula (Messier 42) is one of the closest regions of active, ongoing star formation to Earth. It contains many hundreds of known, newly-formed stars inside, and is located a scant 1,344 light years distant. Image credit: NASA, ESA, M. Robberto (Space Telescope Science Institute/ESA) and the Hubble Space Telescope Orion Treasury Project Team.

revealed with the development of astrophotography. Previously, astronomical objects were viewed optically by a human and sketched; the early astronomers had to be both artists and scientists in order to share their data. But an amateur astronomer — Isaac Roberts — pioneered the technique of astrophotography and applied it to many of the deep-sky nebula. In 1888, he unveiled his masterpiece: a stunning astrophoto of the great nebula in Andromeda, revealing the same spiral structure previously

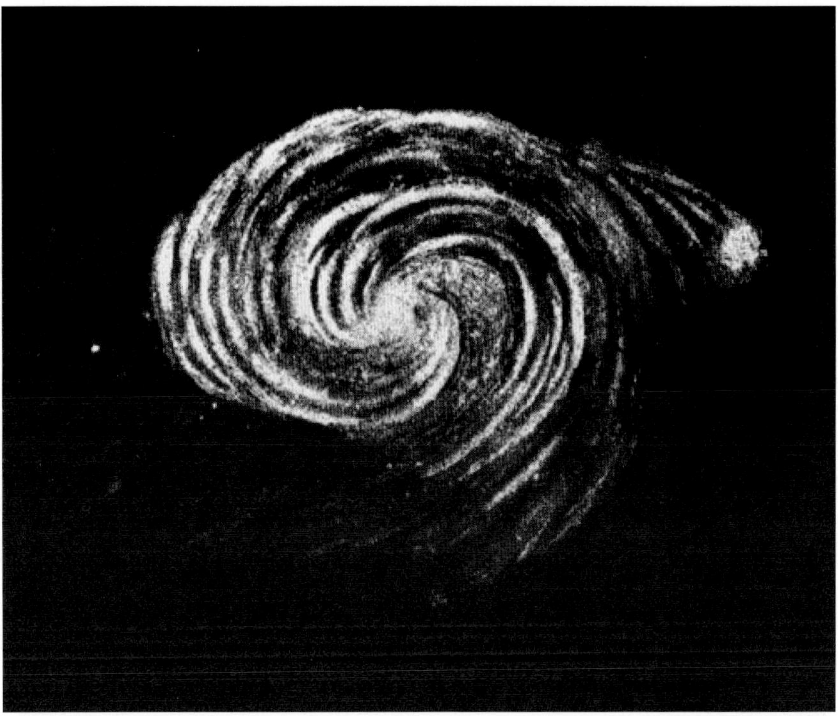

Figure 3.3 This 1845 sketch of the Whirlpool galaxy was the first identification of a deep-sky object as having a spiral structure. Image credit: William Parsons, 3rd Earl of Rosse (Lord Rosse).

seen in numerous other nebulae (Fig. 3.4). Unbeknownst to him, he had taken the first-ever photograph of a galaxy external to our own.

* * *

At the same time that Einstein's General Theory of Relativity was rocking the foundations of fundamental physics, a great debate concerning the nature of these spiral nebulae was dividing astronomers. The other types of nebulae — the open and globular clusters, the supernova remnants, the planetary nebulae and the extended (star-forming) red-and-blue nebulae — were all known to reside within the Milky Way. (At the time, elliptical nebulae were erroneously assumed to be star clusters of some type.) The hotly debated issue was the nature of these numerous spiral nebulae.

Figure 3.4 This is the very first image of the Andromeda galaxy, Messier 31 that reveals its spiral structure. Image credit: Isaac Roberts, (1899). Via *A Selection of Photographs of Stars, Star-clusters and Nebulae*, Volume II, The Universal Press, London.

On one hand, the majority of astronomers thought the best explanation was that these nebulae were proto-stars in the process of forming, also contained within our own Milky Way. On the other hand, a substantial minority contended that these might be "island Universes" in their own right, far beyond the Milky Way itself.

The idea that these nebulae would be proto-stars is not as far-fetched as you might think. Imagine that you start with some matter: a neutral, molecular cloud of gas. If the gas is cool enough, it is going to start to collapse under its own gravity; that much is inevitable. In general, a gas cloud *would not* be perfectly spherical, but rather it will be shortest in one direction compared to all the others. Because of the way that gravitation works, that direction will collapse down the fastest, and because atoms interact with one another, collisions will occur, atoms will stick together, and the gas will begin to emit energy. What we will be left with, in this picture, is a flat, rotating cloud of gas, whose density is highest towards

Figure 3.5 The idea that these spiral nebulae were molecular clouds that collapsed into a disk, began rotating and funneling mass into the center, where they would eventually form stars, was for many years the leading theory as to the nature of spiral nebulae. Images credit (from L-to-R): NASA and The Hubble Heritage Team (STScI/AURA). Acknowledgment: C. R. O'Dell (Vanderbilt University); ESA: C. Carreau; Bill Schoening, Vanessa Harvey/REU program/NOAO/AURA/NSF.

the center. Eventually, it was suspected, at least one star will form at the center, but that these nebulae represented an early stage in the formation of new stars. This was — at least at the time — a completely reasonable explanation for the nature of spiral nebulae (Fig. 3.5).

If these cosmic spirals were, in fact, proto-stars contained within our galaxy, then that would mean that the Milky Way — some 100,000 light years across — encompassed the entirety of the known Universe, with nothing but the vast void of infinity lying beyond. However, if these spirals were "island Universes" — distant, Milky Way-like objects containing billions of stars of their own — then our Universe extended far beyond our own galaxy, going on for at least many millions of light years (and possibly more) in size. Although a huge set of observations, sketches and photographs were taken of these deep-sky objects, a consensus failed to emerge, as the two sides pointed to different pieces of evidence and different interpretations to reach different conclusions. Emotions ran high on both sides of this debate, as the fundamental question of the scale and even the nature of the Universe was at stake!

In 1920, in an effort to resolve the issue, an event known as *The Great Debate* was held, where two renowned astronomers — Harlow Shapley (for the proto-stars side) and Heber Curtis (for the island Universes side) — would present the best arguments and counterarguments on the topic of the scale of the Universe. They took observations and facts that both sides agreed upon, and presented arguments for which interpretation best fit the data. There were six major points of contention between the two factions, enumerated as follows:

1) Observations of Messier 101 (the Pinwheel Galaxy) over the course of many years appeared to show that individual features within this nebula were rotating over time. Shapley contended that this nebula could not be an object even approaching the scale of the Milky Way, as the rotational speeds required would be many times faster than the speed of light, the ultimate speed limit of the Universe. Curtis conceded that *if* those observations were correct, they would disfavor the island Universes picture. However, he countered, the observations were at the very limit of what the best instruments could detect, and that these effects were not observed in the other spirals. Thus, Curtis advocated that the observations themselves could not be trusted (Fig. 3.6).
2) Observations of Messier 31 (the Andromeda galaxy) showed that there are many more objects than normal flaring up in that small region of the sky. They were similar in brightness to the novae that we see in our own Milky Way, except they were *incredibly* dim, and there were more of them seen in this one region than in the rest of the Milky Way *combined*. Curtis estimated that this object must be at least comparable in size to our entire galaxy and *millions* of light years away, placing it far outside the extent of the Milky Way galaxy. Shapley, however, countered that there was a very bright flare-up in 1885 that could not have possibly been a nova, and therefore Curtis' explanation must be flawed.
3) These spiral nebulae were also observed spectroscopically, which means the light coming from them was broken up into individual wavelengths, recorded, and analyzed. The spectra coming from them did not appear to match the spectrum of any known stars, which was puzzling. Shapley contended that this was because these nebulae were not *yet* stars, and therefore should have their own, unique signatures. Curtis, on the other hand, argued that these spirals were, in fact, filled with stars, but that the stars that dominated these island Universes were *not* like the ones nearby us in the Milky Way. On the contrary, he argued, these were dominated by stars that were hotter,

388 ASTRONOMY: A. VAN MAANEN

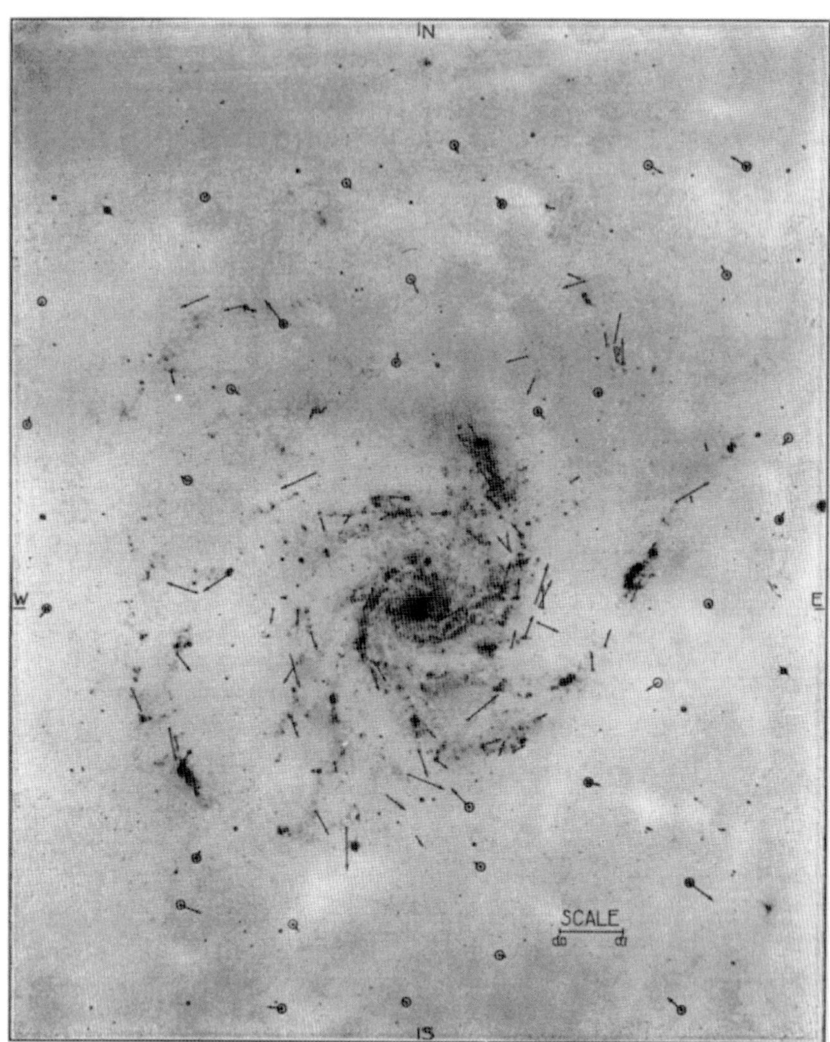

INTERNAL MOTIONS IN MESSIER 101

The arrows indicate the direction and magnitude of the mean annual motions. Their scale (0".1) is indicated on the plate. The scale of the nebula is 1 mm. = 10".5. The comparison stars are enclosed in circles.

Figure 3.6 A claimed observation of internal motions of the stars within Messier 101, if true, would disprove the "island Universes" hypothesis. Image credit: A. Van Maanen, (1916). *Preliminary Evidence of Internal Motion in the Spiral Nebula Messier 101*. Proceedings of the National Academy of Sciences of the United States of America, 2(7), p. 388.

bluer and brighter than the average stars we can see, and were furthermore located in environment very different from the stars we saw. Therefore, it is no surprise that their spectra would be skewed compared to what we are used to observing.

4) A very contentious observation was that there were no spiral nebulae observed in the plane of the Milky Way. This was an especially difficult observation for Shapley to explain, because there are far *more* stars in the plane of the Milky Way than anywhere else in the sky. Curtis advanced the argument that these spiral nebulae are actually everywhere in the sky, but because they are so much more distant than the objects within our galaxy, the plane of the Milky Way *blocks the light* from the spirals that happen to be behind it. Shapley was forced to contend that there must be something about the plane of the Milky Way that disfavors proto-stars from forming there. In perhaps a stroke of brilliance, he argued that the Milky Way itself was not only larger than was previously suspected, but that our Sun was located far from its center, and that there was a vast amount of light-blocking dust *behind the visible stars* that was preventing us from seeing these nebulae. If only infrared astronomy had been pioneered back then, perhaps they would have learned they were *both* correct: the light-blocking dust does obscure the spiral nebulae, which exist in abundance beyond the plane of the Milky Way! (Fig. 3.7)

5) It was pointed out that the starlight from all the known stars in our night sky, if viewed from the great distances that Curtis contended these nebulae were located, would be far *too dim* to account for our observations. In other words, if the spiral nebulae truly were "island Universes" and made out of the same stars we see in our own night sky, they would appear much fainter than they actually do. Shapley pounced on this point, asserting that the only explanation was that these spiral nebulae were not collections of stars located at supremely great distances. Curtis was forced to resort to the same argument he used for the third point: that these spiral nebulae were filled with stars, but that the stars that dominated these distant, island Universes were not representative of the stars found nearby our location in space.

6) Finally, the last observation was that the speeds of most of these spirals had been measured. And while there were a few, such as

Figure 3.7 Looking through the plane of the Milky Way, infrared astronomy is able to find numerous spiral and elliptical nebulae, like Maffei 1 and Maffei 2, shown here. Image credit: NASA/JPL-Caltech/Wide-field Infrared Survey Explorer (WISE) Team.

Bode's Nebula (Messier 81) that were moving at just a few kilometers per second, typical of objects within the Milky Way, the vast majority of them were moving incredibly fast: many hundreds or even over a *thousand* kilometers per second. With only a few exceptions, they were moving directly *away* from us (Fig. 3.8). Neither side had a compelling explanation to deliver at the time, the extraordinary length of the debate perhaps having taken its toll on the two participants.

In the table is a list of the spiral nebulae observed. As far as possible their velocities are given, although in many cases they are only rough provisional values.

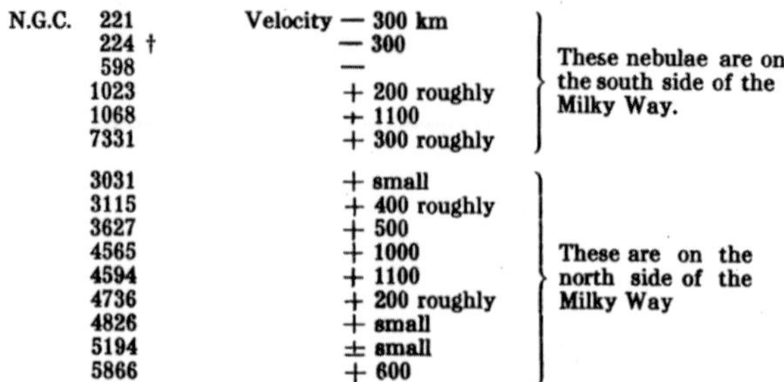

N.G.C.	Velocity	
221	— 300 km	These nebulae are on the south side of the Milky Way.
224 †	— 300	
598	—	
1023	+ 200 roughly	
1068	+ 1100	
7331	+ 300 roughly	
3031	+ small	These are on the north side of the Milky Way
3115	+ 400 roughly	
3627	+ 500	
4565	+ 1000	
4594	+ 1100	
4736	+ 200 roughly	
4826	+ small	
5194	± small	
5866	+ 600	

Figure 3.8 There was no accounting for these incredibly large velocities if these objects were located within our galaxy, as unless the galaxy was much more massive than we thought, they would be gravitationally unbound from us. Image credit: Vesto M. Slipher, (1915). *Spectrographic Observations of Nebulae*. Popular Astronomy, 23, pp. 21–24.

It is not important who was declared the *winner* of the debate (although it was Shapley, for the curious), but rather to understand the merits, challenges and arguments of both sides. In science, it is not the popularity of ideas that leads to a scientific consensus, but rather the power of a theory to both explain the full suite of what is observed and also to predict new, testable phenomena. The decisive observations would wind up coming from an unexpected source.

* * *

Going all the way back to ancient times, it was long thought that the stars in the heavens were fixed points of light. Occasionally, a catastrophic event like a nova or supernova would create a temporarily brightened object, but these are extraordinarily rare, and only a few have been visible to the naked eye in all of human history. While it is true that the vast majority of stars appear to be unchanging in their position and brightness in the sky, this is not true of them all. In 1596, David Fabricius saw what he believed to be a nova, as he saw a point of light brighten in the sky in August and then fade completely from view by the end of October. But much to his surprise, the point of light reappeared again in 1609. No nova had ever reappeared before; what Fabricius had discovered was not a nova at all, but Mira, the first intrinsically variable star!

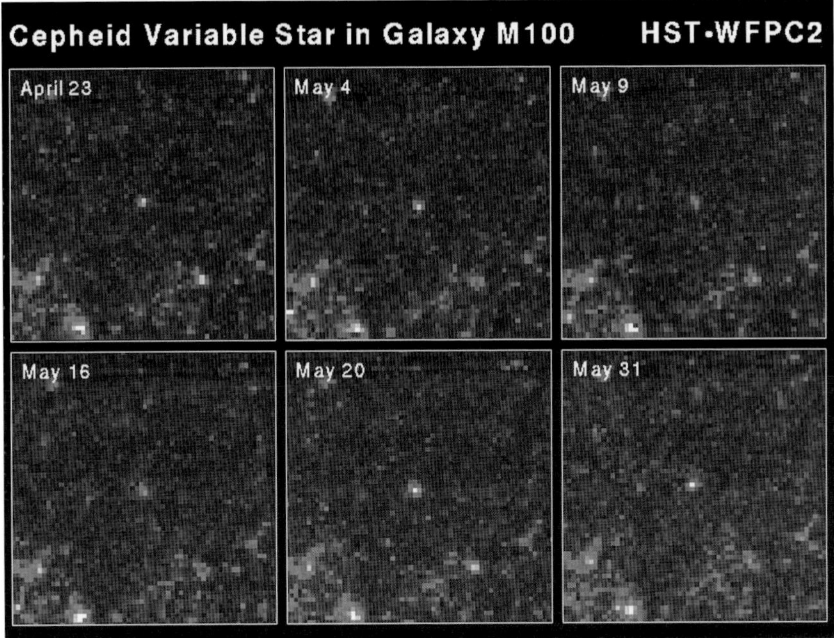

Figure 3.9 This variable star, observed by the Hubble Space Telescope, doubles in brightness and then fades again over a period of 51.3 days. Image credit: Dr. Wendy L. Freedman, Observatories of the Carnegie Institution of Washington, and NASA/ESA.

Variable stars were originally thought to be extremely rare, as it took nearly two centuries for the count of them to finally reach 10, but the number of discovered variables skyrocketed once the technique of astrophotography was developed. By being able to directly compare a star's apparent brightness over periods of days, weeks, months or even years, both the amount of variation and the period of variability could be measured quite accurately (Fig. 3.9).

In the early 1890s, a young woman named Henrietta Leavitt attended the Society for the Collegiate Instruction for Women, now known as Radcliffe College. In 1893, she was hired by the Harvard College Observatory to measure and catalogue the brightness of stars from the observatory's collection of photographic plates. In particular, she was cataloguing stars found in the Small Magellanic Cloud, and over the next two decades, found over 1,000 variables which she catalogued into numerous different classes of variable star.

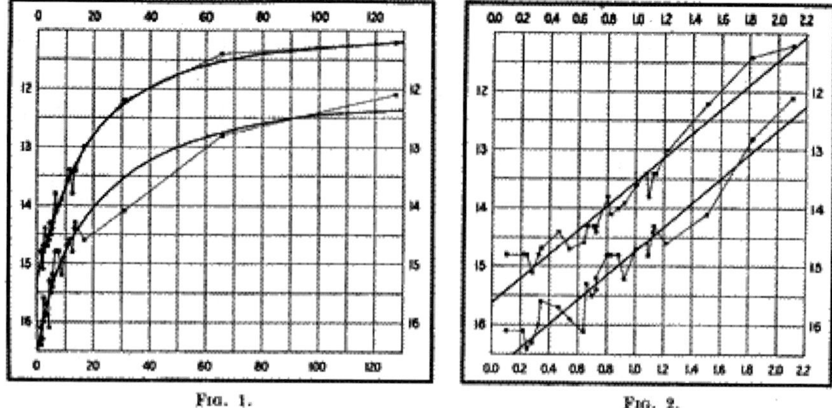

Figure 3.10 The correlation between the periods (on the x-axis) and the magnitudes (on the y-axis) provided the first method of accurately learning the intrinsic brightness of an object, despite being unable to measure it. Both figures show the same data, but on a linear scale at left and on a logarithmic scale at right. Image credit: Harvard College Observatory Circular 173, Edward C. Pickering, with data and figures provided by Henrietta Leavitt, March 3, 1912.

But one particular class — the Cepheid variables — exhibited some unusual behavior, which Leavitt noticed. When she looked at 25 of the brightest Cepheids, they took longer periods of time than the other stars to complete each full cycle of variability: to reach their maximum brightness, dim to a minimum value, and then return to maximum again. While all the stars varied by approximately the same amount (in terms of visual magnitude), the ones with the highest *average* brightness took months to cycle from bright-to-dim-to-bright again. As the average brightness of the observed stars decreased, so did the period of the stars' variability; the dimmer a star was, the faster its brightness varied, down to a minimum of just over a single day. In fact, she found that there was a well-defined correlation between how bright a Cepheid appeared on average and the period of time it takes to pulse (Fig. 3.10).

This relationship is known today as the Period–Luminosity Relationship, and this discovery carried along with it some tremendous implications. Because all the stars in the Small Magellanic Cloud are at approximately the same distance from Earth, the differences in brightness correspond to differences in how *intrinsically luminous* each of these stars are. And if there is a relationship between a star's period and its luminosity, this meant that if you measured a Cepheid variable star's period, you would know how intrinsically luminous it was. If you then measured its apparent brightness, because you know how brightness and distance are related, you could figure out **how far away** the star was located.

This is an amazing achievement, because all you need to do is find and measure the properties of one of these Cepheid variable stars *anywhere in the Universe*, and you can immediately know how far away it is from you! We call these types of objects "standard candles," because if you know how intrinsically bright a light-emitting object is, and then you measure its apparent brightness, you can figure out how far away it is from you. Thanks to Henrietta Leavitt's work on Cepheid variable stars, we now had a standard candle to measure the vast distances across the cosmos. (And as time went on, many other standard candles for astronomy were discovered as well.) For the first time, we had hope for figuring out just how distant the farthest objects in the Universe might be (Fig. 3.11).

* * *

In 1917, the 100-inch (2.54 meter) Hooker Telescope atop Mt. Wilson became operational, finally surpassing the Leviathan as the world's largest telescope. Two years later, Edwin Hubble — a scientist passionate about the physical nature of the spiral nebulae — was hired by the observatory, and began taking repeated surveys of these objects. Although most of the scientists at Mt. Wilson sided with Shapley in the aftermath of the Great Debate, Hubble remained unconvinced, and was particularly interested in undertaking observations of novae in these spirals.

Along with his assistant, Milton Humason, Hubble embarked on a research program where he catalogued and observed these nebulae over time, looking for flare-ups and any transient, changing behavior or phenomena. As Heber Curtis had noted in the Great Debate, multiple objects which appeared to be novae had been observed in these spirals, and at a far greater rate than were observed in other parts of the sky. In fact, Curtis himself first observed novae in these spiral nebulae in 1917, noting that they were — on average — 10 magnitudes fainter than the novae in other parts of the sky. 10 magnitudes is a *huge* number, corresponding to a factor of 10,000 in brightness and therefore, if the

Figure 3.11 Just like a candle whose intrinsic brightness is known, we can measure the apparent brightness of a known "standard candle" in the Universe and compare that with the known intrinsic brightness to calculate the object's distance from us. Image credit: NASA/JPL-Caltech.

brightness/distance relationship we knew was still valid, these objects would be a hundred times farther away than any of the other novae we had previously observed!

Over the course of the early 1920s, Hubble and Humason observed many different known spirals, including Messier 31, also known as the Great Andromeda Nebula. On a photographic plate dated October 6, 1923, Hubble had noted the position of three previously observed novae: the first at the very outskirts of Andromeda, a second one closer towards the center and the third one in between the two.

And then the unexpected happened: a fourth nova was observed. The shocking part about this observation, according to Hubble's data, is that the fourth nova erupted *in the exact same location as the first one*! Why was this so unexpected? Because nova typically take thousands of years before one can be seen to recur. Even in the most extreme modern case — that of RS Ophiuchi, where a white dwarf constantly siphons mass from a red giant star — it takes decades before a nova can recharge. Yet Hubble's data showed that this star brightened to a maximum for a second time just *31 days* after the first observation! (Fig. 3.12)

In other words, due to the short timescales involved, this could not be a nova, after all; Hubble excitedly realized this had to be a variable star! He crossed out the "N" on his photographic plate, which was used to denote a nova, and wrote "VAR!" This turned out not to just be any type of variable star, but in fact the same class of Cepheid variable that Henrietta Leavitt had catalogued just over a decade prior.

Because Hubble was able to measure the period of this variable star, he immediately knew what its intrinsic brightness was, thanks to Leavitt's work. And thanks to the power of the 100-inch Hooker Telescope, he was able to accurately measure the *apparent brightness* of this variable star at its peak, exactly the observation he would need to figure out the distance to it. As with anything, if you know an object's intrinsic brightness and you measure its apparent brightness, you can figure out how far away it is using a simple formula: the brightness falls off as the inverse-distance squared! (In other words, if an object is twice as far away, it is just one-fourth as bright; if it is 10 times as far away, it is only one-hundredth as bright.) Based on what Hubble observed, he concluded that the Great Andromeda Nebula was not just outside the Milky Way, but was actually closer to a **million light years** away, the first accurate estimate of an object beyond our own galaxy! It was later discovered that there were actually two types of Cepheids, one class that Leavitt found her period-luminosity correlation for, and a second type, that Hubble observed in Andromeda. Because this information was unknown to Hubble, his estimate was actually low by a little more than a factor of two. Taking this into account, the modern value — 2.2 million light years — showcases just how high-quality Hubble's data and observations actually were.

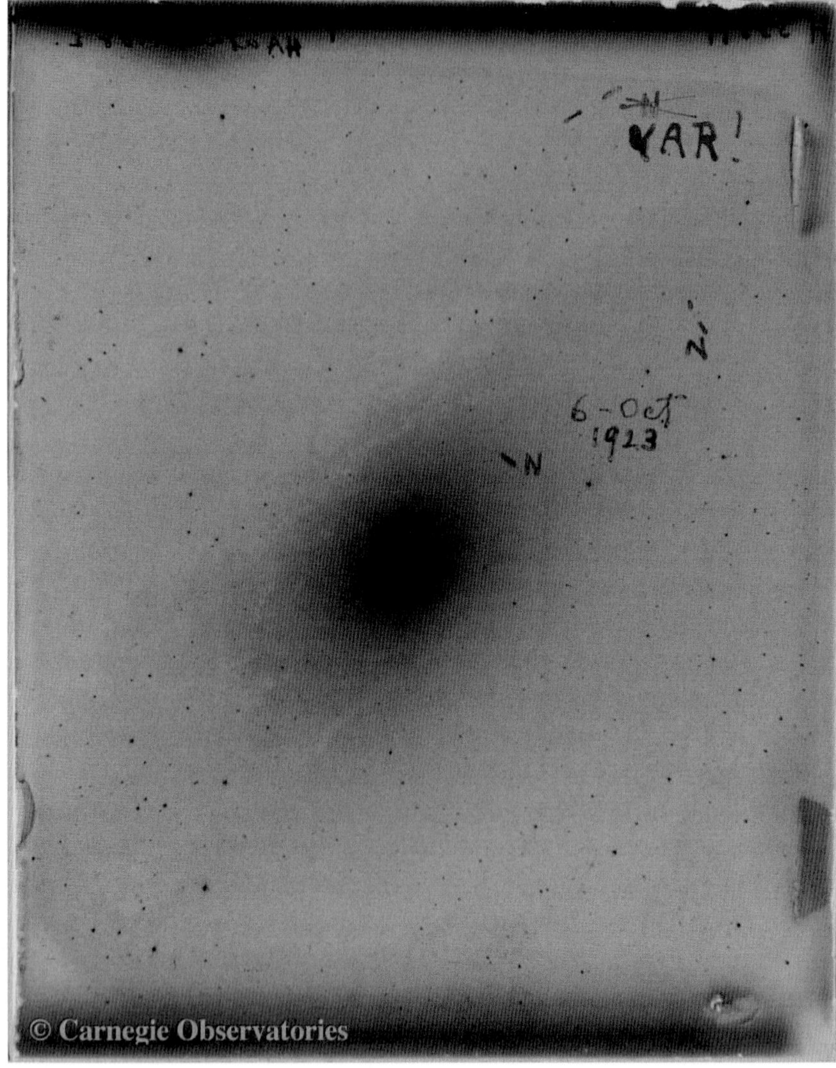

Figure 3.12 Edwin Hubble's photographic plate from the night of October 5–6, 1923. The realization that the object observed was not a nova but rather a variable star meant that its distance could be calculated, and hence the distance to the spiral nebula that hosted it could be uncovered. Image credit: © Carnegie Observatories.

But Hubble did not stop with Andromeda. Once he realized that he could determine the distance to these spiral nebulae by observing Cepheid variable stars within them, he set out to measure these variable stars in as many spirals as possible. Over the remainder of the decade, Hubble was able to precisely determine the distances to more than 20 of these spiral nebulae, discovering in the process that not only was every single one of them outside of the Milky Way, but that Andromeda was actually one of the *closest* to us!

This, on its own, would have etched Hubble's name in stone in the history books. In one fell swoop, he had just definitively settled the Great Debate, and showed that these spiral nebulae were not proto-stars at all, but were island Universes unto themselves, similar in size and scope to our own Milky Way. But Hubble went even farther than that, and changed our picture of the Universe more dramatically than anyone could have imagined (Fig. 3.13).

* * *

Back in 1912, an astronomer named Vesto Slipher was studying the Andromeda Nebula, and had decided to tackle the problem of measuring the spectrum of this object. It had been known since the time of Newton that if you disperse any type of light through a prism, you can decompose that light into the individual wavelengths that make it up, producing a spectrum. And when you measure the spectrum of light from any source, such as the Sun, you will find that, in addition to the deep, long-wavelength reds, the vibrant, short-wavelength violets, and all the visible light in between, there are also dark, black "bands" that appear to be gaps (of varying intensity) in this spectrum. It was quickly realized that individual atoms — such as hydrogen, helium, oxygen, etc. — were found to emit and absorb only very particular wavelengths, characteristic of the election transitions specific to these atoms themselves. So when you see absorption features that correspond to wavelengths of 6563 Å (red), 4861 Å (cyan), 4341 Å (blue), and 4102 Å (violet), that indicates a telltale sign of neutral hydrogen. In particular, that is a sign of neutral hydrogen *in between* the light source and you.

But there is a little more to this story. If you see those four characteristic absorption lines at those exact four wavelengths, that tells you that neutral hydrogen is present between you and the light source, and also that the hydrogen *is not moving* with respect to your frame-of-reference. Think about the sound a police siren makes: you can tell whether it is approaching you or moving away from you based on the pitch of the sound you

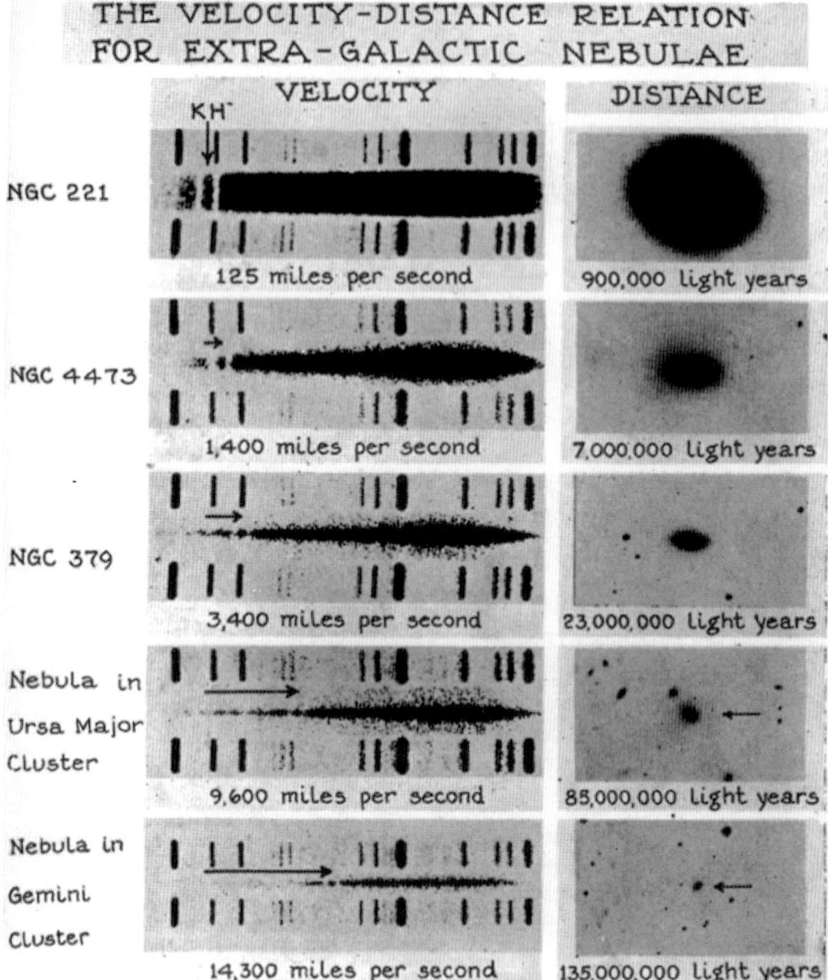

Figure 3.13 This image is Plate VIII from Hubble's book, showing images of five nebulae (i.e., galaxies) with their velocities and distances, calculated from the measurements of Cepheid variable stars. Image credit: Edwin Hubble, *The Realm of the Nebulae* (1936), reprinted 1958 by Dover Publications, Inc.

hear. A higher-pitched sound means that you and the siren are approaching one another, while a lower-pitched sound means you are moving away from one another. What is going on is that the observed wavelength of the sound waves — which determines the pitch you hear — depends on how frequently the sound waves actually pass by your ears. This phenomenon is known as the *Doppler Effect*. When a sound-emitting object moves towards you, the sound waves are more compressed in your direction, and pass by your ears more frequently, so you hear a higher-pitched (and shorter-wavelength) noise. When it moves away from you, the sound waves are more rarefied in your direction, pass by your ears less frequently, and as a result you hear a lower-pitched (and longer-wavelength) sound.

Well, the same thing is true of light, with one interesting caveat that makes it different from sound. If I take a cloud of hydrogen gas that is moving towards you, you will still see those same four absorption features, but their wavelengths will be compressed in your direction, pass by your eyes more frequently, and as a result you will see a higher-frequency, shorter-wavelength, *bluer* color to all of those absorption lines: a *blueshift*. Conversely, if a cloud of hydrogen gas is moving away from you, those same absorption features have their wavelengths rarefied in your direction, pass by your eyes less frequently, and you will see a lower-frequency, longer-wavelength, *redder* color to those features: a *redshift*. The caveat for light is that, in addition to velocities — whether a light-emitting/light-absorbing object is moving relative to you — redshifts and blueshifts can be caused by the Universe itself either expanding or contracting, as permitted by General Relativity. Keep this in mind as we return to Slipher and his work (Fig. 3.14).

Slipher was a pioneer in astronomical spectroscopy, having used the technique to measure the elemental composition of planetary atmospheres. When he turned his spectroscopic techniques towards the Andromeda Nebula in 1912, he was able to take very precise measurements of absorption features from the very center of the nebula itself. What he found was evidence of many of the absorption lines we find in our own Sun (and a few that we did not), but those features were blueshifted. Not by a small amount, either, but by a greater amount than *any other astronomical source* ever discovered at the time. After calculating what its velocity must be, Slipher himself had the following to say:

> "The magnitude of this velocity, which is the greatest hitherto observed, raises the question whether the velocity-like displacement might not be due to some other cause, but I believe we have at present no other interpretation for it."

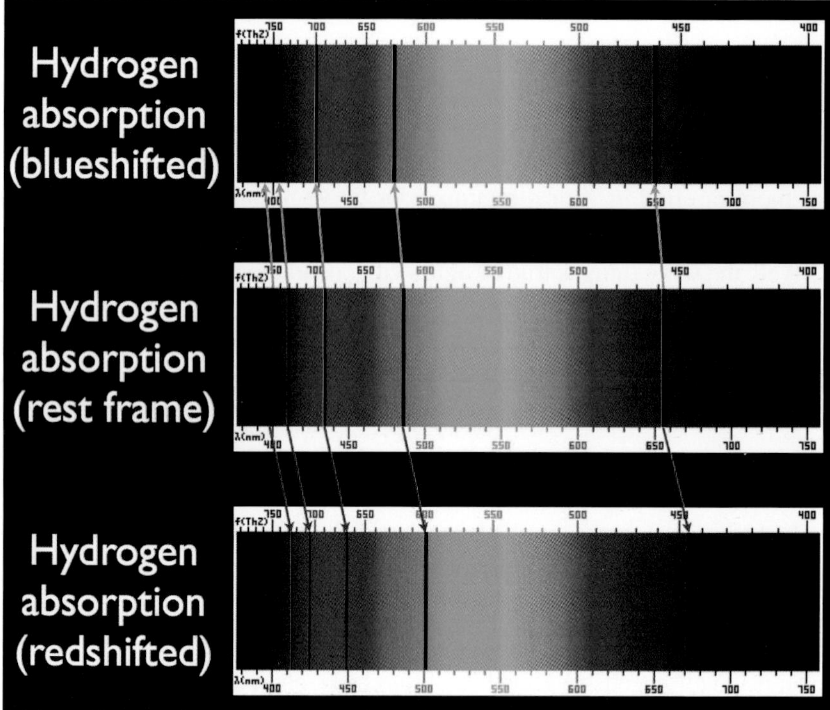

Figure 3.14 Hydrogen absorption spectra superimposed over the visible light spectrum, shown as the human eye perceives it. If you see the characteristic spacing of a particular element or set of elements between emission/absorption lines, you can identify both what elements are present and what their redshift (or blueshift) is. Note how, in the case of a large enough redshift, lines that are ultraviolet (and hence invisible) in the rest frame can be shifted into the visible portion of the spectrum. Image credit: E. Siegel.

Over the next few years, Slipher went on to measure the spectra of 15 of these spiral nebulae, finding that there were two others with slight blueshifts, but that the overwhelming majority were actually *redshifted*, many by a significantly greater amount than even any of the observed blueshifts. (See the table in Fig. 3.8.) This was such a remarkable observation that it was, all on its own, the sixth and final point of 1920's Great Debate!

* * *

When Hubble made his 1923 discovery that the Andromeda nebula was actually the Andromeda *galaxy*, far beyond our own Milky Way, he immediately turned his attention towards measuring variable stars in the other known cosmic spirals. But it was not merely to catalogue the distances to these objects, or to confirm that these other nebulae were also galaxies unto themselves. Hubble was keenly aware of both Einstein's theory and of Slipher's work measuring the redshift of many of these nebulae, and with his new discovery, he concocted a plan. To bring it to fruition, he set out to measure both the distances to and inferred velocities of as many of these spirals as possible.

Together with Humason, they measured the periods and brightnesses of variable stars in more than 20 of these spiral galaxies, as well as their red-or-blueshifts using Slipher's spectroscopic methods. The period-luminosity relationship for Cepheids allowed him to derive the distances to each of the galaxies, and the red-or-blueshift gave either the recession-or-approach speed. This came with a caveat as well: to measure these velocities accurately, the motion of the Earth around the Sun — as well as the Sun's motion through the galaxy — had to be accounted for. It is of paramount importance, in every scientific field and in astronomy in particular, to remember to account for all the effects that could potentially bias your results! By 1929, Hubble had enough data to publish a paper on the relationship between the velocities and distances of these extra-galactic nebulae, and what he found was astounding in both its simplicity and its power (Fig. 3.15).

On average, according to his findings, the *farther away* a galaxy was from us, the *faster* it was *receding away* from us! Not only were these spirals their own "island Universes" containing billions of stars all their own, but they followed a very simple, straightforward relationship: the recession velocity of a distant galaxy was directly proportional to its measured distance, something that would only be expected *if the Universe were expanding*! With one powerful swoop, Hubble had demolished Einstein's dream of a static Universe, and Einstein immediately declared his *ad hoc* cosmological constant to be his "greatest blunder." The relationship between the recession velocity of a galaxy and its distance has been known ever since as **Hubble's Law**, in honor of its discoverer. In fact, the constant of proportionality — which is the slope of the best-fit line that relates an object's distance to its inferred recession speed — is known as the Hubble constant, and tells us the rate at which the Universe is expanding. Not only had he determined the structure of the Universe and the spacetime that described its expanding nature, he devised and utilized a method to measure the rate of expansion itself, a method still in use today!

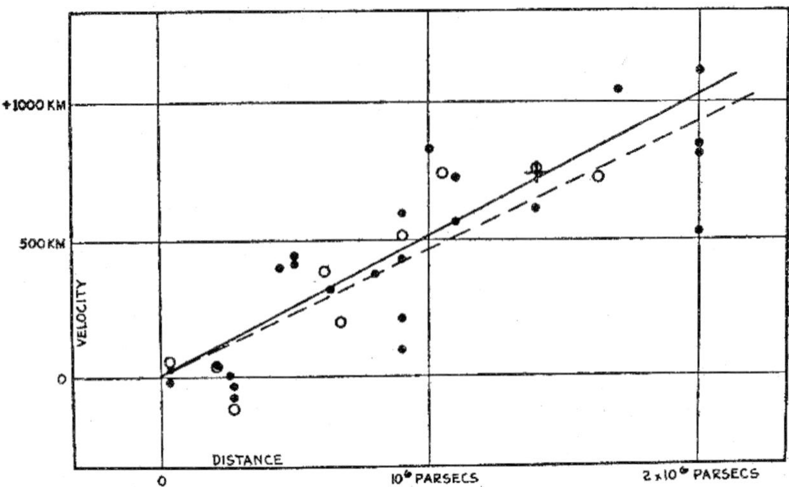

Figure 3.15 Although Hubble was well aware that more data — particularly more distant and higher-velocity data — was needed to determine the exact relationship, it was clear that objects that were more distant from us had higher recessional velocities. Image credit: Edwin Hubble, (1929). *A Relation between Distance and Radial Velocity among Extra-Galactic Nebulae*. Proceedings of the National Academy of Sciences of the United States of America,15(3), pp. 168–173.

There is an interesting historical side-note concerning this discovery. Much later, it came to light that Georges Lemaître, using some of Hubble and Humason's data, independently reached this same conclusion two years prior to Hubble's paper, back in 1927. But Lemaître's discovery of the redshift-distance relation, complete with a calculation of what is now known as "the Hubble constant," was not published in English even upon the paper's translation in 1931. Why not? Lemaître, living and writing in Belgium, and hence, in French, omitted that critical portion of the paper when he submitted it to the Royal Society in England, writing instead in a letter to the editor, "I did not find advisable to reprint the provisional discussion of radial velocities which is clearly of no actual interest, and also the geometrical note, which could be replaced by a small bibliography of ancient and new papers on the subject."

* * *

How does the expansion of the Universe work? Let us remind ourselves how General Relativity functions: the entire Universe is described by *spacetime*, which contains all of the matter and energy present within it. In turn, the presence of that matter and energy determines both how spacetime is curved at any particular instant and also how it evolves as time goes on. There are only a few exact solutions known in General Relativity, but one of them — the one independently arrived at by Alexander Friedmann, Georges Lemaître, Howard Percy Robertson and Arthur Geoffrey Walker — was particularly interesting. What they found was that if you had spacetime that contained the same energy density at all locations (i.e., was homogeneous), regardless of what the value of that energy density was, and regardless of whether that energy was in the form of matter, radiation or anything else, that Universe was *either* going to be expanding or contracting at exactly the same rate in all directions.

This was a remarkable theoretical discovery, and one of the few unique solutions known in General Relativity. The idea that spacetime was its own dynamical entity, and that neither space nor time by itself was complete, was still very new, having only been developed in 1907 by Hermann Minkowski. By showing that Einstein's *Special* Relativity could be expressed as a four-dimensional geometric theory, with three dimensions of space and one dimension of time, Minkowski (a former teacher of Einstein's) changed the way we think about the Universe. In his own words, here is how he expressed that:

> "The views of space and time which I wish to lay before you have sprung from the soil of experimental physics, and therein lies their strength. They are radical. Henceforth space by itself, and time by itself, are doomed to fade away into mere shadows, and only a kind of union of the two will preserve an independent reality."

When Einstein later developed *General* Relativity, the notion that the presence and location of matter-and-energy determined the structure, curvature and evolution of spacetime took center stage. What Friedmann, Lemaître, Robertson and Walker told us was that if you filled spacetime with a uniform amount of matter, radiation or other form of energy, that spacetime was going to either expand or contract in a predictable, well-defined way. Let us try to picture what that looks like.

Imagine the surface of a balloon. I know a balloon is only two-dimensional whereas we have three spatial dimensions in our Universe, but this is just to help you get familiar with thinking about spacetime. Imagine that this balloon is taut, but only slightly inflated, with coins glued uniformly all over the balloon's surface. Each coin represents a galaxy; the surface of the balloon represents spacetime. Pick any coin you like, and imagine that your chosen coin is our Milky Way. Every other coin you see is a galaxy in the night sky. The coins closest to our own represent the closest galaxies, while the coins farther away are fainter, more distant ones.

Now, General Relativity tells us that the balloon — our spacetime — will either expand or contract; it would be *unstable* to simply remain the way it is under Einstein's laws. What happens, then, when we inflate the balloon? The coins themselves do not change, they simply remain the same galaxies they always were. But what of the distance between the coins? Notice how they *all* expand away from one another? In fact, the coins that are closest to us appear to recede from us *slowly*, while the coins that are more distant appear to recede more quickly. In reality, it is not that these galaxies are traveling through spacetime, speeding away from us, but rather that *spacetime itself* is expanding, and as a result, we see these galaxies redshifting away from us! One of the amazing realizations that comes out of this picture is that every galaxy in the Universe would see *exactly the same thing* that we do: nearby galaxies receding away from it, with the apparent recession speed increasing the farther away a galaxy is from it. (Fig. 3.16)

The way the redshift works in an expanding Universe is a little more subtle than a Doppler redshift, which is what you observe when an object moves away from you. In the expanding Universe, we can imagine that light of a specific frequency — and hence of a well-defined wavelength — is emitted. The wavelength of that light is defined by a distance: how far, if you held up a ruler, it would take to move from one wave crest to the next. But if the very fabric of spacetime itself is stretching, then the wavelength stretches along with it, just as a drawing on the fabric of a

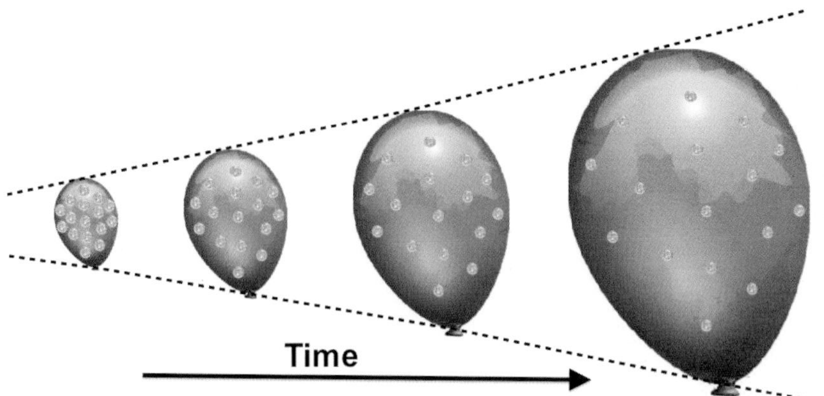

Figure 3.16 As time goes on in an expanding Universe, the fabric of space itself — represented by the balloon — expands and stretches. As a result, the individual galaxies — represented by the coins — find themselves expanding away from one another. Note that there is no single point on the balloon's surface that is the "center," but rather all coins see the same thing: all the other coins recede from them, with the more distant coins receding faster. Image credit: E. Siegel.

balloon would expand as you stretched the fabric. And the longer a light wave has to travel before it reaches you, the more time it spends being stretched by the expanding fabric of space, which means the redder it will be by time it reaches your eye (or your detector's eye). But this redshift *is not* due to the fact that these galaxies are moving away from us, but rather that the spacetime between these galaxies and our own is expanding while the light travels through space from its original location to ours (Fig. 3.17).

A lot of people have difficulty imagining space as the *surface* of a balloon, since our Universe has three dimensions of space instead of just the two dimensions of a balloon's surface. Consider the following three dimensional visualization of spacetime: instead of an expanding sheet, imagine a spherical ball of unbaked dough, with raisins distributed evenly throughout the inside of the dough ball. What happens when the dough bakes? It leavens evenly throughout, and all of the raisins inside expand uniformly away from one another, just like the coins on the balloon, except *in three dimensions* this time! If your galaxy were a raisin, and all you could see were the other raisins (galaxies), it would appear that the

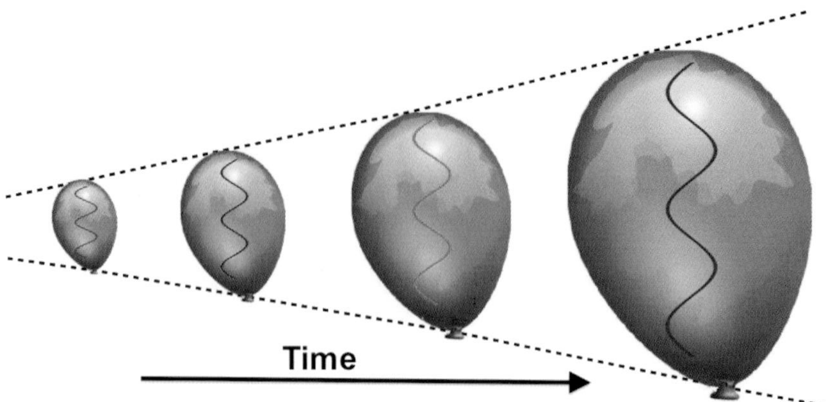

Figure 3.17 Not only does the density go down as the Universe expands, but the wavelength of light gets stretched — and hence redshifted — as well, to longer wavelengths and lower energies. Image credit: E. Siegel.

ones closest to you recede at a slow rate, while the ones farther away move away from you more quickly. In reality, none of these objects are actually speeding away from you, but rather the spacetime that defines the entire Universe — whether you picture it as a balloon's surface, baking dough, or as a purely mathematical construct — is *expanding*.

* * *

The very astute among you may have objected to this picture by now, as the Universe we have is *not* exactly uniform everywhere in space! After all, we have collections of hundreds of billions of stars — galaxies — in some locations, clusters that contain hundreds or even thousands of galaxies clumped together in some places, and great voids in space where there is not even a single star to be found for millions of light years. That is hardly a shining example of what we expect "uniform" to look like! In addition, the "straight-line" relationship observed by Hubble himself is only an approximation; most galaxies *do not* fall exactly on that line, and there are even a handful of galaxies that we do, in fact, observe moving *towards* us, with a blueshift (instead of a redshift) to their light. Finally, you will also notice that — even if we take a precisely measured modern graph of the expansion of the Universe — the straightforward relationship between distance and redshift (or inferred recession velocity, as we

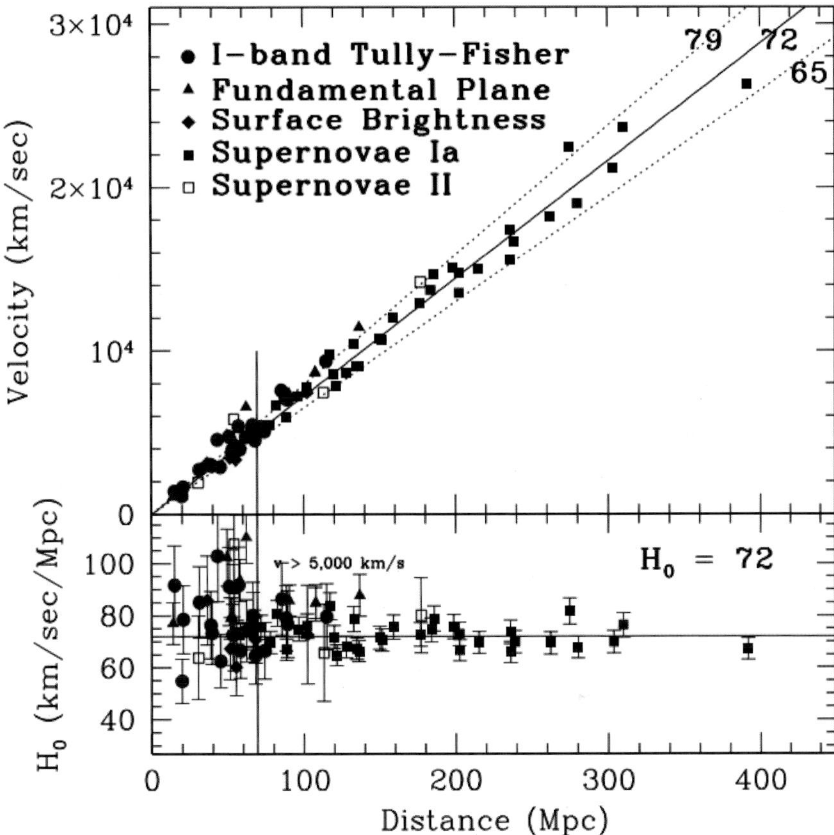

Figure 3.18 Observational results from the Hubble telescope. Note how the distances are hundreds of times greater than those of Hubble's original diagram. Image credit: Wendy L. Freedman *et al.* (2001). Final Results From The Hubble Space Telescope Key Project To Measure The Hubble Constant. *The Astrophysical Journal*, 553: 47–72.

sometimes refer to it) is still only true as an approximation, not an exact relation for the galaxies we observe in our Universe (Fig. 3.18).

Yes, it is true that practically all the points on Hubble's graph — and even in our best modern data sets measuring the expansion rate — *do not* fall exactly on the best-fit line described by Hubble's law. This is not only okay, though, it is *expected.* The reason for this is simple: the Universe is *not* perfectly uniform, and could never have been, even in the distant past! If it were, we would have a tremendous problem: there would be no gravitational imperfections, which are the very things that allow us to form stars, galaxies, clusters and any other type of structure in the Universe. Our Universe would simply have been a smooth, uniformly expanding bath of matter-and-energy, something that never would have led to the existence of stars, planets, human beings and books, among other things.

So what is going on with the *actual* Universe, then? Each galaxy (or island Universe) that we look out at is moving — relative to us — due to *two* factors: the expansion of the Universe *and* the gravitational pull of all the other masses around it. The expansion of the Universe explains why, *on average*, the galaxies are redshifting away from us. But the fact that the Universe is not quite perfectly uniform, especially on small scales, means that individual galaxies are going to be moving around with some additional velocity superimposed on top of the overall Hubble expansion. We have a name for this additional motion: **peculiar velocity**.

For most galaxies like the Milky Way, clumped together in relatively small groups, these peculiar velocities will be on the order of a few hundred kilometers-per-second, while for large clusters of thousands of galaxies, peculiar velocities can be ten times as large, up to *thousands* of kilometers-per-second, or 1–3% the speed of light! When we measure any one individual galaxy, what we are seeing is a combination of both effects: Hubble expansion and peculiar velocity; it is impossible from just a single galaxy to tell how much of its redshift (or blueshift) is due to each component. The way we disentangle this is to measure the redshifts of a great many galaxies over large distances, and piece together what the true Hubble expansion rate actually is. Once we subtract that out, the leftover piece is each galaxy's peculiar velocity, and that gives us a window into exactly how *in*homogeneous, or non-uniform, today's Universe actually is (Fig. 3.19).

Of course, individual variable stars are only discernible — even with today's powerful telescopes — for galaxies that are relatively nearby in the Universe. But astronomers have discovered many more relationships between intrinsic brightness properties found in galaxies and easily observable quantities. For just a few examples, spiral galaxies obey the **Tully–Fisher relation**, which links the intrinsic brightness of a galaxy with the speed of its rotation; elliptical galaxies obey (depending on their orientation) the **Faber–Jackson relation** or the **fundamental plane**,

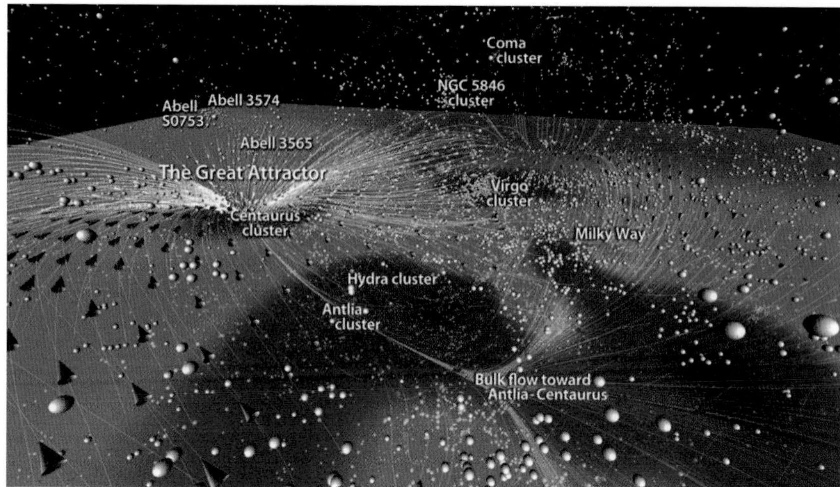

Figure 3.19 Peculiar velocity of galaxies. The scale of this map is approximately 300 million light years; regions with positive (attractive) peculiar velocities are in red, while those with negative (repulsive) peculiar velocities are in blue. Image credit: Helene M. Courtois, Daniel Pomarede, R. Brent Tully, Yehuda Hoffman and Denis Courtois, (2013). Supplemental material from their work Cosmography of the Local Universe. *The Astronomical Journal*, 146, 69.

where the intrinsic luminosity is related to the central velocity dispersion (observable as a spectral line width) of the galaxy; the fact that galaxies are made up of individual stars means that the brightness of different areas of the galaxy will exhibit **surface brightness fluctuations**, where a galaxy's distance can be determined based on the magnitude of the fluctuations in brightness away from the mean; and finally, **type Ia supernovae** exhibit similarities in their peak intrinsic brightness, which means if we measure the brightness curve of such a supernova over time, we can determine the distance to the galaxy where it occurs. In all cases, we have been able to measure individual stars in *some* relatively nearby galaxies that exhibit these relationships, and use those galaxies — along with the relationships mentioned earlier — to help figure out how far away even more distant galaxies are. The new technique(s) can then be applied to galaxies where individual stars are not measurable. The net result is that we have been able to measure the Hubble constant, and hence the expansion rate of the Universe, out to distances of well over a billion light years. The key finding in all of this? Hubble's Law, or the relationship between a galaxy's observed redshift and its distance from us, is incontrovertibly valid.

* * *

So the Universe is governed by General Relativity, linking matter and energy with spacetime, and not only contains a huge number of galaxies separated by millions of light years, but — on the largest scales — the Universe is expanding. As a consequence of this expansion, the distances between galaxies ought to be getting greater as time goes on, so long as they are not gravitationally bound to one another.

But what does this mean for our Universe? What did it mean for the past history of our Universe (the question of where it came from) as well as its future (what will eventually become of the planets, stars and galaxies that make it up)? There are a few different possibilities that are allowed in General Relativity, but which one would be correct? While it is true that there are often multiple mathematical solutions to equations, we only have *one* physical Universe! It would take a way of physically testing the Universe as a whole to determine which one of many reasonable ideas best represents the Universe we inhabit.

Chapter 4

The Great Leap Backwards: Theories On Where It All Came From

With the revolution of General Relativity now firmly entrenched, along with the observation that the Universe was filled with galaxies that were expanding away from one another on the largest scales, we had developed a clear picture of what our Universe looked like today. On small scales, matter was clumped together to form individual stars and planets, star clusters, globular clusters, nebulae, dwarf galaxies and individual spiral and elliptical galaxies. On intermediate scales, galaxies were clumped together in groups and, in some larger cases, clusters, containing as many as a few thousand times the mass of our Milky Way. These were the systems that were gravitationally bound, and so, over time, these individual systems would not expand away from any other system that they were bound to. And on the largest scales, the Universe was mostly uniform. Any galaxy, group, or cluster that lay beyond an individual, gravitationally bound system would inevitably find itself caught up in the Hubble expansion of the Universe (Fig. 4.1).

Over time, these bound systems — the galaxies, groups and clusters — will separate ever farther from one another; even the galaxies that are nearby today will become more distant from us as time goes on. Perhaps, for some of these systems drifting apart through space, gravitation will win out in the future, and more of these galaxies and groups will find one another, merging together. For others, perhaps the expansion of the Universe will win out, and these unbound systems will recede from one another for an eternity. One thing was for certain, though: if the total amount of matter in the Universe were a constant — which is what was

Figure 4.1 This is the "El Gordo" galaxy cluster, ACT-CL J0102-4915, the largest distant galaxy cluster ever discovered, as imaged with the Hubble Space Telescope. Despite its incredible mass, all the mass not bound inside of it still expands away from it, along with the rest of the Universe. Image credit: NASA, ESA, J. Jee (University of California, Davis), J. Hughes (Rutgers University), F. Menanteau (Rutgers University and University of Illinois, Urbana–Champaign), C. Sifon (Leiden Observatory), R. Mandelbum (Carnegie Mellon University), L. Barrientos (Universidad Catolica de Chile), and K. Ng (University of California, Davis).

expected, since energy can neither be created nor destroyed — but the Universe were expanding, then the Universe should be getting progressively less and less dense over time. Since energy density is just the energy of the Universe divided by its volume, then as the Universe expands, its volume increases, and hence its energy density must be dropping.

But will this go on forever? Many physicists, astronomers and cosmologists of the late 1920s, 1930s and 1940s concerned themselves with this pressing question. They wanted to know what the future fate of our Universe would be. You can imagine, in an expanding Universe filled with matter and other forms of energy, that there are two great, cosmic forces struggling against one another. On one hand, there is the Hubble expansion of the Universe, a consequence of having an isotropic, homogeneous Universe governed by General Relativity. The expansion causes galaxies, groups and clusters to increase their separation distances relative to one another. On the other hand, these are all massive objects, and General Relativity is our universal theory of gravitation, meaning that these galaxies, groups and clusters will attract one another. Cosmologists were then compelled to consider three different scenarios, which correspond to three different fates for the Universe:

(1) The expansion rate could be too large for the amount of matter and energy in the Universe. In the cosmic struggle between the initial, speedy rush of matter and radiation compelling things to fly apart and the long-lasting efforts of gravity to bring it all back together, the initial expansion dominates. As time goes on, the Universe will continue expanding indefinitely. While a small percentage of gravitationally bound systems will find one another and merge, the expansion will defeat gravitational attraction on the largest scales, and the Universe will eventually become cold and desolate. This equates to expansion defeating gravitation in great cosmic struggle.
(2) The amount of matter and energy could be very large relative to the expansion rate, which would mean that even though the Universe is expanding *now*, gravitational attraction will continue to slow down the expansion rate until all the galaxies reach some maximum separation distance from one another. At this point, the expansion will *reverse*, and gravitational contraction will ensue, causing the Universe to recollapse. This equates to gravitation winning the cosmic battle over expansion.
(3) Finally, you could imagine a case that was right on the border between the first two options, where a tiny bit more matter — just one more atom, perhaps — would cause it to recollapse, but the Universe just barely lacks the necessary matter to make it so. In this scenario, the expansion rate will asymptote to zero more quickly than in any other expand-eternally case, but it will never turn around. This case, right between the two extremes, is known as either a *critical Universe* or as the "Goldilocks" case, where the expansion rate and the amount of matter-and-energy in the Universe balance one another perfectly. This is what it would look like if neither expansion nor gravitation were favored in the great cosmic conflict between the two (Fig. 4.2).

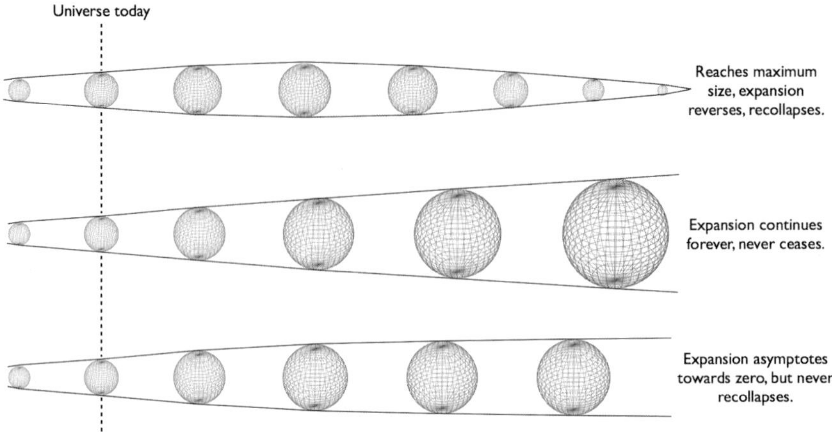

Figure 4.2 The three possible fates of the Universe, as envisioned in the early years after the discovery of the expansion of space. It could either recollapse (top), expand forever (middle), or exist on the border between those two possible cases (bottom). Image credit: E. Siegel.

These different fates — and their observational consequences — were first worked out by Georges Lemaître, and then expanded upon by Howard Robertson and Richard Tolman. The race to measure the both the redshift-distance relationship and how galaxies clumped and clustered together out to arbitrarily large distances, which would allow us to discern which of these three possibilities best described our Universe, *was on.*

* * *

But there was an equally important consideration in the *opposite* direction of time: what was the Universe like in the distant past? If the Universe is expanding now and matter-and-energy are conserved, then yes, of course the Universe will become less dense in the future. Remember that matter density is simply mass-per-unit-volume, and if the Universe is expanding, then the volume will increase, while the mass will remain the same. Well, that means that in the past, the matter density had to be much higher. Presumably, we had the same amount of mass in the Universe, but the volume was *smaller*, as the Universe must have expanded in order to reach its present size.

But matter is not the only thing in the Universe; for one thing, there is also radiation, or energy in the form of photons. More simply, we know it as light! Unlike matter, where the amount of energy it contains is primarily defined by its mass, photons are completely *massless*, and so their energy is defined by their *wavelength*, or the distance between two successive crests of a light wave. When the Universe expands, the energy content of matter (i.e., its mass) remains unchanged, but the energy content of radiation (i.e., its wavelength) *does* change. Let's take a deeper look at how this happens.

Imagine that you have a box that is exactly one wavelength in length, with a single wave of one wavelength spanning the entire box. If you now stretch that box, or increase its length, what happens to the wave inside? You might think that the wave could retain its original wavelength while the box expands, so that it simply now takes multiple wavelengths to fill the box. But this is not how waves work at all, due to the fact that you need either an integer (1, 2, 3, etc.) or half-integer ($\frac{1}{2}$, $\frac{3}{2}$, $\frac{5}{2}$, etc.) number of waves to span that box! This is true for other waves that you might be more familiar with. If you pluck a guitar string that is pinned at two ends, the entire string between those ends vibrates, and if you change where the string is pinned (by sliding your finger up and down the fretboard), the size of the vibrating part (and hence the wavelength of the sound) changes. If you have an open pipe of a certain length (e.g., wind chimes, a pipe organ, a trombone, etc.), it can only play a specific set of notes, and that is due to the fact that only certain specific wavelengths can exist inside that pipe. In fact, in the case of a trombone, you can slide the pipe to change its length while you are blowing air through it, causing the wavelength of the sound waves to change accordingly: lengthening (and becoming lower-pitched) if you lengthen the trombone, shortening (and becoming higher-pitched) if you shorten the trombone.

What does this mean when we try to apply this principle to light — or any other form of radiation — in the context of the expanding Universe? As space itself expands (or contracts, as the case may be), the wavelengths of the individual photons passing through that space get stretched (or compressed, if the Universe is contracting) in direct proportion to the scale of the Universe! As the Universe continues to expand, the wavelengths of the photons (or light-waves) get longer as they travel

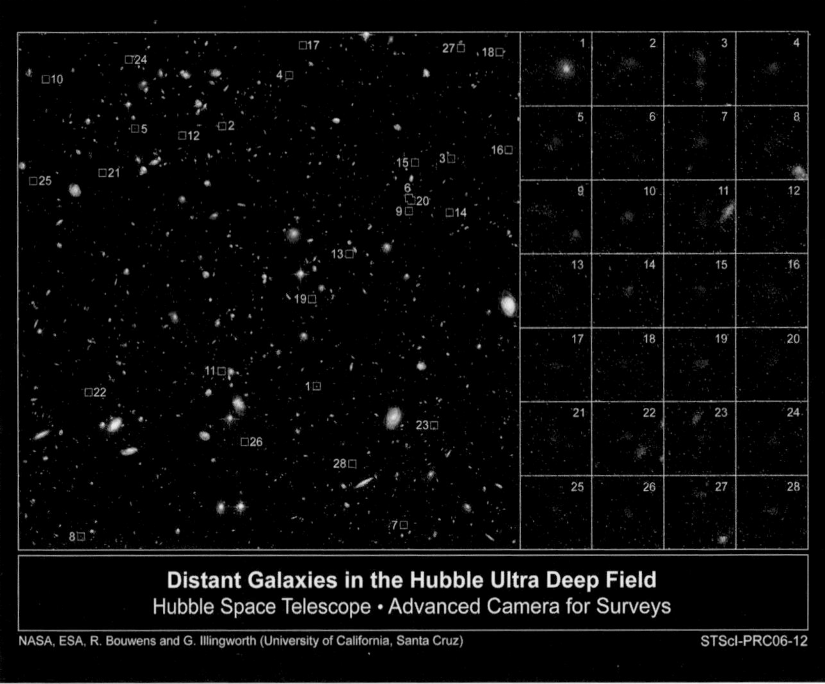

Figure 4.3 The faint, red galaxies highlighted in the numbered boxes are not *intrinsically* red, but rather have had their light — which, when emitted, had the same spectral properties as any nearby galaxy — shifted far towards the red by the expansion of the Universe. Image credit: NASA, ESA, R. Bouwens and G. Illingworth (University of California, Santa Cruz).

through the expanding space, and become shifted to longer-wavelengths and lower energies as a result, as we saw back in Fig. 3.17. This is the reason *why* distant galaxies appear redshifted: the light emitted and absorbed by their stars and atoms is no different than the light coming from the stars and atoms in our own galaxy. But as that light travels through space — which is expanding — the wavelength of the traveling light expands, too! By time it reaches our eyes, it is longer in wavelength, and hence, redshifted (Fig. 4.3).

Now that we understand how this works, let us apply this same principle to light in the Universe's distant past. If space has been expanding, that means that the Universe was *smaller* in the past, and that

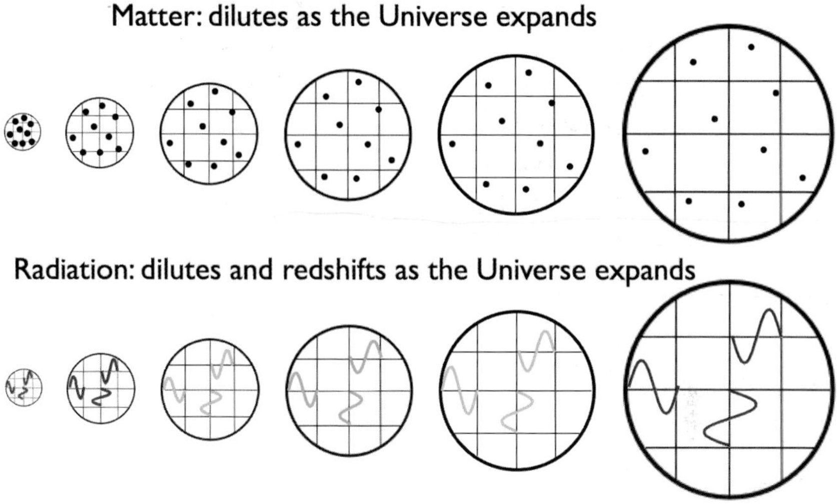

Matter: dilutes as the Universe expands

Radiation: dilutes and redshifts as the Universe expands

Figure 4.4 Both matter and radiation's number densities are higher in the past, but radiation gets an extra boost to its energy density owing to the fact that its wavelength was shorter, which translates to higher energies for each individual photon. Image credit: E. Siegel.

all the matter and radiation was closer together and denser. While massive particles like protons, neutrons and electrons still have the same respective masses regardless of density, when the Universe was smaller, the wavelength of the radiation was *also* smaller, meaning that the radiation was of a *higher energy*. Remember, particles of light with shorter wavelengths are more energetic than those with longer ones. When the Universe was half the scale (in all directions) compared to today, the number density of particles (both matter and radiation) was eight times higher than it is today, but each photon had half the wavelength it has today. This means that each individual photon had *double the energy* compared to what it presently has today, so the total photon energy density was *sixteen times* what it is right now! When the Universe was just 10% the scale it is today, the matter density was 1,000 times higher, but the energy density in radiation was *10,000 times higher*, since the number of photons-per-unit-volume was 1,000 times greater, but each photon had 10 times the energy. This rapid rise in energy describes how radiation worked in the Universe's distant past (Fig. 4.4).

* * *

In the 1940s, theorist George Gamow — a prominent advocate of Lemaître's work — was thinking about this property of radiation in the young Universe along with his student, Ralph Alpher. By this point, it was not only understood that all the known matter in the Universe was made of atoms, but that atoms themselves were made of positively charged atomic nuclei and negatively charged electrons, bound to one another through the electromagnetic force. In addition, atomic nuclei themselves were made up of positively charged protons and uncharged neutrons, which could be broken apart under the right conditions.

If this is what makes up the matter in our Universe, then we can combine this with our picture of expanding space to arrive at a Universe that is getting larger and less dense as we move forward in time. If we extrapolate backwards in time instead, we expect to find that it was denser, hotter and smaller in the past. Why would it have been hotter? Because the wavelengths of all forms of radiation, if we extrapolate backwards, were smaller, meaning the photons existing back then were higher in energy. At some sufficiently early time, Gamow realized, the energy of each photon would have been so high that each time it ran into a neutral atom, it would knock an electron clean off of it, creating an ionized (positive) nucleus and a free (negative) electron. In fact, the Universe was so dense back then that even if another electron came along immediately to meet up with the ionized nucleus, another photon would blast the atom apart practically instantly. In other words, by noticing the effect of the expanding Universe on radiation, it was easy to show that there was a time in the Universe's distant past where it was *too hot and too dense* to form neutral atoms!

This was a remarkable conjecture, extrapolating backwards billions of years in time. Once Gamow realized this, there was no reason not to go back even *farther*. While temperatures in the thousands-of-kelvin range would be enough to ionize every atom in the Universe, producing a hot, dense sea of plasma, when the Universe was young enough that the temperature reached the billions-of-kelvin range, even atomic nuclei would be destroyed by the high-energy radiation bath, leaving behind only protons and neutrons. And if protons and neutrons (or electrons) were made of still smaller, more fundamental particles, at early enough

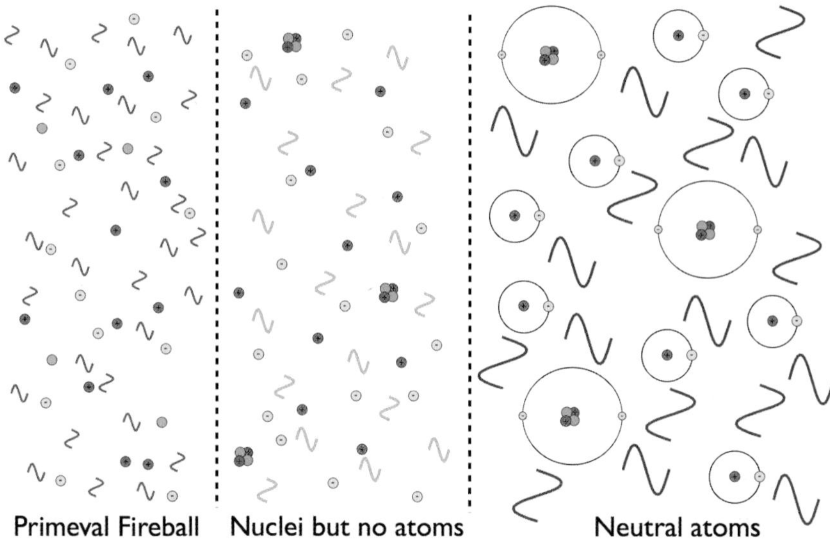

Figure 4.5 At late times, the energy of the Universe's radiation is low enough that neutral atoms can stably form. But at earlier times, when the radiation was energetic enough, any neutral atoms that formed would be immediately blasted apart, creating a sea of ionized plasma. Even earlier than that, if the radiation energy was sufficiently high, even atomic nuclei would be blasted apart, creating an era where protons, neutrons, electrons and photons were all unbound, with radiation as the dominant form of energy. Image credit: E. Siegel.

times, when the Universe was hot enough and dense enough, and the radiation had sufficiently short wavelengths, even they would be blasted apart, too.

Gamow called this state the **Primeval Fireball**, where radiation ruled the early Universe, and the matter than makes up you, me, and everything we know of could not have existed the way it does today (Fig. 4.5).

* * *

The implications of this idea were far-reaching and profound, and the way to figure them out was to imagine the earliest possible state of this hot, dense, radiation-dominated, rapidly expanding Universe, and then to step forward in time, obeying the laws of physics the entire time. Very early on, all you would have was a "primordial soup" of radiation plasma, made up of photons and every fundamental particle of matter (and antimatter) imaginable, all whizzing around at speeds approaching the speed of light. Collisions between particles would happen on an almost continuous basis, and energies would be so high that new particle–antiparticle pairs would create and annihilate at will. But through all this, the hot, dense Universe is expanding and cooling at an incredibly rapid pace.

By time just a single second has passed, the Universe will have cooled to a temperature of "only" about 11 billion kelvin, cool enough that matter–antimatter pairs can no longer be formed. Because matter and antimatter are oppositely charged and annihilate one another on contact, and all the galaxies in our Universe are dominated by matter (and not antimatter), all the antimatter particles have annihilated away with the vast majority of the matter particles. What is left over, by this time, is a small amount of matter, maybe one-part-in-a-billion, amidst a bath of radiation particles, which includes photons and neutrinos (Fig. 4.6). (If you are wondering how we know the Universe is matter-dominated and how it came to have more matter than antimatter, a full discussion can be found in Chapter 7.)

The matter particles that are left over include protons, neutrons and electrons, although none of them can bind together because the temperature is still too hot. But what *can* happen is that high-energy protons can collide with electrons, and if they have enough energy, they can convert themselves into neutrons and neutrinos. And the reverse can happen: neutrons and neutrinos can collide, and they can convert into protons and electrons. When temperatures in the Universe are high enough, these two reactions happen at roughly equal rates, so we have a Universe whose matter content is made up of 50% protons and 50% neutrons (along with the exact number of electrons necessary to keep the Universe electrically neutral by balancing the number of protons) to start. This makes sense, because protons and neutrons are *almost* exactly the same mass as each other, and hence have almost the same exact energy content as one another: a neutron at rest is just 0.138% more massive than a proton at rest.

This sets up two incredibly interesting things. First, when the temperature drops substantially *below* that mass difference, which happens after about one-to-three seconds, it becomes more difficult to collide a proton with an electron and produce a neutron-and-neutrino than the other way around.

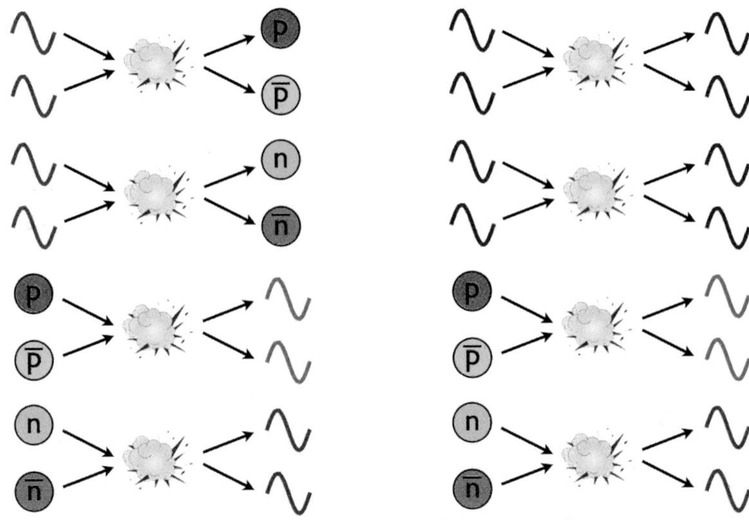

Figure 4.6 In the very hot Universe, with energies above a certain threshold (defined by $E = mc^2$ for each particle), radiation can spontaneously create pairs of particles and antiparticles, which can then spontaneously annihilate as well. When the energy of the radiation drops below the threshold for particle creation, only annihilation occurs. Note that particle–antiparticle pairs of a specific mass always produce photons of a specific energy, but only photons of that critical energy *or greater* can create the particle–antiparticle pairs in question. Image credit: E. Siegel.

Why is that? As the temperature of the Universe drops due to the expansion and cooling of the Universe, so does the kinetic energy of each particle. Proton–electron pairs simply no longer have enough energy to create the extra mass for a neutron when they collide with one another. But you can still turn neutrons-and-neutrinos into protons-and-electrons for a time, until the energy drops enough that neutrinos stop effectively interacting with neutrons. This means that the original 50/50 split between protons and neutrons, by time our Universe is a few seconds old, has become about an 85/15 split, with nearly six times as many protons as neutrons (Fig. 4.7).

But there is a second remarkable thing that happens: even though the temperature of the Universe has now dropped sufficiently to prevent

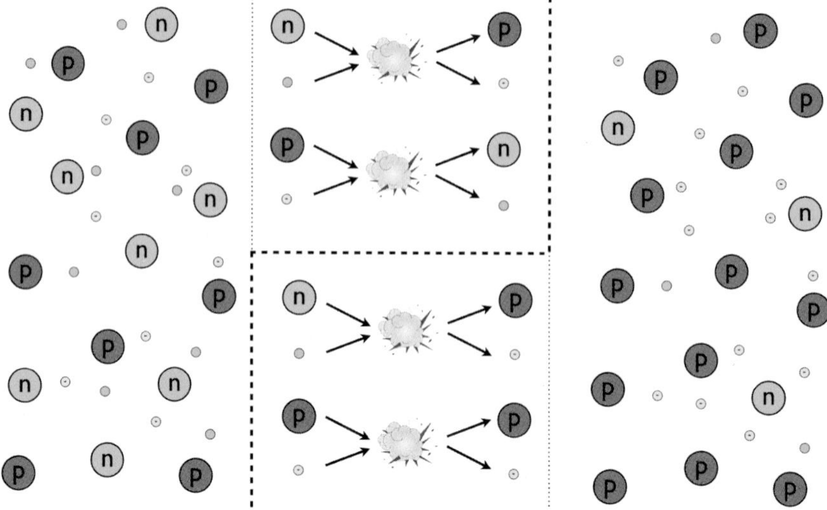

Figure 4.7 When the Universe is under a second old, there are approximately as many protons as neutrons, as protons (in red) and electrons (in yellow) can combine to form neutrons (in green) and neutrinos (in blue) just as easily as neutrons and neutrinos can combine to form protons and electron. But over the first few seconds, the temperature drops, meaning that while neutron–neutrino interactions can still easily produce protons and electrons, the proton–electron interactions no longer have enough energy to create a neutron–neutrino pair when they collide. This winds up creating an excess of protons to neutrons in approximately a 5:1 ratio. Image credit: E. Siegel.

neutrons-and-neutrinos from converting into protons-and-electrons (and vice versa), it is still too hot to fuse those protons and neutrons together. Yes, the temperatures and densities are high enough that nuclear fusion can occur, but the intense bath of radiation presents a problem known as the *deuterium bottleneck*. Forming deuterium — a nucleus consisting of one proton and one neutron bound together — is the first step in all nuclear fusion chain reactions that lead to the formation of heavier elements. By combining one proton with one neutron, you can form deuterium, which is about 0.2% lighter than an individual neutron-and-proton, each on their own. But when you are immersed in a sea of high-energy radiation, if you get struck by a particle whose energy is *higher* than the binding energy of that deuterium nucleus you just formed, that deuterium will get blasted back apart into a proton and a neutron. Even though the *average* energy of radiation is much lower than this binding energy, as long as your typical deuteron gets destroyed faster than it gets created (and remember, there are more than a *billion* photons for every proton in the Universe), you will continue to effectively have a Universe filled with free protons and neutrons.

The protons are not harmed by the inability to fuse into anything heavier; they can wait. But a free neutron is *unstable*! With a mean lifetime of just under 15 minutes, the neutron is by far the longest-lived simple, unstable particle. [By "simple" particle, we mean either a lepton (e.g., electron, muon, tau), meson (a combination of a quark and antiquark, like a pion) or a baryon (a combination of three quarks, e.g., protons and neutrons). Unstable composite nuclei, such as tritium (a nucleus consisting of one proton and two neutrons), can live for much longer. The muon, the second longest-lived "simple" particle, has a mean lifetime of only 2.2 *micro*seconds!] Even though it took only about three seconds for the Universe to go from a 50/50 split of protons and neutrons to an 85/15 split, it will take more than three *minutes* for the radiation to cool enough so that it will not blast deuterium apart (back into a proton and neutron) immediately after its formation. During the time that the neutrons are free, a substantial fraction of them will decay: a free neutron decays into a proton, an electron and a neutrino (or more specifically, an electron antineutrino). By the time deuterium can be stably formed by interacting protons and neutrons, the matter in the Universe is nearly 88% protons, with just a smidge over 12% in the form of neutrons (Fig. 4.8).

You might be wondering why it is so important to trace out just what fraction of the Universe is in the form of protons and what fraction is in neutrons right about now. After all, it seems like a small detail amidst a sea of expanding, ultra-hot radiation. We need to remember that protons and neutrons are the building blocks of *all atomic nuclei*; what we learn from thinking through this process is exactly which elements existed (and in what quantities) before the formation of the very first star! With this in mind, let us continue.

When the Universe is able to form stable deuterium — which happens when the Universe cools to "only" about 800 million kelvin — protons and neutrons combine to do so at an incredibly rapid pace. After waiting

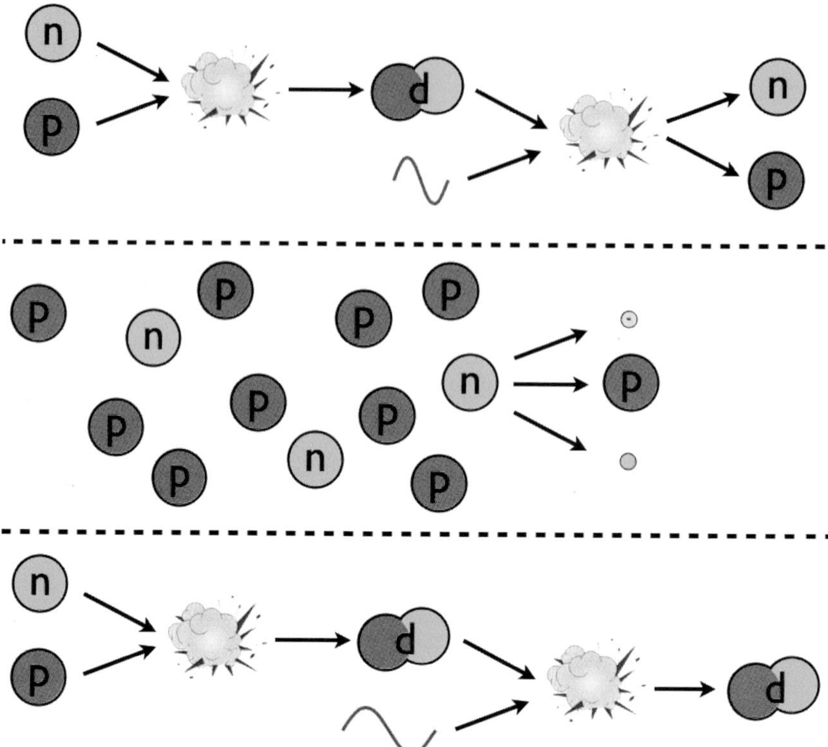

Figure 4.8 When protons and neutrons first attempt to fuse together, the radiation in the young Universe is *too hot*, immediately blasting apart any deuterium that forms (top). For slightly over the first three minutes of the Universe, the radiation keeps the Universe filled with free protons and neutrons, and some fraction of the neutrons spontaneously decay during this time (middle). Only after the Universe cools so that the radiation is low enough in energy that deuterium can stably form does it do so (bottom), overcoming the deuterium bottleneck. At last, nuclear fusion can proceed. Image credit: E. Siegel.

around for nearly four minutes, the free neutrons disappear incredibly rapidly, fusing with protons to form deuterium. But the Universe does not stop there! The Universe is still hot and dense enough that some deuterium nuclei find another proton, forming an isotope of helium known as helium-3, with two protons and one neutron. Other deuterium atoms fuse together with a neutron, becoming a quasi-stable isotope of hydrogen known as tritium, with one proton and two neutrons. Both

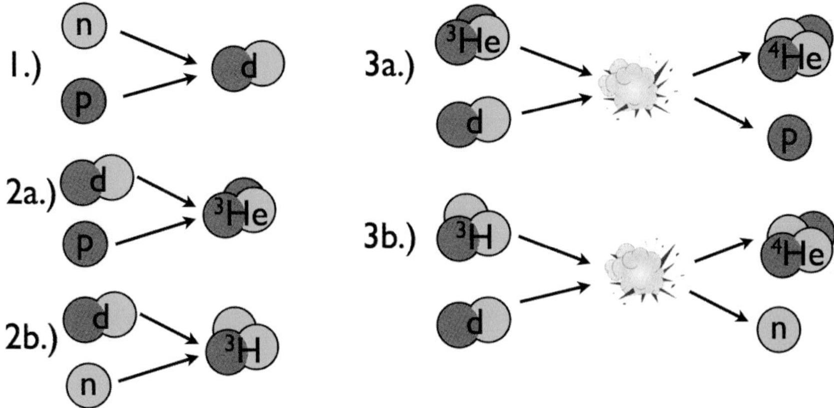

Figure 4.9 This is the major mechanism that the lightest atomic components form heavier elements through in the early Universe. In the first stage, protons and neutrons finally combine to form deuterium, which practically *all* of the neutrons do spontaneously. In the second stage, deuterons combine with either free protons or, more rarely, any leftover neutrons, to form the first nucleus with a mass of three: helium-3 or tritium (hydrogen-3), respectively. Finally, those mass-3 nuclei can combine with another deuteron, giving rise to helium-4 and either a free proton or neutron. Image credit: E. Siegel.

helium-3 and tritium can interact with another deuterium nucleus, forming helium-4 (two protons and two neutrons) and spitting out either a proton or a neutron, respectively, which go back to the beginning of the chain (Fig. 4.9).

But what about elements *heavier* than helium-4? We could try adding a proton to it, forming lithium-5, or another neutron, forming helium-5. Although this can happen, they live less than 10^{-21} seconds apiece, decaying back into helium-4; there is no nucleus with a combination of 5 protons-and-neutrons that is stable. We could also try combining two helium-4 nuclei together, forming a nucleus of beryllium-8. This can happen too, but fares little better, living just under 10^{-16} seconds before decaying back into helium-4. The tiny increase in lifetime does not give beryllium-8 the opportunity to form much of anything heavier that might be stable (like adding another neutron to form beryllium-9) before it disappears. Because it had to wait around for nearly four minutes for nuclear fusion to proceed, the Universe is too cool and diffuse to fuse any significant amount of elements heavier than helium. While practically all of the neutrons that are left by this time wind up becoming part of helium-4 nuclei, only trace amounts of lithium and beryllium (the third and fourth elements in the periodic table) are able to form, with not a single atom heavier than those.

The process that fuses protons and neutrons into helium completes in mere seconds, and leaves us with a Universe that's about 75–76% protons (i.e., hydrogen nuclei) and 24–25% helium-4 nuclei, by mass. (That's the equivalent of about 92% protons and 8% helium-4, by number.) A small amount of deuterium and helium-3 — about 0.001% each — will also remain, along with about 0.0000001% lithium. (The vast majority of beryllium formed is beryllium-7, which decays into lithium-7 with a half-life of 53 days.) With temperatures and energies now low enough, none of these atomic nuclei will be destroyed, nor will any new atomic nuclei be created, not for millions of years. This entire process — the formation of the lightest elements in the Universe — is known as Big Bang Nucleosynthesis, and its predictions are in spectacular agreement with our best modern observations. The ratios of elements found in the Universe will remain unchanged for millions of years, until the first stars form (Fig. 4.10).

* * *

With the Universe now consisting of hydrogen and helium nuclei along with electrons and radiation, you might suspect that the electrons will find these atomic nuclei to form neutral atoms. But just as nuclear reactions (like atomic bombs) are far more powerful than chemical reactions (which are powered by electron bonds and transitions), the energy required to ionize an atom is nearly a factor of a *million* lower than the energy needed to split apart an atomic nucleus. The Universe is expanding according to the laws of General Relativity as time goes on, which means the radiation within it continues to redshift and lose energy. In the meantime, the nuclei and electrons spread out as the Universe's density drops, but for hundreds of thousands of years things remain too energetic to form neutral atoms. The atomic nuclei are totally stable, the electrons are totally stable, but every time they attempt to bind together, a photon above the ionization threshold comes along and liberates the electron. For all this time, the Universe remains 100% ionized.

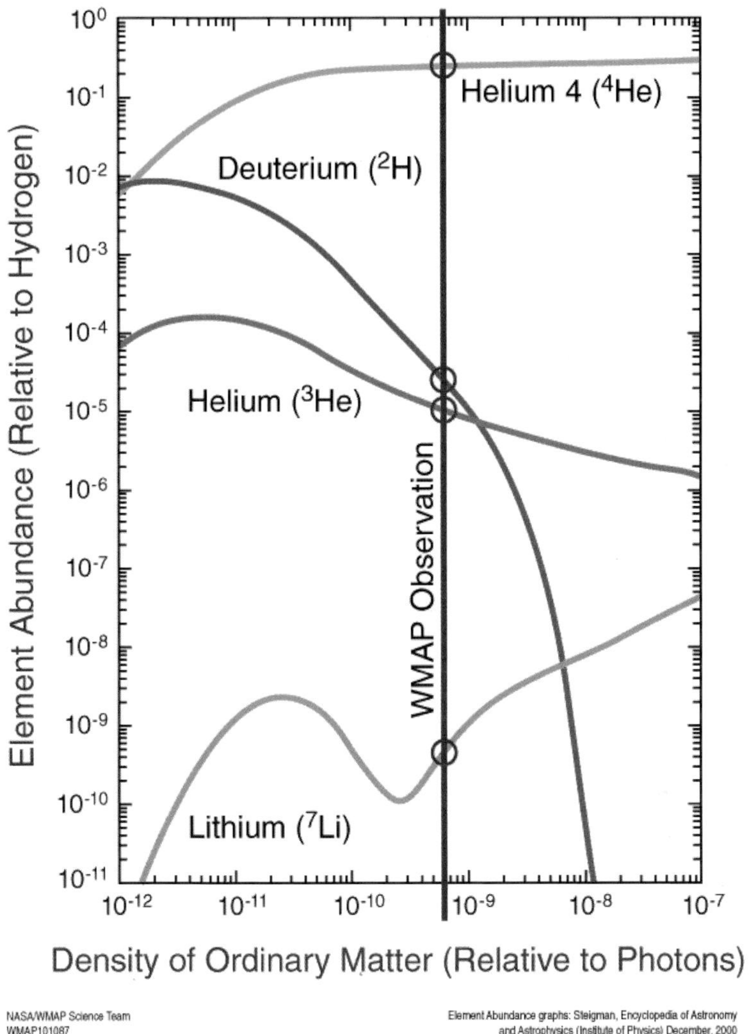

Figure 4.10 This is the best measurement of the present abundances of all of the light elements in the Universe, compared with the theoretical predictions from Big Bang Nucleosynthesis. The circles show the points where theory and observation match. Image credit: NASA/WMAP science team.

With photons outnumbering electrons by something like a billion-to-one, the wavelength of the Universe's radiation needs to stretch by incredible amounts in order for neutral atoms to form. Instead of hundreds of millions of kelvin, the Universe needs to cool down to temperatures of only about three *thousand* kelvin, which takes more than 300,000 years! During this time, collisions between electrons and photons are plentiful, with each photon bouncing off a free electron (through a process known as Thomson scattering) each chance it gets. Given that photons move at the speed of light, and that free electrons are found everywhere throughout the Universe at these early times, you can imagine each photon is like a pinball, bouncing off of electrons continuously and from every direction. As the Universe expands and becomes both cooler and more diffuse over time, the frequency of electron collisions that each photon experiences drops from trillions upon trillions of times per second all the way down to just a handful of times per second. The Universe, once denser than the core of the Sun, now has electrons and nuclei separated not by tiny, atomic-scale distances, but by millimeters.

During the first few hundred thousand years of the Universe, electrons and nuclei had been finding one another copiously, forming atoms like neutral hydrogen pretty much at will. The problem was that almost instantaneously — in less than a billionth of a second — a photon with enough energy to knock that electron back off would come along and smack into that atom, reionizing them immediately. By the time 300,000 years have passed, the vast majority of the photons around are too low in energy to have that effect. Remember that atoms can only absorb photons of a very specific wavelength: those are the wavelengths that can allow an electron within the atom to transition from one energy level to another. Even though photons outnumber electrons and nuclei by about a-billion-to-one at this point, the vast majority of photons do not have enough energy to ionize an atom. If all electrons and nuclei needed to do was wait for the background temperature to fall to a low enough value in order to become stable atoms, they could have done it long before this.

But another phenomenon occurs, preventing the formation of neutral atoms even when photons lose their ionizing energies: every time an electron and a nucleus find one another to form a neutral atom, they also *emit* a photon. And that emitted photon will have the right amount of energy — when it runs into *another* neutral atom — to ionize the one it runs into. This brings up a puzzle: how, if every time you make a neutral atom, the photon emitted from that process goes ahead and ionizes a different neutral atom, can the Universe ever become neutral at all?

The expansion of the Universe helps a little bit; occasionally, a photon emitted when an atom becomes neutral will be redshifted (thanks to the expansion of space) enough so that it can no longer be absorbed by another atom. Although this does happen, it is not the reason the Universe's atoms become neutral. In fact, if this were the process we relied on for the Universe to form neutral atoms, it would have taken tens of *millions* of years, not hundreds of thousands. Instead, we owe the formation of neutral atoms to a curiosity of atomic physics.

Atoms do not simply transition from a free electron and an ionized nucleus to their neutral, lowest-energy (or ground) state. There are many different energy levels in an atom, and the largest energy differences — where the most energetic photons get emitted and absorbed — come from transitions into and out of the ground state. A transition into the ground state will emit a powerful ultraviolet photon, which can then be absorbed by another neutral atom already in the ground state, exciting it, which either allows it to be easily ionized or, if that does not happen, it will drop back down into the ground state, emitting another ultraviolet photon. So long as these ultraviolet photons fill the Universe, no neutral atom is safe for long, and the net number of neutral atoms will not increase. At least, that is what *would* happen 100% of the time, if it were not for the fact that there are different types of atomic orbitals, with different configurations. The ground state (n = 1) orbital is always spherical, while the excited states can either be spherical or have a preferred axis, dependent upon what type of excited state the electron happens to be in. A transition from a non-spherical state to the ground state will always emit one ultraviolet photon, which gives us the problem we have just talked about. But if the electron happens to be in an excited state with a spherical configuration (e.g., the 2s, 3s, 4s, etc., orbital), the rules of quantum mechanics forbid it from emitting just one photon containing all the energy. Instead it emits two photons, each with half the energy, as it transitions to the ground state!

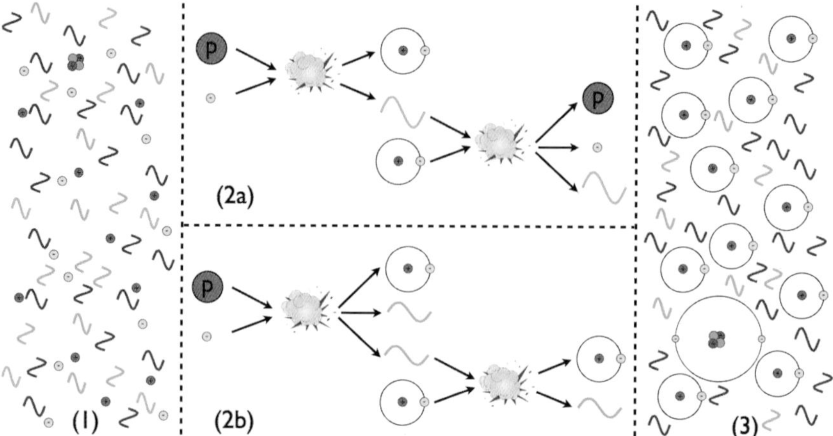

Figure 4.11 For hundreds of thousands of years, there are enough energetic photons to ionize atoms as soon as they become neutral (panel 1). When they do become neutral, most of the time, they emit an ionizing photon, which will be absorbed by another atom, causing that new atom to ionize (panel 2a), which results in no net neutral atoms. It is only due to a rare two-photon transition — a property of the hydrogen atom itself — that we increase the number of neutral hydrogen atoms, and simultaneously decrease the number of ionizing photons (panel 2b). In the end, we are left with not only a neutral Universe, but one without any photons energetic enough to ionize atoms (panel 3). Image credit: E. Siegel.

The probability of having two such photons strike a neutral, ground-state atom simultaneously is so small that the odds are minuscule that it happened *even once* in our Universe while it was undergoing the process of forming neutral hydrogen. It is this rare, two-photon transition into the ground state that allows the Universe to form neutral atoms so quickly; after approximately 380,000 years, it "only" takes about 117,000 years to go from a Universe that is 100% ionized, made up of free electrons and nuclei, to one that is 100% neutral, made up exclusively of stable, neutral atoms (Fig. 4.11).

Once the Universe becomes neutral, all the photons left over from the Big Bang no longer have free charged particles to interact with and scatter off of. Even though they outnumber protons, neutrons and electrons by billions-to-one, they no longer have the correct energies to interact with them in their neutral configurations. All they can do is travel in a straight line, affected only by gravitation and the expansion of the Universe, until

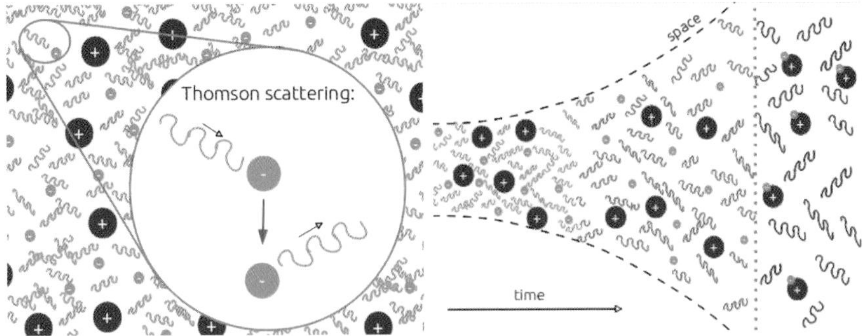

Figure 4.12 When the Universe is ionized, photons scatter off of charged particles (especially electrons) with an incredibly high frequency, via a process known as *Thomson scattering*. Once the Universe becomes neutral, however, the photons no longer collide with them, and simply travel in a straight line at the speed of light, a process known as *free-streaming*. The surface where the photons last scattered is marked by the vertical dotted line. Image credit: Amanda Yoho (Case Western Reserve University), used with permission.

something comes along with the right properties to interact with them (Fig. 4.12).

* * *

Gamow, of course, did not know all the specifics of this. In the 1940s, when this was first being worked out, no one had seen any evidence for either a leftover bath of radiation, nor was the physics of nuclear fusion well-understood. Many of the caveats of atomic physics necessary to understand neutral atoms had yet to be discovered, including the energy differences between spherical and non-spherical excited electron states. And yet, Gamow still knew enough to make two tremendous, albeit relatively general, predictions:

1) The Universe started off, before *any* stars ever formed, with more than just hydrogen. Heavier elements existed thanks to the nuclear physics that occurred during the first few minutes of the Universe, forming deuterium, helium and possibly heavier elements than that. The only things that the initial ratios of elements should depend on are the laws of nuclear physics and the number of

protons, neutrons and photons that the Universe had at that early time. If we came to understand the laws that governed nuclear interactions, the only thing that the early abundance of elements should depend on is one parameter: the ratio of baryons (protons and neutrons combined) to photons. Since there are many elements and isotopes to observe, it should be easy to check whether these predictions are correct or not.

2) The Universe became neutral — and hence transparent to the radiation from this primeval fireball — when the Universe was a few hundred thousand years old. Since all of this radiation was colliding and interacting with charged particles copiously during this early stage, it should exist with a very particular energy spectrum: a **blackbody spectrum**. Once the Universe becomes neutral, this radiation should be mostly unaffected by everything except the expansion of the Universe, which will cause it to redshift significantly. The net result is a prediction that, today, there should be a blackbody spectrum of radiation left over from this early time, emerging from the surface of last scattering, and having cooled by this point to be only a few kelvin above absolute zero.

All of this could be predicted using only the laws of physics (gravitational, electromagnetic and nuclear) known at the time, and applied to an expanding, isotropic and homogeneous Universe. Gamow's big idea to consider what happened to the matter in the Universe — both through electromagnetic and nuclear reactions — as the Universe expanded and cooled from an initial, arbitrarily hot and dense state. If this picture of the Universe was correct, then we should, in principle, be able to detect these two observable signatures. If we could find a sample of gas from an early enough time, that neither formed stars nor was enriched by material ejected from prior generations of stars, we could measure the abundance of the elements found within, and see whether they matched the predictions of Gamow's theory or not (Fig. 4.13).

And if we could search sensitively enough, at the right energies, we should be able to find this blackbody radiation that peaked somewhere in the microwave portion of the electromagnetic spectrum: the leftover glow from the earliest times in our Universe.

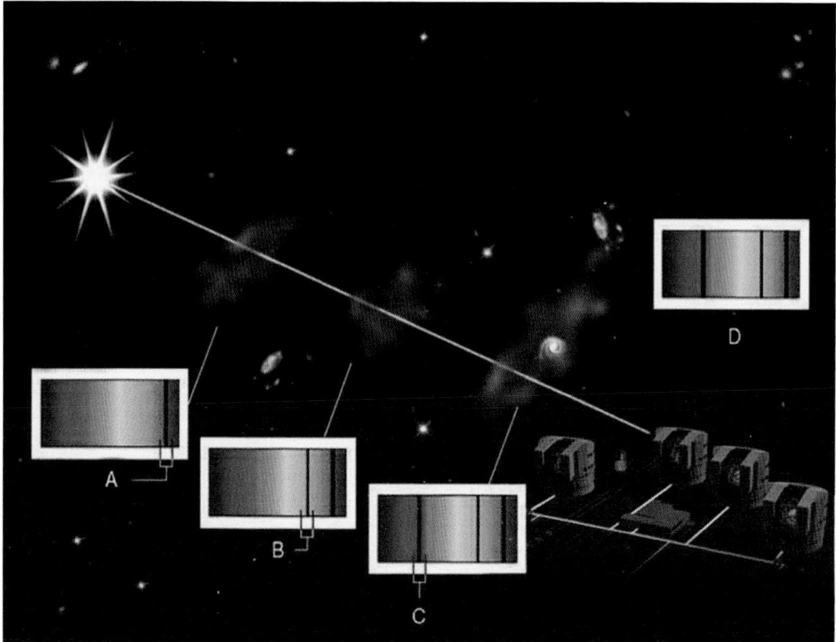

Figure 4.13 If light from a very distant source happened to pass through a molecular gas cloud that had not yet formed any stars within it, we would be able to measure the concentration of pristine elements and isotopes by looking at the spectrum of absorption lines. Image credit: European Southern Observatory (ESO).

This was a remarkable idea, and Gamow, along with Ralph Alpher, Robert Herman, and — in one famous paper — Hans Bethe (pronounced "Beta," and added to the Alpher and Gamow paper without his knowledge, so that Gamow could have the author list make the pun "Alpher Bethe Gamow") spent large portions of the 1940s and 1950s working out the details of these predictions. But this was not the only big idea bandied about concerning the earliest times in the Universe.

* * *

The key observation that every idea about the history of the Universe had to take into account was the redshift–distance relation discovered and developed by Hubble and others. The earliest alternative interpretation was that rather than an expanding Universe, the galaxies could be receding from us as the result of some type of explosion. Indeed, if you take a collection of matter at rest and give it a large amount of energy, the faster-moving particles will, over time, wind up farther away from the source of the explosion than the slower-moving ones. In principle, there is no reason why this could not be true, though if it were the case, it would mean both that Einstein's General Relativity was incorrect, and also that the Milky Way was suspiciously located at (or remarkably near) the very center of this initial, spherical explosion.

This, of course, runs counter to the central idea that the Universe is spatially homogeneous (the same at all locations) and isotropic (the same in all directions) on average on the largest scales. This is sometimes called the Copernican principle, but is also known as Einstein's cosmological principle or just the cosmological principle, a term coined by Edward Arthur Milne, who developed this alternative a little further in the 1930s. Milne's model is consistent with Special (but not General) Relativity, and assumes a spherical Universe that's isotropic but *not* homogeneous, that increases in density as we move radially away from our location. His central argument was that if the Universe appeared homogeneous to one observer, then a moving observer — by the principle of relativity — would see a different density in one direction than in another. Only if the Universe were isotropic, non-homogeneous and spherical could a coordinate transformation be made to reproduce the observed Universe. This argument only holds true in a flat spacetime without the laws of General Relativity governing it. However, the successes of Einstein's theory of gravitation are inconsistent with this alternative picture, and have made this an uninteresting avenue to subsequently pursue.

Another alternative dates all the way back to 1929, proposed by Fritz Zwicky. Rather than a cosmological origin of redshifts, due to the expansion of the Universe, Zwicky put forth the proposal that over great distances, light simply *lost energy* over the great distances it traveled before it reached our eyes. An object that was twice as far away as another would have its wavelength lengthen by twice the amount neither because it was moving away twice as fast (which would be due to a Doppler shift) nor because the Universe expanded for twice as long before the light reached our eyes (which would be due to cosmological expansion), but because the light got twice as *tired* on its journey. This alternative explanation — a *Tired-Light* cosmology — was an interesting non-cosmological option that would explain Hubble's Law, as more distant objects would simply be more redshifted due to light intrinsically losing energy as it traveled through the cosmos. The only known way for light to lose energy in this manner is to have it interact with other particles, causing it to scatter, and any sort of scattering would blur distant object by an amount that was greater than the observational limits of the day. An interesting consequence of any Tired-Light cosmology would be that distant objects would not be subject to any time-dilation effects that *are* present in the standard picture. Distant supernovae would brighten and dim in the same amount of time as nearby ones, the intrinsic surface brightness of galaxies would *not* appear to change with redshift (and they do in the standard picture, since they were closer to us in the past, when the light was emitted), and the spectrum of any distant galactic or stellar radiation would be shifted away from a blackbody spectrum. All of these are in conflict with observation, and so while Tired-Light might be a fun alternative to consider, it never was a viable candidate for describing our Universe (Fig. 4.14).

Plenty of other alternatives have been proposed throughout the years, including the Gödel Universe, which is a solution to General Relativity where the Universe has a global rotation to it, and the mathematical curiosity of also having closed time-like curves, which means that time travel would be possible if this described our Universe. The *plasma Universe* (or plasma cosmology) held that, on large scales in the Universe, electromagnetic forces were more important than gravity, that the clustering of galaxies was due to large electric currents that caused them to align and intersect, and that electromagnetic forces gave spiral structures to galaxies and triggered the collapse of gas clouds to cause the formation of stars. There was also one alternative focused on objects known as quasars, which is an acronym for **q**uasi-**s**tellar **r**adio **s**ources (QSRS). When quasars were first discovered in reasonably large numbers, they only appeared at what seemed to be regularly spaced redshifts, whereas the standard picture predicts that they ought to exist smoothly at all redshifts. The idea of *redshift quantization* took hold among a few cosmologists, who claimed that these quasars had intrinsic redshifts, rather

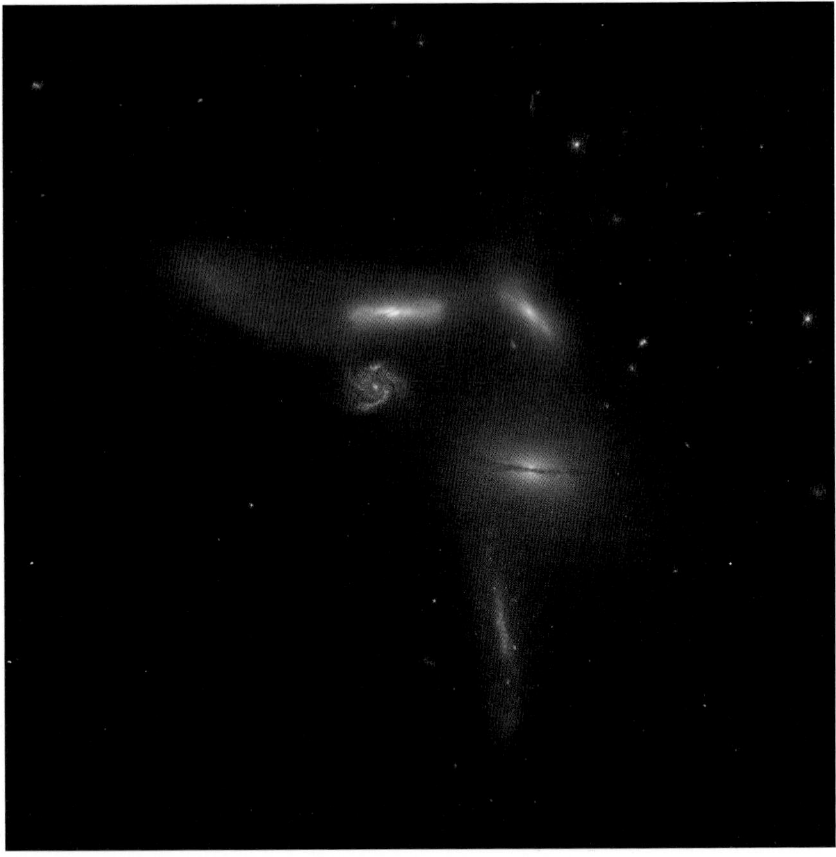

Figure 4. 14 This group of galaxies is known as Seyfert's Sextet, which actually consists of four galaxies in a group, one of which is significantly gravitationally disrupted, and one galaxy — the face-on spiral — that is some five times as far away as all the others. The fact that the more distant galaxy is not at all blurred relative to the closer ones, and that events in it do experience time dilation, rule out the Tired-Light description of our Universe. Image credit: Hubble Legacy Archive, NASA & ESA, with processing by Judy Schmidt.

than appearing at cosmologically tremendous distances. Although these alternatives were all worth considering at various points in time, they have all been thoroughly ruled out: observational limits on the global rotation of the Universe show that it is negligibly small; plasmas turn out to be important in a number of astrophysical processes but cannot be responsible for *any* of the above claims made by plasma cosmology; and the subsequent discovery of tens of thousands of quasars show a continuous, smooth distribution, with absolutely no redshift quantization present.

* * *

But there was one major contemporary competitor to Gamow's big idea, one that deserved to be taken just as seriously as his prediction of a primeval atom: the Steady-State model of the Universe. In addition to a Universe that was isotropic and homogeneous in space — or the same at all locations — the Steady-State model held that the Universe was also the same at all locations *in time*, which its proponents called the "perfect" cosmological principle. This idea came along with a very different interpretation for the past and future histories of the Universe, as well as a very different set of predictions for what we would find with improved technology.

Gamow made a great leap to suggest that observing an expanding Universe today necessarily implied an arbitrarily hot, dense state in the past. Instead, building upon an older idea of James Jeans, the British scientists Fred Hoyle, Hermann Bondi and Thomas Gold pointed out that instead of Hubble's Law implying that the matter density in the Universe dilutes over time, the density could be kept constant if new matter were created over time *in between* the distant galaxies as they expanded away from one another. The rate of matter creation needed to keep the density constant would actually be *tiny*, something like one hydrogen atom per cubic meter every ten billion years! If matter were spontaneously created at this meager rate, then each and every location in the Universe — on average — would *always* contain the same numbers of galaxies, the same populations of stars, and the same abundance of elements, no matter where or when we looked (Fig. 4.15).

A Steady-State Universe would have no beginning and no end, either in space or in time. Instead of a hot, dense primeval fireball giving way to a cooler, less dense state amidst the rise of heavy nuclei and neutral atoms, the Steady-State theory predicted that heavy elements arose exclusively through nuclear processes taking place in stars, and that the only background radiation that exists would originate from reflected (or absorbed and re-emitted) starlight. In addition, there would be very

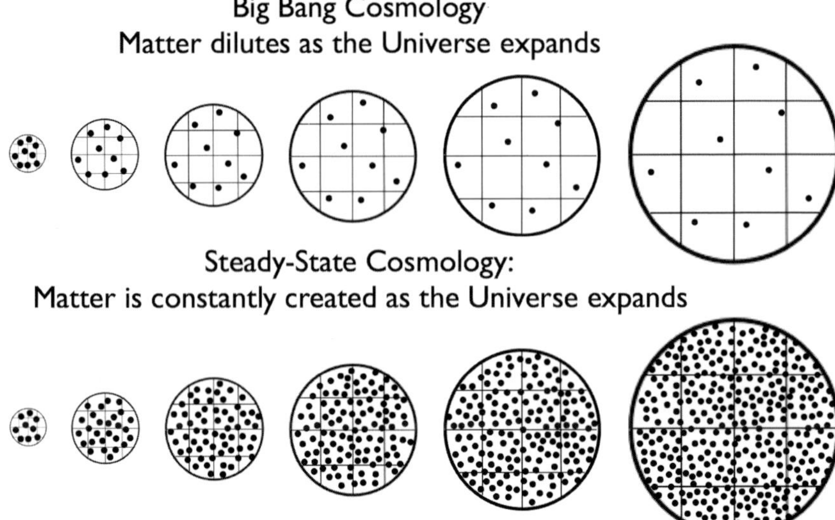

Figure 4.15 While an expanding Universe would seem to imply that the density of matter would dilute over time, it is also conceivable that new quantities of matter could get created as the Universe expands, leading to a Universe that was not only homogeneous (the same everywhere) in space, but the same *at all times* as well. This was the idea of a Steady-State Universe, requiring a very low but consistent rate of matter creation as the only new piece of physics. Image credit: E. Siegel.

different predictions of galaxy counts at large distances (since, in Gamow's model, the Universe was denser in the past, but not in the Steady-State model), the background temperature of the Universe would be uniform (rather than increasing with increasing redshift), and there would be slight deviations from Hubble's redshift-distance relation at extreme distances.

The straightforwardness of measuring these predictions — as well as the predictions of Gamow's idea — made these two models the leading scientific candidates to describe the Universe we live in. The two camps often snarked at one another from across the Atlantic, with Fred Hoyle deriding Gamow's primeval fireball idea on BBC radio in 1949, claiming that the Steady-State model's matter creation hypothesis

> "[…] replaces a hypothesis that lies concealed in the older theories, which assume, as I have already said, that the whole of the matter in the Universe

> was created in one 'Big Bang' at a particular time in the remote past. On scientific grounds this Big Bang hypothesis is much the less palatable of the two. For it is an irrational process that cannot be described in scientific terms."

The term of "the Big Bang" — coined by the idea's biggest detractor — would continue to live on as Gamow's greatest scientific legacy.

Chapter 5

An Element-ary Story: How The Stars Gave Life To The Universe

One of the most hotly contested points of the two main competing theories of the cosmos — the Big Bang and the Steady-State models — was about the origin of the heavier elements that we find not only here on Earth, but across the Universe as well. We know that atoms come in around 90 naturally occurring types here on Earth, with the number of protons in each nucleus determining what type of atom you have. We also have neutrons and electrons along with the protons, where the neutrons and protons are bound together in the atom's nucleus, and the electrons orbit the nuclei. The early stages of the Big Bang predicted that only free protons, neutrons and electrons should have existed, with heavier elements forming through nuclear reactions, while the matter creation hypothesis of the Steady-State theory only adds individual particles to the Universe. Were the heavy elements that are now so abundant created in the early stages of the hot Big Bang, as Gamow contended? Or were they created in stars, as favored by Hoyle and his contemporaries? To answer this, let us first take a look at how we know what elements actually exist in the Universe (Fig. 5.1).

* * *

It might seem crazy, but a mere century ago, we did not even know what elements the Sun and the stars were made out of. The prevailing assumption — believe it or not — was that the Sun, and therefore *all* stars, were made out of the same elements that the Earth is, and in

1 H																	2 He
3 Li	4 Be											5 B	6 C	7 N	8 O	9 F	10 Ne
11 Na	12 Mg											13 Al	14 Si	15 P	16 S	17 Cl	18 Ar
19 K	20 Ca	21 Sc	22 Ti	23 V	24 Cr	25 Mn	26 Fe	27 Co	28 Ni	29 Cu	30 Zn	31 Ga	32 Ge	33 As	34 Se	35 Br	36 Kr
37 Rb	38 Sr	39 Y	40 Zr	41 Nb	42 Mo	43 Tc	44 Ru	45 Rh	46 Pd	47 Ag	48 Cd	49 In	50 Sn	51 Sb	52 Te	53 I	54 Xe
55 Cs	56 Ba		72 Hf	73 Ta	74 W	75 Re	76 Os	77 Ir	78 Pt	79 Au	80 Hg	81 Tl	82 Pb	83 Bi	84 Po	85 At	86 Rn
87 Fr	88 Ra		104 Rf	105 Db	106 Sg	107 Bh	108 Hs	109 Mt	110 Ds	111 Rg	112 Cn	113 Uut	114 Fl	115 Uup	116 Lv	117 Uus	118 Uuo

Lanthanides	57 La	58 Ce	59 Pr	60 Nd	61 Pm	62 Sm	63 Eu	64 Gd	65 Tb	66 Dy	67 Ho	68 Er	69 Tm	70 Yb	71 Lu
Actinides	89 Ac	90 Th	91 Pa	92 U	93 Np	94 Pu	95 Am	96 Cm	97 Bk	98 Cf	99 Es	100 Fm	101 Md	102 No	103 Lr

Figure 5.1 Understanding the cosmic origin of all the elements heavier than hydrogen can give us a powerful window into the Universe's past, as well as insight into our own origins. Image credit: Wikimedia Commons user Cepheus.

roughly the same proportion. The reasoning, although flawed, was pretty straightforward, and based on the simple physics-and-chemistry of atoms.

Every element in the periodic table (which *was* well-understood back then) has a characteristic spectrum to it, for both emission and absorption. When a neutral atom is heated up, the electrons in it transition to a higher energy state; as the electrons then drop down to lower energy states, they emit light of very particular wavelengths. Based on the number of protons and neutrons in the nucleus, each atom has its own particular *emission* spectrum, based on the physics of electrons in orbit around their particular atomic nucleus. On the other hand, the very act of heating up an atom — such as by a multi-spectral light source — causes an atom to absorb energy at those very same particular wavelengths: its *absorption* spectrum. If we consider an object like our Sun, which is perhaps the most obvious example of a multi-spectral light source, and we break it down into all the different individual wavelengths possible, we could figure out what elements are present in its outermost layers from the absorption features that we see in the solar spectrum.

This technique of breaking up the light from an object into individual wavelengths for further study is known as spectroscopy. When we do this

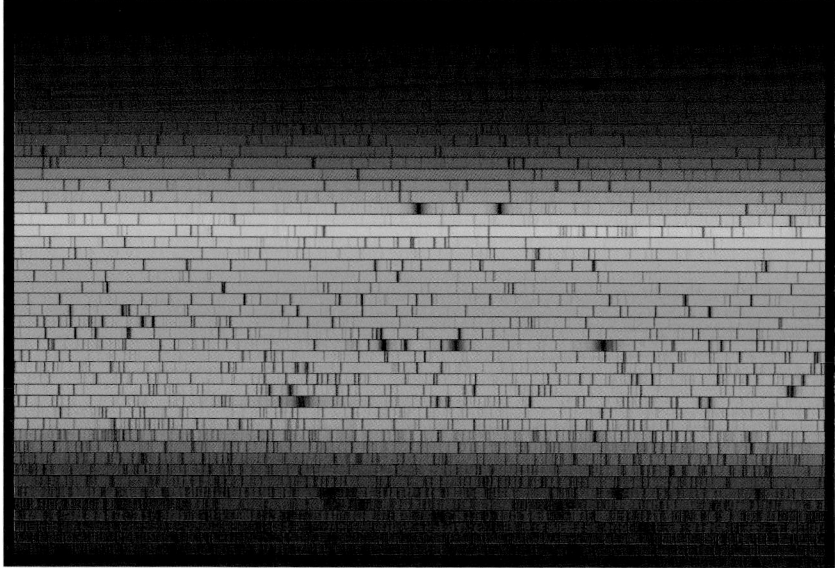

Figure 5.2 The solar spectrum shows a significant number of features, each corresponding to absorption properties of a unique element in the periodic table. The darkest, widest line occurs at 6,563 Å, in the red portion of the spectrum, while the entire chart shows the wavelength range from 7,000 Å (at the upper left) to 4,000 Å (at the lower right). Image credit: Nigel A. Sharp, NOAO/NSO/Kitt Peak FTS/AURA/NSF.

to the Sun, to no one's surprise, we find absorption features that are consistent with the same elements that we find here on Earth. But there is a notable feature that appears in the Sun's spectrum: some lines are *darker* and *wider* than others, while others are narrower and weaker. In particular, one absorption feature stands out more than any other in our Sun, in the middle of the red part of the spectrum. At a wavelength of 6,563 Ångströms (Å), where an Ångström is 10^{-10} meters, the darkest, widest and therefore *strongest* absorption feature in the entire visible spectrum of sunlight stands out against all the others (Fig. 5.2).

There are two factors that determine the strength (or weakness) of each of these lines, one of which is obvious and another which is not. The obvious one is that the more of a particular element you have, the stronger the absorption feature is going to be. The particular wavelength of 6,563 Å, if you were wondering, corresponds to the strongest visible line of the hydrogen atom, known as the Balmer-alpha (or Hα) line. It makes a lot of sense, intuitively, that the more of an element you have, the better it will be at absorbing light. But the second factor that must be understood to be able to successfully predict the strengths of these lines is a little more subtle: the level of *ionization* of these atoms.

Each atom, with its own specific atomic nucleus and its own unique pattern of electron shells and orbitals, has an explicitly determined spectrum when it is in its neutral configuration, or when it has *all* of its electrons. But each atom will *ionize*, or lose one (or more) electron(s), at different energies particular to the number of protons in that atom's nucleus. So while it takes 13.6 electron-Volts (eV) of energy to knock an electron (its only one) off of hydrogen, it takes a hefty 24.6 eV to knock one off of helium, but only 5.2 eV to knock one off of lithium. To knock a second electron off of helium (to make it doubly ionized) takes an additional 54.4 eV of energy, while to remove a second one from lithium takes a whopping 73.0 eV. So dependent on the energy that each atom is bathed in, it will exist in a particular ionized state (neutral, singly ionized, doubly ionized, etc.), and in that state will have its own, unique spectrum of absorption lines (Fig. 5.3).

Back before we understood how stars worked, we were still able to observe these absorption lines and classify stars by the relative strengths of their absorption. Stars were originally divided into classes (Secchi classes) according to the strength of their hydrogen lines, followed by

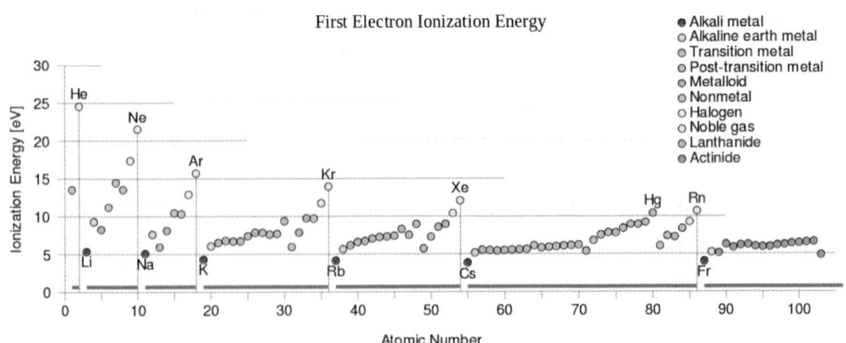

Figure 5.3 There is a periodic structure to ionization energy, with the first elements in each row the easiest to ionize, and heavier elements easier to ionize than the lighter ones in the same group. Noble gases and halogens are the most difficult of the atoms to ionize. Image credit: Wikimedia Commons user Sponk, under a c.c.-by-s.a.-3.0 license.

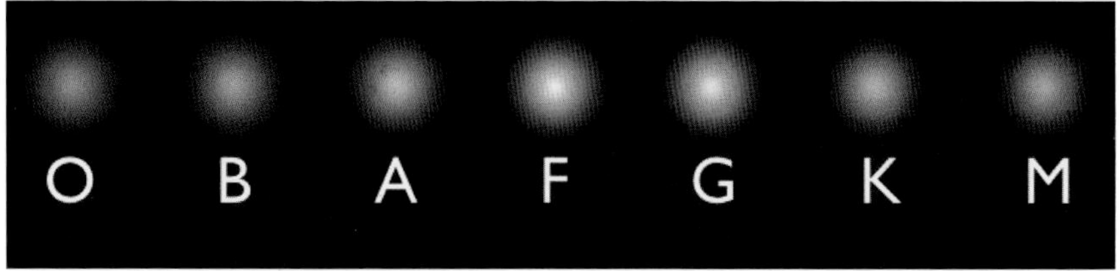

Figure 5.4 The seven major star classes, organized by their colors. It turns out that these colors also correspond to a star's surface temperature, and so O-stars are the hottest, while M-stars are the coolest. Image credit: E. Siegel.

metallic lines, complex band structure, carbon lines and finally their emission lines. These were eventually broken up into finer delineations denoted by the letters A through Q, which were later consolidated into just seven types (ABFGKMO) and reordered by their color (OBAFGKM), from blue-to-red. We still use this ordering today, with many different clever acronyms abounding to help people remember the order, ranging from the classic, "**O**h **B**e **A** **F**ine **G**irl, **K**iss **M**e," to the scatalogical, "**O**ats, **B**ran **A**nd **F**iber **G**et **K**ids **M**oving," to one that highlights a student's fear of failure, "**O**h **B**oy, **A**n '**F**' **G**rade **K**ills **M**e," to one in memory of the classic television show, The Golden Girls, "**O**ld **B**ea **A**rthur **F**ound **G**old **K**nocking **M**cClanahan." (Fig. 5.4.)

But it was not until 1925 — thanks to the Ph.D. dissertation of Cecilia Payne, which was called "undoubtedly the most brilliant Ph.D. thesis ever written in astronomy" by her contemporary, Otto Struve — that we understood *why* this was the case.

Why do the Sun's absorption features appear the way they do? There is an underlying cause for the phenomenon of how absorption lines and the colors of stars are related that was only uncovered by Payne herself: the underlying temperature of the stars. For each atom, *energy* is the only thing that determines its ionization state. This means that if we place any atom in an environment at a particular *temperature*, we will get different relative levels of ionization, and therefore different (but well-known) absorption features. Different lines will begin to appear with certain strengths as we turn up the temperature, and then as temperatures continue to increase, those lines will start to disappear, eventually being replaced by others. Remember, absorption in an atom happens because an electron

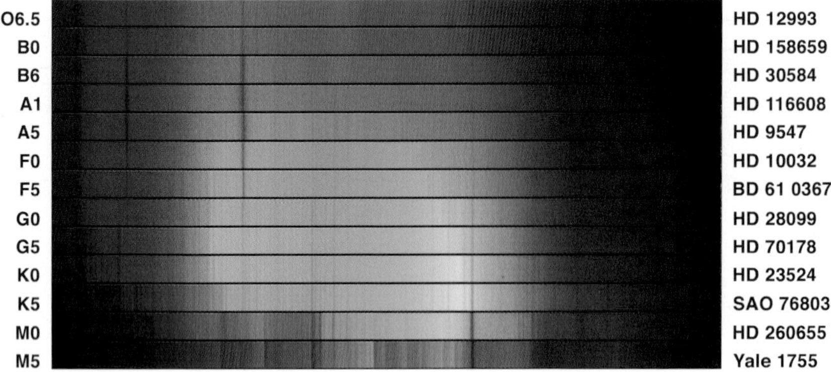

Figure 5.5 O-stars, the hottest of all stars, actually have *weaker* absorption lines in many cases, because the surface temperatures are great enough that most of the atoms at its surface are at too great of an energy to display the characteristic atomic transitions that result in absorption. Image credit: NOAO/AURA/NSF, modified to illustrate the stars that demonstrate this phenomenon.

is absorbing a photon and transitioning to a higher-energy state; if the electrons are not in the right configuration to be able to absorb that photon — because the atoms have *either* have too much or too little energy — the transition, and hence the absorption, simply does not occur. In fact, in the most extreme case, all you would have is a completely ionized nucleus, which is incapable of absorption lines altogether! The appearance and disappearance of different spectral features can be seen in stars of different colors, as shown in Fig. 5.5.

But this means, by measuring the relative ionization of atoms as well as the color of a star, you can learn what the intrinsic temperature of that star is. As it turns out, each spectral class corresponds to a range of temperatures, and so within each class, we break these seven letters up into numbers — 0 through 9 — which span the range from hottest to coolest, respectively. Once you know a star's temperature and also have its spectrum, you can finally take that long-awaited step to figure out exactly what it is made out of.

The giant revolution that came along with Cecilia Payne's work was learning that, sure, the elements found on the Sun were pretty much the same as the elements on Earth, but with two major exceptions. Helium and

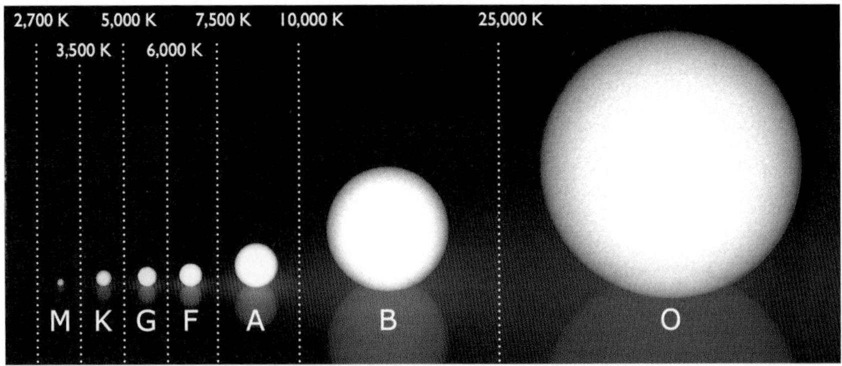

Figure 5.6 The (modern) Morgan–Keenan spectral classification system, with the temperature range of each star class shown above it, in kelvin. O-stars are shown so much larger than the others, which decrease in size, to represent how much more massive a typical (main-sequence) O-star is, which can be as great as 260 times the mass of our Sun. M-stars, by contrast, may be as little as 8% the mass of our Sun, which is a G2-class star. Image credit: Wikimedia Commons user LucasVB.

hydrogen were both *vastly* more abundant on the Sun than they are on Earth, with helium abundances many thousands of times richer and hydrogen, most spectacularly, about **one million** times as abundant on the Sun than it is on Earth. The reason that the hydrogen absorption lines on the Sun were so strong is both because the Sun is at the right temperature to have hydrogen in an Hα-absorbing state and because it is primarily *made out of hydrogen*! Combining our observations with Cecilia Payne's discovery, we learn that hydrogen is the most common element in not just our Sun, but in practically all the stars we've ever observed (Fig. 5.6).

* * *

So if the Sun and the stars are made out of hydrogen, what is it, exactly, that *powers* them? After all, our Sun is the biggest, most massive thing in our Solar System, coming in at about 2×10^{30} kg, or about 300,000 times the mass of Earth. And yet, *un*like Earth, the Sun is made out of about 71% hydrogen and 27% helium, with just a tiny amount of heavier elements like oxygen (1%), carbon (0.4%), iron (0.1%) and others. Yet, despite being composed primarily out of the lightest two elements in the Universe, the Sun manages to put out an *incredible* amount of power: 4×10^{26} watts, or about 10^{16} times as much as a fully operational nuclear power plant. To put that in perspective, if you covered the entire land area of our planet with nothing but nuclear power plants running at full capacity, it would take more than 80,000 Earths to equal the power output of the Sun.

The question of just what it is that powers the Sun was one of the biggest mysteries at the start of the 20th Century. From the work of Darwin, it was evident that the Earth needed at least hundreds of millions of years for evolution to produce the diversity of life we see today, and from contemporary geologists, the Earth had apparently been around for at least a couple of billion years. But what type of power source could be that energetic for that long a period of time? Lord Kelvin — the famed scientist who discovered absolute zero — considered three possibilities: one, that the Sun was burning some type of fuel; two, that the Sun was feeding on material from within the Solar System; and three, that the Sun generated its energy from its own gravity (Fig. 5.7).

The first possibility, that the Sun burned some type of fuel source, made a lot of sense. Given that the Sun is *mostly* made up of hydrogen, and how easily hydrogen combusts here on Earth, it seems very straightforward that burning such a giant store of hydrogen could provide a tremendous amount of energy. Indeed, if the Sun were made entirely out of hydrogen, and we considered that hydrogen fuel combusted the exact same way that it does here on Earth, there would be enough fuel for the Sun to produce that incredible amount of power — 4×10^{26} watts — for tens of thousands of years. Unfortunately, even though that is quite long when compared to, say, a human lifetime, it is not nearly long enough to account for the long history of life, Earth, or our Solar System. Kelvin, therefore, was able to rule this first option out.

The second possibility was a little more intriguing. While it would not be possible to sustain the Sun's power output from whatever hydrogen atoms were presently in there, it *could* be possible, in principle, to continuously add some type of fuel to the Sun to keep it burning. It was well-known that comets and asteroids abound in our Solar System, and so long as there was enough new (unburned) fuel being added to the Sun at a roughly steady rate, its lifetime could be extended by the addition of masses such as these. However, you could not add an *arbitrary*

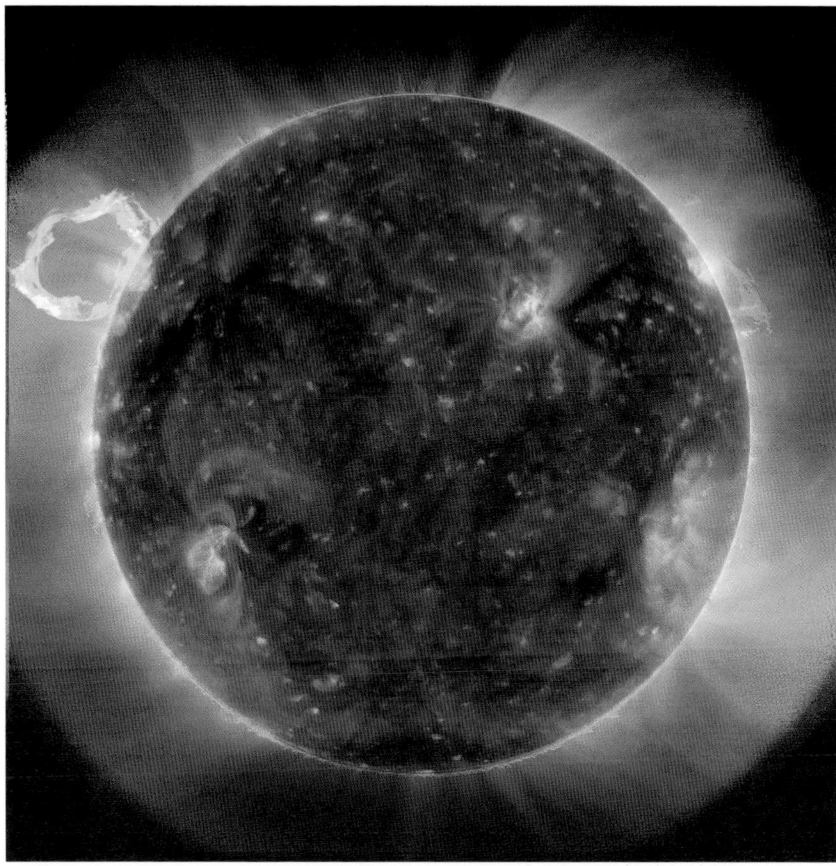

Figure 5.7 This is a multiwavelength ultraviolet image taken by NASA's Solar Dynamics Observatory, shown in false color, with reds tracing cooler regions and blues and greens tracing hotter ones. The Sun emits a tremendous amount of power, 4×10^{26} watts, but the source of that power was a mystery until the process of nuclear fusion was discovered in the 20th century. Image credit: NASA/Goddard/SDO AIA Team.

amount of mass, because at some point, the increasing mass of the Sun would slightly change the orbits of the planets, which had been observed to incredible precision since the 16th century and the time of Tycho Brahe. A simple calculation showed that even just adding the small amount of mass necessary to fuel the Sun over the past few centuries would have a measurable effect, and that the steady, observed elliptical orbits definitively ruled this option out.

That left only the third possibility: that the energy released as the Sun was gravitationally contracting over time was responsible. In our common experience, a ball raised to a certain height on Earth and then released will pick up speed and kinetic energy as it falls, with that kinetic energy converted to heat (and deformation) when it collides with the Earth's surface and comes to rest. Well, that same type of initial energy — gravitational potential energy — causes molecular clouds of gas to heat up as they contract and become denser. Moreover, because these objects are now much smaller (and more densely-packed) than they were back when they were diffuse gas clouds, it will take a long time for them to radiate all of that heat energy away through their surface. Kelvin was the foremost expert in the world on how the mechanics of how this would happen, and the *Kelvin–Helmholtz mechanism* is named after his work on this subject. For an object such as the Sun, Kelvin calculated, its lifetime for emitting as much energy as it does would be on the order of tens of millions of years: somewhere between 20 and 100 million years to be more precise.

As long as this timescale was, it was not quite long enough, posing a huge problem for scientists! It meant that, as the 19th century drew to a close, there was an unresolvable tension between the sciences of biology and geology, which argued for an Earth that was at least *billions* of years old, and astrophysics, which placed an upper limit on the age of the Sun at around 100 million years. Darwin was flummoxed, and when he wrote about the tension of evolution via natural selection with the source of the Sun's energy, simply stated:

> "The case at present must remain inexplicable; and may be truly urged as a valid argument against the views here entertained."

So who was right? In a sense, everyone was; Darwin was correct about the age of the Earth, while Kelvin was correct that none of the three conceivable mechanisms could account for a Sun as old as Darwin and geology required. But what neither man fathomed was that there was an entirely new type of fuel at work inside the cores of stars like our Sun, a discovery that would change our understanding of the Universe forever (Fig. 5.8).

Figure 5.8 There are objects that emit light based off of the Kelvin–Helmholtz mechanism, but they are not stars like our Sun (at left), but rather are white dwarfs (at right), stellar remnants that emit less than 0.01% of the light that Sun-like stars do. White dwarfs will shine for many trillions of years, contracting under their own gravity, but only live that long because of their low luminosities and their small surface areas, which restrict them to only slowly radiate their energy away. Image credit: NASA, ESA; created by: G. Bacon (STScI).

* * *

Even before Cecilia Payne discovered what the stars were made out of, it was recognized that there was a tremendous variety among stars as far as what we were seeing. While a cursory glance at the night sky might leave you believing that the stars are all white (varying only in brightness), very dark skies or telescopes reveal that they come in a huge variety of colors, from red to orange to yellow to white to blue! And while a star's color is an *intrinsic* property to that particular star itself (our Sun, for example, is white), the apparent brightness of a star depends on both how intrinsically bright it is as well as *how far away* it happens to be. Once we discovered how to account for a star's distance — first through parallax measurements, and later using other methods, too — it became possible to simultaneously know the intrinsic color and intrinsic brightness (or magnitude) of a large number of stars relatively easily.

When we first measured, determined, and studied in tandem these two properties — color and magnitude — of stars side-by-side, we found three surprising things:

1) That stars of specific colors *only* exist at particular brightnesses,
2) That the overwhelming majority of stars followed a very particular relationship between their color and their magnitude, and
3) That every cluster of stars had its own unique characteristics when it came to the colors and magnitudes of the stars present within it.

Let us first take a look at what we see when we look at all the stars within our sight in general.

When you graph their colors and their magnitudes together, you find what appears to be a simple general relationship: the *bluer* a star tends to be, the *brighter* it tends to be as well, while the *redder* a star tends to be, the *dimmer* it appears. The most prominent feature of this color–magnitude diagram (or Hertzsprung–Russell diagram) is a characteristic curved line snaking from the upper-left down to the lower-right. It is known as the **main sequence**, and the vast majority of stars in the Universe land on that very line (Fig. 5.9).

But what determines what color-and-brightness a star will be, and what about all the stars that *are not* on that line? One of the most wonderful things about scientific questions like these — questions about the very nature of the Universe — is that they can be answered simply by looking at the Universe and listening to what it tells us about itself. If we want to learn more about how stars form, burn through their fuel and evolve, we need something accessible in the night sky that can teach us the critical information. Luckily, in order to figure it out, we have hundreds of such objects right here in our own galaxy: a variety of star clusters.

The reason star clusters are such a great place to look for clues to this puzzle is because stars do not form in isolation, but rather in clusters, where thousands (or more) of individual stars are all created at roughly

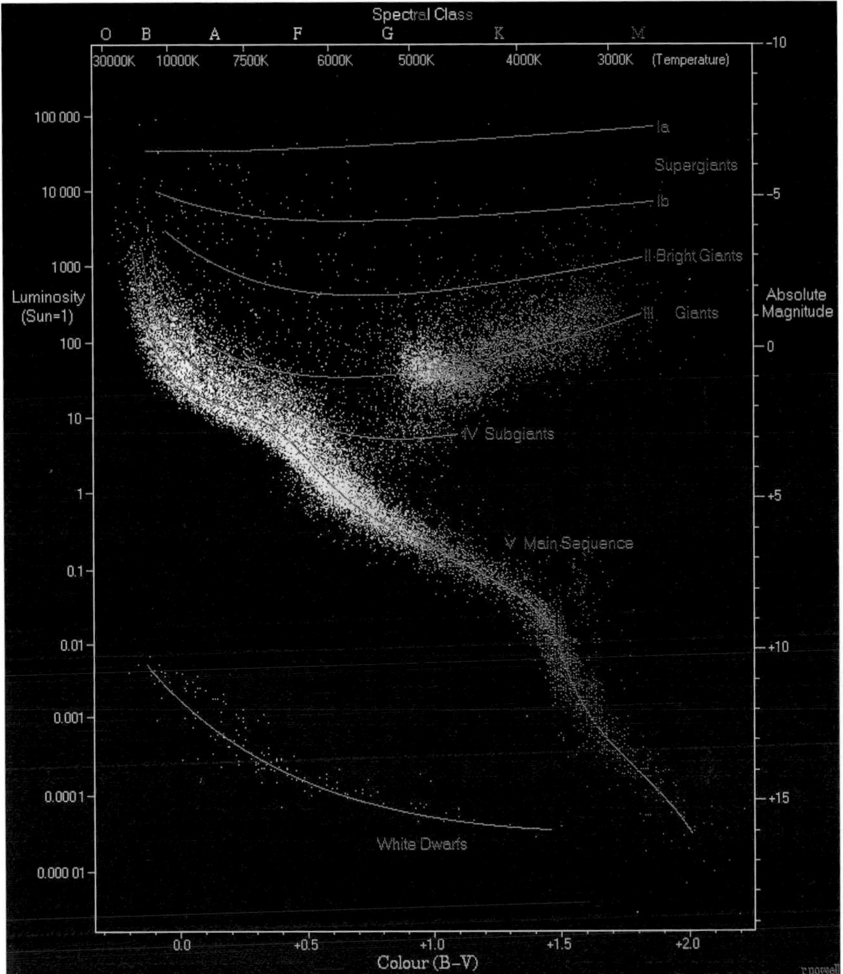

Figure 5.9 The color–magnitude relationship between stars, shown here as the Hertzsprung–Russell diagram, teaches us that most stars lie along the main sequence, running from lower right to upper left, and is defined by stars undergoing the process of fusing hydrogen into helium in their cores. Image credit: Richard Powell, under c.c.-by-s.a.-2.5.

the same time. So when we look at a cluster, we are getting a snapshot of a wide spectrum of stars — different colors, sizes, brightnesses and masses — as they exist after aging for an amount of time unique to that particular cluster! And when we make these measurements, and see how

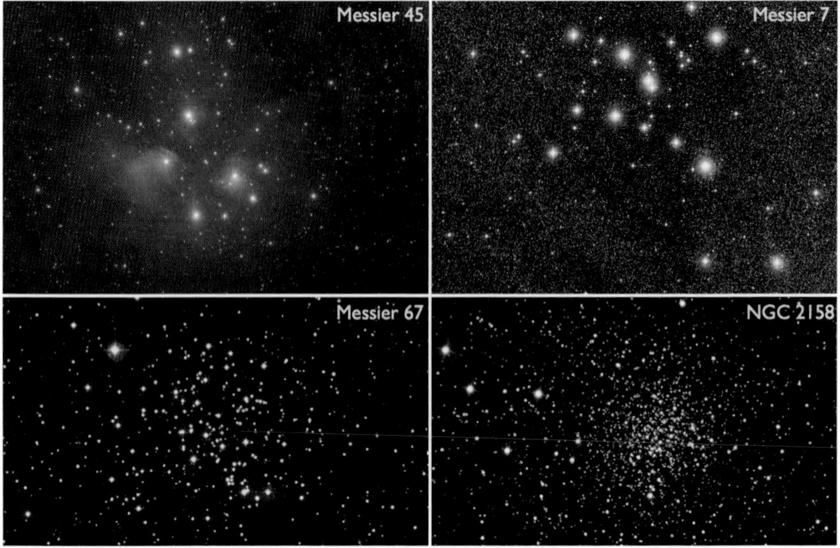

Figure 5.10 A variety of star clusters, with different populations inside. Clockwise from top left, they are Messier 45 (the Pleiades), Messier 7, NGC 2158 and Messier 67, with estimated ages of 80 million years, 200 million years, one billion years, and four billion years, respectively. Note how common bright, blue stars are in younger clusters, and how even the brightest stars appear dimmer and redder the older a cluster gets. Images credit (clockwise from top left): Boris Stromar of Wikimedia Commons under c.c.-by-s.a.-3.0; European Southern Observatory (ESO); Adam Block/Mount Lemmon SkyCenter/ University of Arizona; Digitized Sky Survey 2/STScI/WikiSky.

different classes of stars evolve and change after a certain amount of time has gone by, we can learn a tremendous amount about the stars that make up our Universe (Fig. 5.10).

The youngest clusters — the ones that still have gas-and-dust condensing to form new stars — are home to the brightest and bluest stars on the main sequence. In addition, the redder stars will initially appear brighter (and a bit redder) than they will at later moments in time. For the absolute youngest clusters, these main sequence stars (and the red ones just above the main sequence) are the *only* stars we see. As we look at progressively older and older clusters, what we find is that the dimmer, redder stars begin to settle down onto the main sequence, while the bluest stars begin to *leave* it! When a typical star leaves the main sequence, it increases in brightness and becomes slightly redder in color, also increasing its radius in the process. What is happening to these stars is that they are transforming into subgiant stars, over time becoming *much* brighter and redder as they transition into true red giant stars. Giant stars often vary in color over time from red to blue and back again, but all giant stars eventually end their lives in either a catastrophic supernova explosion or — for the giants that evolved from about B3 class and cooler main sequence stars — by blowing off their outer layers and settling down into white dwarf stars, which have brightnesses of only a tiny fraction of the main sequence stars they started from. Occasionally, older star clusters also contain bright, blue main sequence stars that you would naïvely think would be too massive to still exist. But they *do* exist, thanks to two (or more) redder, lower-brightness stars *merging* and creating a brighter main sequence star known as a *blue straggler* (Fig. 5.11).

With this knowledge of stellar evolution firmly in place, the stage was set to uncover exactly what the mechanism was that powered the life-and-death of these stars, and to finally explain why they shine exactly as they do.

* * *

It seems ironic that the secret to understanding what powers the stars — some of the largest single objects in the Universe — lies in some of the smallest building blocks of matter. Yet that is exactly where to look to solve this mystery. When you examine conventional fuel sources — things like hydrogen gas, oil, coal or hydrocarbons — there is energy stored in the bonds between atoms. By combining these bound atoms with oxygen in the presence of heat, they will happily rearrange themselves into a more stable configuration, releasing energy in the process. The energy released is on the order of a few eV (electron-Volts) for each atom involved, something that holds true for all chemical reactions, which conventional fuel sources are examples of.

But deeper inside each atom, beneath the electrons orbiting them, lies the atomic nucleus: a combination of protons and neutrons bound together. And while binding an electron to a nucleus releases a few eV of energy, binding either a proton or neutron to a pre-existing nucleus (or even a lone

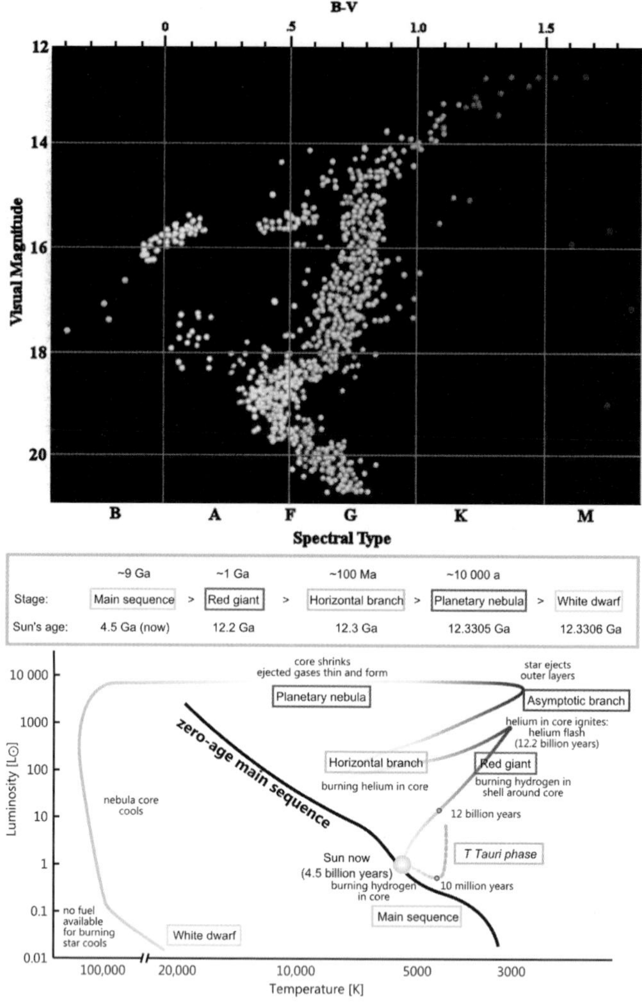

Figure 5.11 The Hertzsprung–Russell diagram for an individual cluster (Messier 3, top) shows a "turn-off" from the main sequence dependent on its age, with progressively dimmer, redder and lower-mass stars becoming red giants as a cluster ages. Occasionally, two lower mass stars will merge, producing the blue straggler population that's upwards and to the left of the main sequence turn-off. An individual star like our Sun (bottom) will follow a specific life-cycle path, spending most of its life on the main sequence soon after forming, then expanding and cooling as it becomes a red giant, eventually blowing off its outer layers while its core contracts down to a white dwarf. Images credit: R. J. Hall of Wikimedia Commons, under c.c.-by-s.a.-1.0 (top); Szczureq of Wikimedia Commons, under c.c.-by-s.a.-4.0 international (bottom).

proton) releases a few **MeV** (Mega-electron-Volts, where 1 MeV = 1,000,000 eV) of energy! Put simply, the strong nuclear force — the force that holds protons and neutrons together in complex nuclei — allows for a tremendous amount of energy to be released by forming bonds between these tiny particles.

It stood to reason that if the stars were powered by an energy source such as this, one that was *millions* of times more efficient than conventional, chemical-based fuels, that they might live *millions* of times longer than Kelvin's "conventional-fuel-source" estimate for the lifetime of the stars and the Sun. But there was something else that came along with this idea that delighted proponents of the Steady-State Model, as well as all detractors of Gamow's Primeval Fireball idea. You see, the stars were made primarily out of hydrogen and helium, the lightest elements in the Universe, and yet we know that the Universe is full of significantly heavier elements that had to come from *somewhere*. The Primeval Fireball model may do a satisfactory job of creating the lightest elements, but the temperatures and densities of the early Universe were simply insufficient for creating elements heavier than lithium in any substantial abundance. Even though elements heavier than helium make up only about 2% of all the elements in the Universe, that is a very, *very* important 2% from our point-of-view! But there *is* a place where temperatures and densities could rise to the necessary levels to fuse lighter nuclei into heavier ones: the cores of stars (Fig. 5.12).

Steady-State proponent Fred Hoyle, together with collaborators Geoffrey Burbidge, Margaret Burbidge and Willie Fowler, put forth a breathtaking paper — simply known as B^2FH (after its authors) — in 1957, detailing how nuclear fusion ought to work in the hearts of stars. Above a certain density and temperature threshold, protons from hydrogen nuclei in the core of massive enough stars — of at least 8% the mass of our Sun — could fuse together, first into deuterium and then quickly into helium-3 and then helium-4, releasing a tremendous amount of energy: 28 MeV for every helium-4 nucleus that is produced. And the release of this energy through fusion reactions in stellar cores would not only explain why the Sun shines, it would explain why *all the stars on the main sequence* shine as they do.

Figure 5.12 Inside the cores of stars, it is neither a chemical nor a gravitational process that results in the release of energy, but a *nuclear* one, which enables the formation of heavier elements from lighter ones. Image credit: NASA.

In the core of a star like our Sun, the temperatures reach up to around 15 million K, with gravitational pressures so intense that the density of the plasma is 13 times greater than solid lead. As massive as our Sun is, it contains a total of around a whopping 10^{57} protons, a little less than 10% of which are in the core at any given time. Under all that pressure and at such a high temperature, protons in the Sun's core have tremendous amounts of kinetic energy, moving at a significant percentage of the speed of light. The collision rate is tremendous, as they bounce off of other protons and nuclei, with every particle undergoing billions of interactions every second.

With all these collisions and interactions, you can then calculate how many protons have enough energy to take that first step in the nuclear chain reaction, and fuse into deuterium. **The answer is *exactly zero*.** Of all the proton pairs colliding in the sun's core, not a single one has sufficient energy to fuse into anything heavier, meaning temperature and density can't be telling the whole story. Then how does nuclear fusion occur? Through the power of quantum mechanics! The individual protons in a star's core might not have enough energy to overcome the repulsive force caused by their electric charges, but there's always a chance that these particles can undergo *quantum tunneling*, and wind up in a more stable bound state (e.g., deuterium) that causes the release of this fusion energy. Even though the probability of quantum tunneling is very small for any particular proton-proton interaction — somewhere on the order of 1-in-10^{28}, or the same as your odds of winning the Powerball lottery three times *in a row* with buying exactly three tickets — the fact that there are so many interactions in the core happening continuously means that a whopping 4×10^{38} protons fuse into helium *every second* in our Sun. And this process, of nuclear fusion fueled by quantum physics, is what is responsible for powering all the main sequence stars in the Universe (Fig. 5.13).

In a very low-mass star, the volume of the fusion-producing core is small, and so fusion proceeds at a leisurely pace, causing these stars to be cool, red in color, and dim, or low in luminosity. If your star is more massive, the volume of the core is going to be larger, with higher temperatures, densities, and rates (and probabilities) of fusion. The more massive your hydrogen-burning star is, the hotter, bluer and brighter your star is going to be, explaining why bluer stars are also the more luminous ones. But, perhaps counterintuitively, the more massive, luminous and blue your star is, the *shorter* its lifetime will be. You see, a star that is twice as massive as another might have twice as much hydrogen fuel to burn, but it burns through the fuel in its core approximately *eight* times as fast, while one that is *ten* times as massive burns through its fuel around a *thousand* times as quickly. Over very long timescales (many tens of billions of years), the spent (helium) fuel in a star's core can convect out while unburned hydrogen convects in, allowing the longest-lived stars to burn through 100% of their hydrogen fuel. But for higher-mass stars, including stars like the Sun, when the core runs out of hydrogen, that's the end of their time on the main sequence. This epiphany led to a spectacular prediction on the part of Fred Hoyle.

* * *

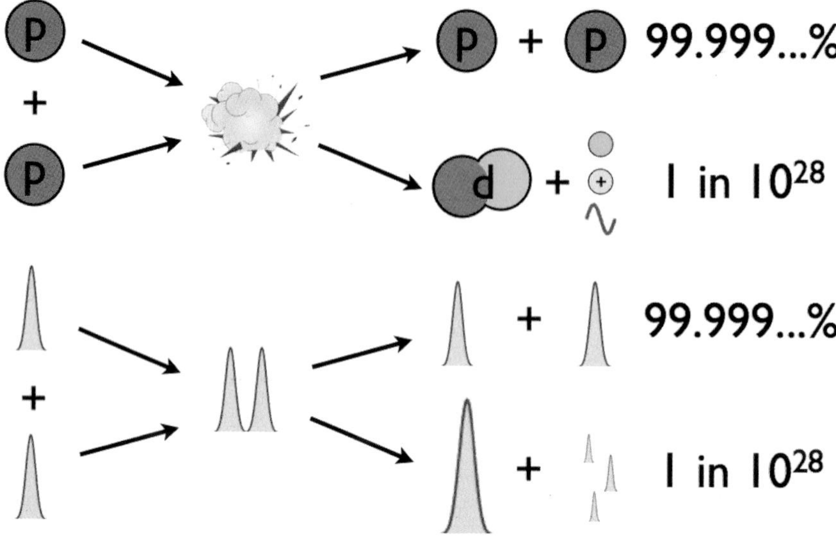

Figure 5.13 At the extreme temperatures in the core of our Sun, collisions between bare hydrogen nuclei (protons) are common, and occur at extremely high energies. However, these energies are insufficient to overcome the repulsive electric force on their own. It is only due to the laws of quantum physics, and the fact that each proton is a quantum particle with a probabilistic function describing its location, that enables the two wavefunctions to overlap ever so slightly. In virtually all the interactions, no fusion will occur, and the protons simply scatter off of one another. But in 1-in-10^{28} interactions, a fusion reaction does occur, forming a heavier element — like deuterium in this figure — and releasing energy in the process. Image credit: E. Siegel.

The only reason a hydrogen-burning star does not collapse under its own gravity is the incredible radiation pressure resulting from the nuclear fusion happening in the core. Yet, this process *only* creates the element helium, and judging from the sheer number and diversity of elements we see on Earth, there *must*, Hoyle reasoned, be a way to create heavier elements in these stars. At the temperatures and densities present in the core of a hydrogen-burning star, there would be no way to move up past helium-4; a proton could not fuse with it because there is no stable nucleus with a mass of 5, and another helium-4 could not fuse with it because there's no stable nucleus of mass 8. Any nucleus temporarily formed with these masses decays back to helium-4 in a minuscule fraction of a second. But when you run out of hydrogen fuel in your core, the

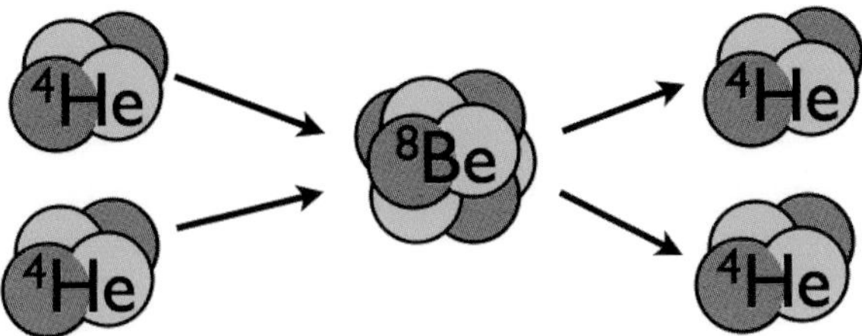

Figure 5.14 When temperatures reach a high enough value, two helium-4 nuclei can occasionally form beryllium-8, albeit for an incredibly short amount of time: only 10^{-17} seconds. After that brief amount of time, it decays back to two helium-4 nuclei, with no net gain, loss or release of energy, and no stable, heavier elements formed. On its own, this pathway would be a dead-end for the formation of heavier elements. Image credit: E. Siegel.

radiation pressure begins to drop, and suddenly the core of the star begins to collapse under its own gravity.

When you have a very large number of particles together in a small space — like the core of a star — there is a lot of energy stored in the gravitational field between the particles. Unless you both collapse it very slowly and provide a very efficient way for that energy to escape, the process of collapse is going to cause a tremendous increase in both the temperature and the energies of the particles inside. This is the same principle behind how a diesel engine works — the rapid compression of the fuel inside causes it to ignite — only when the helium-4 reaches a certain threshold, it does not ignite, but rather begins fusing into beryllium-8! This isotope of beryllium is unstable, of course, and decays back to two helium-4 nuclei in approximately 10^{-17} seconds, but Hoyle recognized the importance of this isotope's existence, even if it remained for only the briefest amounts of time (Fig. 5.14).

You see, the reason that nuclear fusion is so effective at releasing energy — the reason it can happen via quantum tunneling and the reason it can power so much — is because the rest mass of the *products* of a fusion reaction is significantly and measurably *smaller* than the mass of the reactants. Hydrogen eventually fuses into helium-4 because helium-4 has the mass equivalent (via the famous $E = mc^2$) of 28 MeV *less* than the four hydrogen nuclei that went into it. Beryllium-8, on the other hand, has almost *exactly* the same mass as two helium-4 nuclei that go into it (the energy difference is less than 0.1 MeV), so there is no reason for it to be energetically favored, and that is why it decays back into helium-4 almost immediately. But that is an important *almost*, reasoned Hoyle, because if you could get *a third* helium-4 in there quickly enough, you could theoretically combine it with the beryllium-8 to form carbon-12, which is heavily energetically favored. But there was an important hurdle to overcome, which led Hoyle to make the most breathtaking prediction of his entire career.

Just like atoms have excited states — where electrons can be in unstable, higher energy configurations that decay down to lower energy states, emitting a photon in the process — nuclei can also have excited states, or a spectrum of configurations where the lowest one, or the ground state, may be stable. The big difference between atomic excited states and nuclear excited states is that the latter are so significantly different in energy from one another that they have *measurable* mass differences (due to $E = mc^2$) between them. Combining three helium-4 nuclei together would not be able to give you carbon-12, as the mass differences between the two systems are too great, with carbon-12 being significantly less massive. But *if*, Hoyle proposed, there were an excited state of carbon-12 that had strictly the same mass as three helium-4 nuclei combined, you could continue fusing elements in the hearts of stars *even after* they ran out of their hydrogen fuel. Since carbon exists, he reasoned, and is also required as a building block to form the plethora of even heavier elements present in stars, on Earth and throughout the Universe, this new excited state — which had never been observed — *must exist*, and must exist at precisely the same mass as three helium-4 nuclei combined! (Fig. 5.15)

This new hypothesized state was dubbed the Hoyle state, and the theoretical new process by which it was formed was known as the **triple-alpha process**, since a helium-4 nucleus is also known as an alpha particle, emitted in some radioactive decays. Hoyle told his collaborator Willie Fowler about this hypothesis in 1952, and Fowler conceded that it was *possible* that such a state existed, and had been missed by nuclear physicists up until that point. It took five years of research, but in 1957, the Hoyle state of carbon-12 was found, and confirmed to have exactly

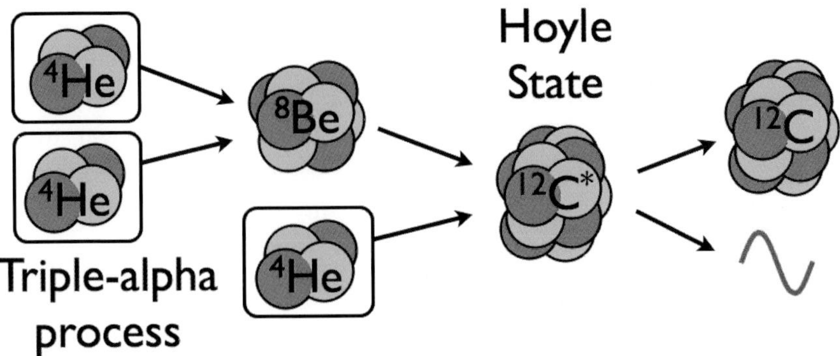

Figure 5.15 The "Hoyle state," as its now known, was a proposed excited state of carbon-12. By reasoning that carbon must come to exist in great abundance, there must have been a way to form it in the core of stars. Beryllium-8 may be unstable, but at high enough densities and energies, it should be possible to get another helium-4 nucleus in there before it decays. If that happens, it can create the excited carbon-12 through what's called the "triple-alpha" process, since a helium-4 nucleus is also an alpha-particle (emitted in some radioactive decays) and it takes three of them to make carbon-12. The excited carbon-12 then decays to normal, stable carbon-12 and emits a very high-energy photon, giving the core of the star an influx of new energy. Image credit: E. Siegel.

the energy necessary to produce carbon in the cores of massive stars that had burned through their hydrogen! The key to unlocking the origin of all the heavy elements in the Universe had just been discovered.

* * *

If the Universe we live in today is filled with the wide variety of atoms we find here on Earth, with naturally occurring elements all the way up to Uranium (element 92) or even Plutonium (element 94), how and where were they made? Even though the elements heavier than helium only make up about 2% of our Universe today, that important 2% makes up over 99% of what we find on Earth! We know these elements must have been made in stars, but not *all* stars are equally responsible for the elements beyond hydrogen and helium in the Universe. It is not surprising that, to solve this puzzle, we need to take a journey through the lives and deaths of stars, from the smallest red dwarfs (the main sequence M-class stars) to the brightest blue (O-class) supergiants. What might surprise you is that by understanding the smaller, cooler stars that *do not* produce the heaviest elements, we can better understand the largest ones that do!

The dividing line between an object becoming a true star and a failed star is defined by whether it reaches a self-sustaining nuclear fusion reaction in its core, where hydrogen fuses into heavier isotopes and eventually into helium-4. This requires densities many times that of solid lead and temperatures of around 4 million K, something that would not be achieved unless you can get about 26,000 Earths worth of mass (mostly hydrogen) together in one place, or about 8% the mass of our Sun. Below that threshold, the individual, positively-charged protons simply would not be able to fuse together, although other light-emitting and energy-releasing reactions can occur. But at-or-above that threshold, the core can begin fusing protons together in a chain reaction, producing helium-4 and releasing a tremendous amount of energy. At the beginning of every true star's life, this hydrogen-burning reaction is the start of the core's climb up the periodic table.

Remember, also, that the *more massive* a star is, the *larger* and *hotter* its core region — where fusion occurs — is as well, meaning that the star will be:

- Bluer in color, due to its higher temperature,
- More luminous, due to the increased rate of fusion, and most importantly,
- Shorter lived, as the increased rate of fusion means it burns through the fuel in its core more quickly.

For the lowest mass stars, in fact, the M-class stars whose masses are only between 8–40% the mass of our Sun, hydrogen fusion occurs so slowly that the entire star has time to convect, moving fused material out of the core and unburned hydrogen from the outer layers into the core. This convection occurs the same way a pot of heated water convects and has the hot water from the bottom rise to the top while the colder water sinks to the bottom where it can be heated. The lowest-mass of these M-class stars can live as long as **20 trillion years**, or more than 1,500 times the current age of the Universe, before running out of hydrogen fuel at long last. When they are all out of hydrogen, the entire star contracts, but with

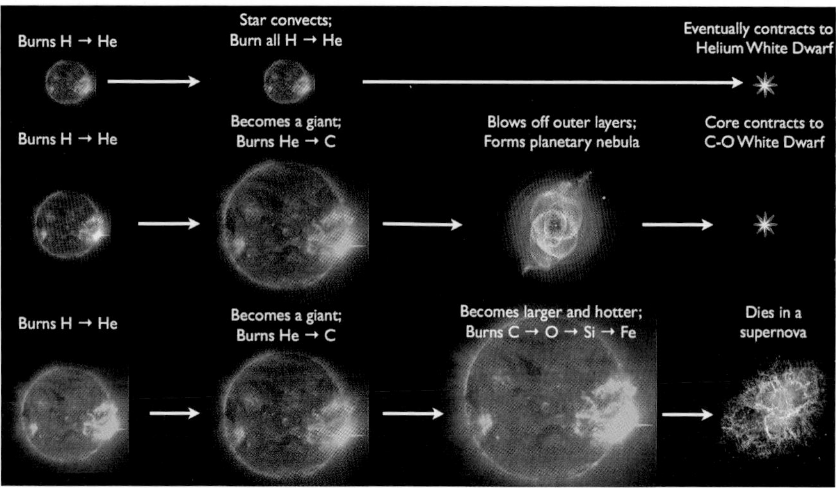

Figure 5.16 In the top row, we see the fate of M-class stars, which will burn hydrogen into helium until the process is complete, but without enough mass to burn helium. They will end their lives as helium white dwarfs. In the middle row, K-class stars up through mid-mass B-class stars will burn hydrogen into helium, then helium into carbon, but will be unable to ignite carbon fusion in their cores. These stars will blow off their outer layers when helium fusion ceases, creating a planetary nebula and contracting down to a carbon-and-oxygen white dwarf. The most massive B-class stars and all O-class stars will burn hydrogen into helium, helium into carbon and then carbon into still heavier elements, culminating in iron. They will end their lives in a type II supernova, leaving behind a neutron star or black hole at their core. Image credit: E. Siegel, using an image of our Sun from the NASA SOHO mission.

insufficient temperatures and densities to ignite the fusion of helium-4 into heavier elements, and so a low-mass white dwarf made purely out of helium will be the result. It will not come to pass for more than a hundred billion years that the first helium white dwarf will be created in our Universe! (Fig. 5.16.)

If, instead, your star contained more than 40% the mass of the Sun, not only do you become blue enough to bump yourself up to become a K-class star (or higher), but your star will burn through its fuel more quickly and with a larger, higher-temperature core. How is this different from an M-class star? In addition to being brighter and bluer, the rapid fuel-burning means that only small amounts of convection have the opportunity take place, and the burned fuel remains in the core. When there is insufficient hydrogen left in the core for the rate of fusion to hold the star up against the gravitational force working to compress it, the star contracts, but this time, the temperatures and densities *do* become high enough to fuse helium-4 into carbon-12 through the triple-alpha process. When that new fusion reaction ignites, there is a tremendous increase in the energy emitted from the star, which causes the star to swell into a red giant, thousands of times the brightness and radius of the original star. In a thin shell just *outside* that helium-fusing region, hydrogen continues to fuse into helium, creating more of the Universe's *second* most common element. But what is most important is happening in a relatively tiny region just about the size of the planet Jupiter: elements heavier than helium are being created in great abundance!

Any remaining hydrogen (i.e., free protons) in the core can fuse together with the large amounts of carbon-12 to create nitrogen and then oxygen. If a proton fuses with carbon-12, it creates the unstable nitrogen-13, which will radioactively decay after a few minutes into carbon-13, which incidentally makes up about 1.1% of all the carbon here on Earth. The temperatures inside are sufficient so that *another* proton added to carbon-13 makes nitrogen-14, which is stable, *or* you could add a helium-4 nucleus to carbon-13, producing oxygen-16 (which is stable) and a **free neutron**. (Remember that some nuclear processes produce free neutrons; this will be important shortly.) For many stars — particularly the lower-mass K-class stars — these are the heaviest elements that can be formed through nuclear fusion, and the end of the line for the star. The core will contract once again when its current main fuel source, helium-4, runs out in sufficient quantities to continue creating carbon-12. Although this causes the temperature and density of the core to further increase, it does not increase *enough* to initiate subsequent stages of nuclear fusion. Instead, the outer layers of the giant star are slowly (over tens-to-hundreds of thousands of years) blown off into interstellar space by intense stellar winds generated by the central region, while the inner core contracts down to a white dwarf made mostly of carbon and/or oxygen. These white dwarfs are roughly the physical size of the planet Earth, but are hundreds of thousands of times as dense, containing anywhere between 20–140% the mass of the Sun.

If your star were heavier, however, it *could* take that next step, where helium-4 nuclei fuse with oxygen to produce neon, which our Sun — a G-class star — will likely do at some point billions of years in the future. Even more massive stars can add helium to neon, producing magnesium (and another free neutron), while those with even higher temperatures can then add helium to magnesium to make silicon, and possibly combining helium with silicon, producing sulphur. In all cases, these stars wind up with a core of mostly carbon and oxygen at the center, where some helium reactions occur, forming heavier elements, but most of the helium fusion occurs in a shell outside of the core, with a shell of hydrogen fusion occurring outside of that. Stars like this, which are thought to include all K, G, F, A, and the lower-mass B-class stars, form a significant amount of carbon, nitrogen and oxygen, and (sometimes) lesser amounts of elements slightly heavier than that, but the vast majority of them get trapped in the stars' cores. The outer layers are mostly made of hydrogen and helium, with only some heavier elements transported outward from the core. During the final helium-burning phases, the incredible flux from the core's radiation pressure will blow the already tenuous outer layers of the star into the interstellar medium, forming a planetary nebula. Eventually, the core will contract down to a carbon/oxygen white dwarf, which will take quadrillions ($\sim 10^{15}$) of years to cool down to the same temperatures as interstellar space (Fig. 5.17).

This might not seem like much of an enrichment to the Universe, since most of what gets returned to the interstellar medium to help form the next generations of stars are hydrogen and helium, with much smaller amounts of still relatively light elements like carbon, nitrogen and oxygen. Of the known stars in the Universe, three out of four were born as the lowest mass (M-class) stars, which return *nothing* to the Universe. Of the remainder, about 99.5% of those will follow the life cycle outlined above: dying in a planetary nebula/white dwarf combination, only adding a relatively small amount of the first sixteen-or-so elements of the periodic table back into the Universe. But there are two extra things that happen that make all of this possible: the **free neutrons** we mentioned earlier, and the *most massive* stars in the Universe.

Unlike protons or any of the other atomic nuclei, free neutrons are special because they have no electric charge. While all the fusion processes

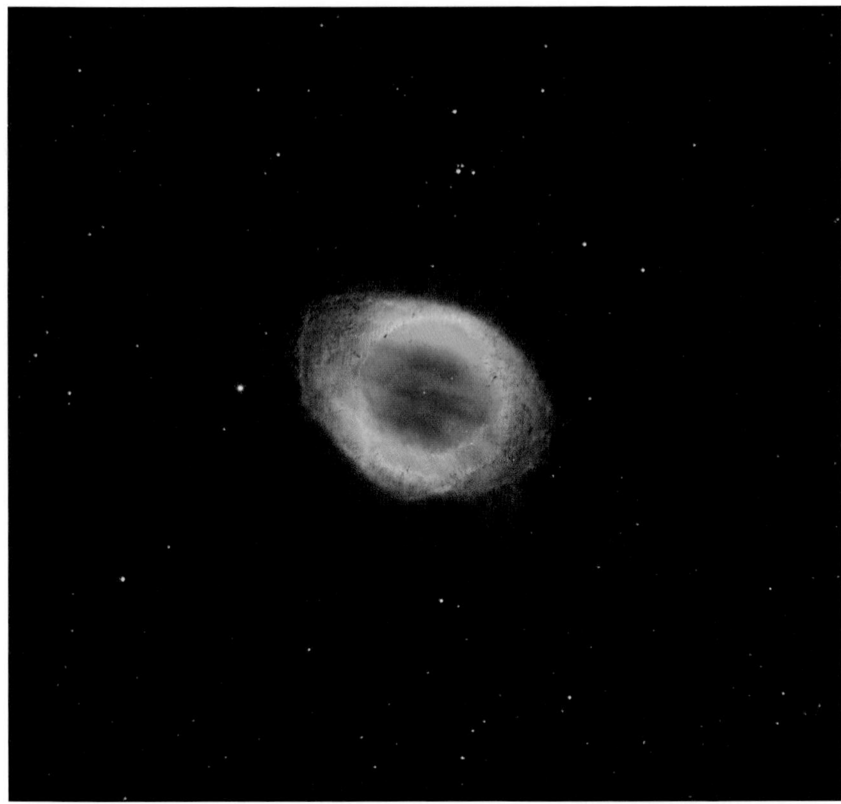

Figure 5.17 This is the Ring Nebula, a typical planetary nebula. The outer layers consist primarily of hydrogen but with significant amounts of elements like helium, carbon, oxygen, silicon and sulphur, which are returning to the interstellar medium where they will help enrich future generations of stars. The inner core of the now-deceased star contracts down to a white dwarf. It is thought that our Sun will suffer a similar fate in roughly seven to eight billion years. Image credit: NASA, ESA, C.R. O'Dell (Vanderbilt University) and D. Thompson (Large Binocular Telescope Observatory).

mentioned earlier require tremendous temperatures and densities to overcome the electrical repulsion between two positively-charged bodies, neutrons can simply "add" to most nuclei with no significant barrier, making them heavier. Typically, if a nucleus absorbs a neutron, it will either be stable, making it a heavier isotope of the same element, or it will be unstable and undergo beta decay (emitting an electron), moving one element *up* the periodic table. The free neutrons we produce in red giant stars can be readily absorbed by most heavier nuclei, allowing us — in principle — to create elements all the way up to bismuth (element 83). The main drawbacks are that the process is relatively slow (its official name is **the s-process**, where "s" stands for slow), as hundreds of years typically pass between neutron captures, and it requires somewhat heavier elements to build upon before it can be effective. This is not good enough to produce the majority of heavier elements, but it does play an important role in the lives of all stars that are massive enough to enjoy a helium-burning phase.

* * *

To make the absolutely heaviest elements in the Universe, as well as the majority of everything heavier than lithium, we need to look to the shortest-lived stars in the Universe. The stars are born hot, bright, massive and blue, representing all of the O-class stars and, perhaps, some of the most massive B-class stars as well. Like all stars heavy enough to do so, they begin their lives by fusing hydrogen into helium in their cores. (They also produce energy through the C–N–O cycle, which uses carbon, nitrogen and oxygen as a catalyst to produce helium from hydrogen. This is a process that happens in small amounts in lower-mass stars like the Sun, but becomes more important the more massive your star is.) Because they are so massive, they burn through a tremendous amount of nuclear fuel at an incredible rate; the most massive O-star known is (as of its discovery in 2010) R136a1, with 265 times the mass of the Sun, 8,700,000 times the Sun's intrinsic brightness, a surface temperature of 53,000 K (more than eight times the Sun's) and an anticipated lifetime of just one or two million years. Just like its lighter counterparts, the core of this star, too, will rapidly run out of fusible hydrogen, and when it does, hydrogen begins burning on a shell located at the star's outer core, the outer layers expand the star into either an ultra-massive red giant or a blue supergiant, and the inner core contracts, heats up, and begins fusing helium into carbon (Fig. 5.18).

So far, there is nothing *too* different from what we have seen before; only the timescales are shorter. The helium-burning stage will be brief — a few hundred thousand to a million years — and then, just like with our other stars, a core will develop filled with mostly

Figure 5.18 This is the super star cluster 30 Doradus, found in the Tarantula Nebula just 160,000 light years away. Containing hundreds of thousands of new stars, the most massive one at the center, R136a1, is the most massive star known in the Universe, coming in at 265 times the mass of the Sun. Image credit: NASA, ESA, F. Paresce (INAF-IASF, Bologna, Italy), R. O'Connell (University of Virginia, Charlottesville) and the Wide Field Camera 3 Science Oversight Committee.

carbon and oxygen, with some neon, magnesium and silicon inside. Just like before, the inner core will become devoid of helium and contract, with helium continuing to burn in a shell outside of this core, and hydrogen in a shell outside of that. Only this time, the star is so massive (and so is its core) that **carbon fusion** ignites under the

Figure 5.19 As the most massive stars burn through their fuel, the inner core continues to contract, heat up and fuse lighter elements into heavier ones, climbing the periodic table rapidly. Image credit: Nicole Rager Fuller/National Science Foundation.

tremendous core temperatures. While hydrogen fusion may have taken place for millions of years, and helium fusion for hundreds of thousands, carbon fusion occurs at a much faster rate, and leads to an even more concentrated central region that begins burning oxygen, followed by one that burns neon, magnesium, and eventually silicon. The end result is an onion-like structure, with each progressively inner layer burning heavier and heavier elements at ever-increasing temperatures (Fig. 5.19).

Just as the hotter, more massive stars burn through their fuel more quickly, leading shorter lives, the reactions that produce progressively heavier elements — because they require higher temperatures — also proceed more quickly. While carbon fusion may last a thousand years or so, silicon fusion ought to only last a few minutes, producing an innermost core made of iron, nickel and cobalt, the three most stable elements in the Universe. So far, every type of fusion reaction we've talked about has given us products that are more tightly bound than the reactants, and thus the process of fusion *releases* energy. But once we hit iron, fusing two elements of iron together would give us a heavier element that is *less* tightly bound than the two iron atoms we started with, which means that fusing iron would absorb energy, rather than release it. The iron, nickel and cobalt that rapidly builds up in the cores of the Universe's most massive stars is a form of ash, in the sense that it can be burned no further.

At the same time, there actually *is* enough available energy in these cores for iron to fuse, but iron fusion comes with a terrible cost for the star. As the iron begins to fuse, the core temperature begins to rapidly drop, which causes the pressure to drop and the core to collapse under its own gravity. This causes a catastrophic, runaway chain reaction: the core collapses, the iron fusion rate increases, the pressure drops, the core collapses faster, etc. In a timespan of just a few seconds, the core collapses down to the minimum size that matter is allowed to be compressed down into. At this moment a number of things happen all at once:

- The runaway fusion reactions produce not only a myriad of heavy elements, but also a tremendous number of free neutrons (and neutrinos) all at once.
- The outer layers of the inner core "rebound" off of this minimally-sized object, imparting a great amount of energy to the layers outside of it.
- The sudden outrush of energy results in both an increased set of fusion reactions and a shock wave affecting all of the outer layers; this is the first stage of a **supernova** event!

The very inner core of this star will wind up in a collapsed, degenerate state as a result of this catastrophe. Either a neutron star will be left at its core — a solid object made entirely of neutrons, about the mass of the Sun but just around 10 km in diameter — or, if the mass is too great, a black hole! (Fig. 5.20.)

But what of the remainder of the star: the outer layers that were filled with everything from hydrogen up through iron, nickel and cobalt? First off, they themselves undergo a runaway fusion reaction, which allows many more of the intermediate elements of the periodic table to be created. But during this supernova, the outer layers are bombarded with

Figure 5.20 The Crab Nebula is the most well-studied supernova remnant in history; it was identified as a supernova in 1054 and became the very first object in Messier's catalogue. The outer layers blown off in the explosion can be clearly seen with an amateur telescope, and are still expanding today, extending for more than ten light years in diameter. Image credit: NASA, ESA, J. Hester and A. Loll (Arizona State University).

unprecedentedly large numbers of neutrons, which send these intermediate elements rapidly scampering up the periodic table. Unlike the slow s-process that occurs in stars experiencing helium fusion, this rapid **r-process** — that occurs in supernovae — is likely to produce *every* element we have ever produced on Earth and more, upwards of Uranium, Plutonium, Curium and even the unstable ones we have created only briefly in the laboratory! Unlike planetary nebulae, these supernova remnants return the *majority* of matter from the progenitor star back into the Universe, including a large fraction of heavy elements.

Every time a new star cluster forms, a little more than one-in-a-thousand new stars will eventually die in a supernova like this. Because these are the most massive stars in the Universe, they do an excellent job of enriching the interstellar medium with carbon and all the heavier elements. Many generations of stars lived and died before enough heavy elements were around to create a Solar System like our own, while the stars that are forming today tend to be even richer in these heavy elements than our own Sun. Practically every heavy element in your body, from the carbon in your muscles to the oxygen in your lungs, from the calcium in your bones to the iron in your blood, *all* of these heavy elements originated in a supermassive star that underwent a supernova explosion. These atoms were returned to the Universe where they participated in future generations of stars, and after billions of years and billions of stars doing exactly that, *our* Sun and Solar System came about.

This is how the heavy elements in the Universe — and everything made out of them — owes their existence to the nuclear reactions happening inside the stars that came before us. In a very real sense, they are the common ancestors of everything on our world today.

Chapter 6

All The Way Back: It Started With A Bang

By the late 1950s, astronomers and physicists alike were sharply divided as to which model of the Universe was more likely to be correct. Of the two main competing theories, proponents of the Steady-State framework — governed by the "Perfect" Cosmological Principle — were able to successfully show how the heavy elements in the Universe originated in the hearts of prior generations of stars. Proponents of the Big Bang were forced to accept that *if* their model was correct, it could only explain the abundance of the *lightest* elements: hydrogen, helium and their isotopes, along with lithium-7 and *nothing* heavier than that. But there was another test to be performed, one that would prove far more discriminating between the two options.

By this time, no serious scientists doubted the fact that the Universe was expanding, nor that gravitation had led to the formation of galaxies and stars, which burned hydrogen into heavier elements as time went on. What was up for debate is where the entirety of all this originated! In a Steady-State Universe, the cosmos had always appeared infinite, full of stars and galaxies that were expanding away from one another. As the galaxies grew apart, new matter must be generated — in the form of hydrogen atoms — that would eventually collapse into galaxies, keeping the Universe electrically neutral, full of new fuel to form new stars and galaxies, and eternally expanding at a constant rate. Whatever photons (particles of light) were around today must have been created by the stars and scattered throughout the Universe, where they could have bounced off of whatever gas clouds were around (Fig. 6.1).

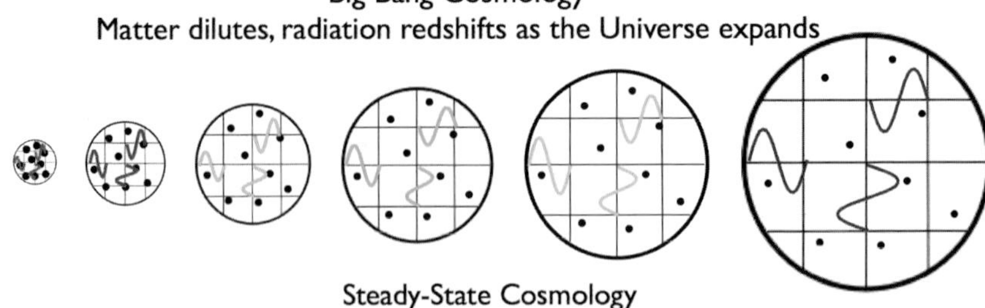

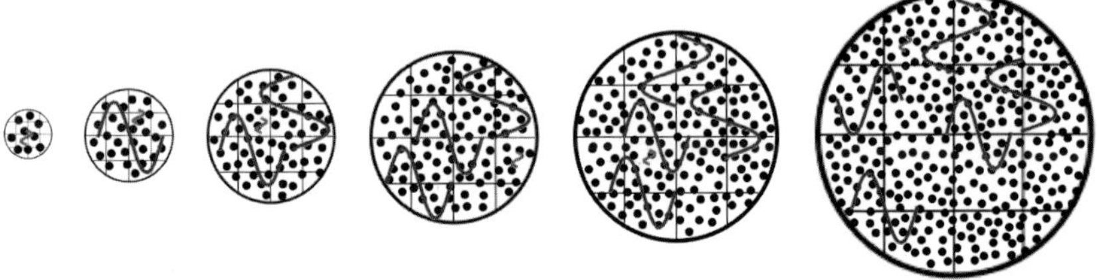

Figure 6.1 Not only will the matter density be different at different times in a Universe governed by the Big Bang compared to the Steady-State theory, but the radiation will be vastly different as well. Even setting the predictions of uniform temperature aside, the Big Bang model predicts that the radiation should have a perfect blackbody spectrum, while the Steady-State model does not. Image credit: E. Siegel.

By contrast, in a Universe obeying the Big Bang model, the cosmos *had a beginning*, and has been expanding and cooling ever since. Initially pristine matter-and-radiation once existed in an incredibly hot, dense state, followed by the Universe cooling through a number of important stages, with the expansion rate slowing down as the energy density dropped over time. In particular, these stages include:

- Whatever composed protons and neutrons cooling enough to form stable protons and neutrons.
- The Universe cooling enough to form the light atomic nuclei without them immediately being blasted apart again.
- The leftover photons cooling enough so that nuclei and electrons could come together and form neutral atoms.
- Gravitation acting on this neutral gas for enough time so that the first stars would form, creating heavy elements.

- And finally, after many generations of stars living and dying, enough heavy elements were created to give rise to rocky planets and the ingredients for life.

As George Gamow, the father of the modern Big Bang theory famously declared, "It took less than an hour to make the atoms, a few hundred million years to make the stars and planets, but five billion years to make man!" But without a surefire measurement to be taken or experiment to be performed, there would be no way to discern between the Big Bang and Steady-State models. There was, however, one thing that could be done for each of these models that *could* install one or the other as the undisputed leader. For the Steady-State model, the observation of spontaneous matter creation would plug the most assailable hole in that theory; for the Big Bang model, evidence that the Universe originated from a hot, dense, *radiation-filled* state would provide the necessary proof. Let us take a look at what the latter scenario would entail.

If the Universe began from a hot, dense, matter-and-radiation-filled state, the radiation from that early time could not have been destroyed, and *must* still be around today. The Universe expanded and cooled tremendously in its early stages, with the radiation blasting nuclei apart in the early moments of the Universe, then bounced off of the ionized nuclei and electrons before the formation of neutral atoms. When neutral atoms did begin to form, that radiation would knock electrons off of hydrogen and helium atoms for many thousands of years, before finally cooling enough that the neutral atoms would wind up in a stable state. But after the formation of neutral atoms, that radiation did not just *disappear*, it ought to still be here! From the early collisions that it underwent, it should be in a state of thermal equilibrium, meaning that it needs to have a very specific type of energy distribution: a **blackbody** distribution, which is an even more idealized energy distribution than the Sun's light possesses (Fig. 6.2).

Only, since *billions* of years must have passed since those neutral atoms first stably formed, with the Universe expanding and cooling throughout that time, those leftover photons from the Big Bang must be tremendously cold and low-in-energy today. We normally measure the energy of photons in units of electron-Volts (eV), where 1 eV is the amount of energy

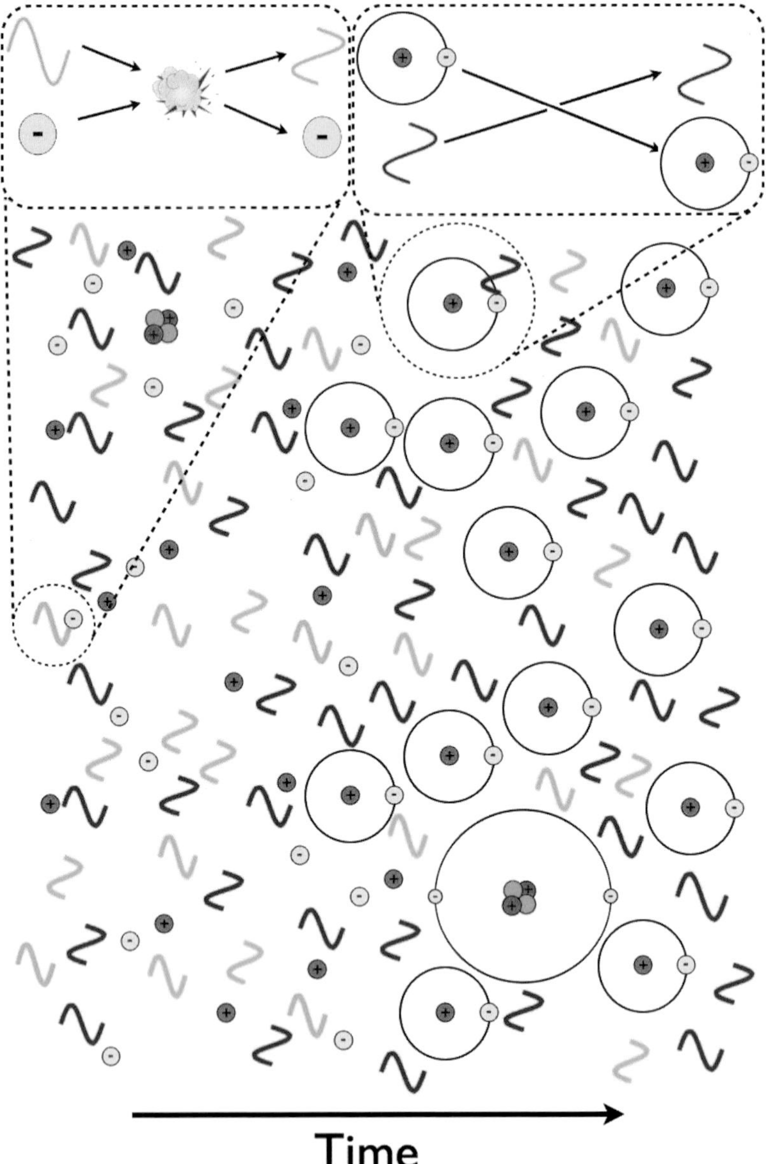

Figure 6.2 Before the Universe forms neutral atoms, photons scatter off of electrons frequently, exchanging energy rapidly and giving the radiation a blackbody energy distribution. After neutral atoms form, the photons no longer interact with them, free-streaming until the present day. Image credit: E. Siegel.

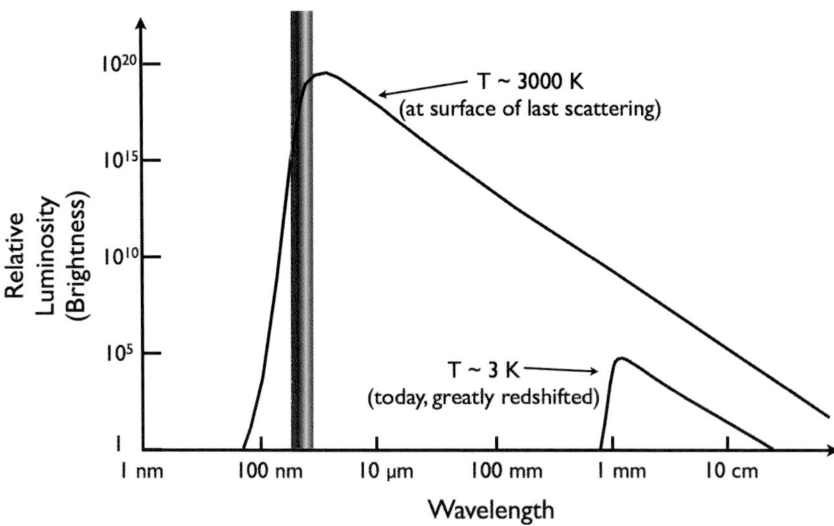

Figure 6.3 The blackbody spectrum of the radiation from the early Universe is shown from when it last scattered off of free electrons (upper curve), just prior to the Universe becoming completely neutral, and that same spectrum as it should appear today, redshifted by a factor of approximately 1,000. Note how the luminosity has not only dropped precipitously, but the peak wavelength of the spectrum has lengthened by that same factor of about 1,000 as well. The portion of the spectrum that corresponds to visible light is shown for comparison. Image credit: E. Siegel.

required to accelerate an electron to an electric potential of one Volt. To blast the lightest atomic nucleus — a deuterium nucleus (with one proton and one neutron) — apart, it requires a photon of 2,200,000 eV of energy, so the Universe must have been *at least* that energetic in the past. To ionize a neutral hydrogen atom requires a photon with 13.6 eV worth of energy, so the Universe must have cooled from an energy above that to one below it in the past as well. Based on the accepted laws of physics, the Universe must have subsequently expanded by at least a factor of close to a thousand in all directions, so that the *typical* photon energy would just be a few hundred *micro*-eVs by this point in time, or (converting to its temperature equivalent) just a few kelvin, or a few degrees above absolute zero (Fig. 6.3).

But if this leftover radiation from the Big Bang, the **primeval fireball**, as Gamow called it, could be detected, it would be a smoking gun that the Big Bang — and *not* the Steady-State theory — was the one that described our Universe.

* * *

Let us imagine what the important components of the Universe were doing back when it was so hot and dense that neutral atoms could not have stably formed. First, there were atomic nuclei: protons and bound states of protons-and-neutrons combined. These are massive, positively charged particles moving relatively *slowly* compared to everything else, due to their heavy mass. Second, there were electrons: negatively charged particles that weigh in at less than 0.1% the amount of even the lightest atomic nucleus, and hence moving much more quickly, but still far below the speed of light. And finally, there were photons: massless particles of radiation that are constantly dropping in energy as the fabric of the Universe itself expands, but always moving at the speed of light. When the Universe is in this very hot and dense state, collisions between photons and charged particles are frequent, with each collision resulting in an exchange of kinetic energy. In short order, the temperatures of *all* the different types of particles: electrons, nuclei and photons alike, acquire the same blackbody distribution, just as the air molecules in a heated room will eventually take on the same temperature if the air in the room is well-mixed.

When the Universe cools enough that neutral atoms form, suddenly the radiation (in the form of photons) has no more free, charged particles to interact with! Neutral atoms, remember, can *only* absorb or emit photons of very specific wavelengths, meaning that the vast majority of the photons (which *are not* of those wavelengths) can do nothing but simply move in a straight line at the speed of light. As the Universe continues to expand, the matter becomes more and more sparse, eventually achieving the low density our Universe currently has: less than one atom per cubic meter on average! By all accounts, that would mean that these photons scattered very efficiently off of the ionized plasma of the young Universe, before neutral atoms could stably form. But once neutral atoms did form, the scattering ceased entirely; all those photons could do was travel unimpeded, with no effects on it save the expansion of the Universe (Fig. 6.4).

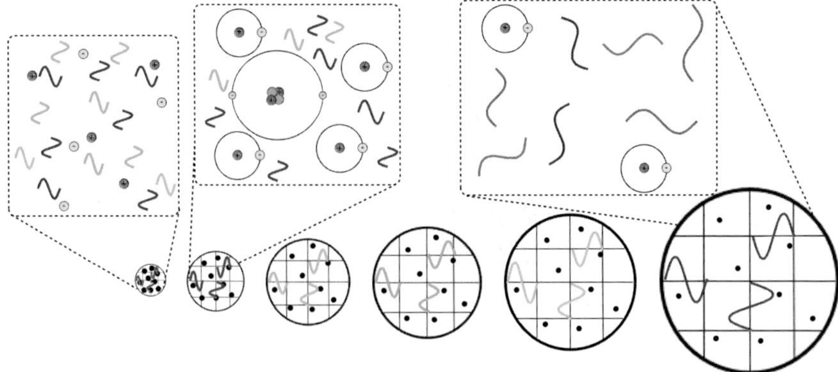

Figure 6.4 After the Universe's atoms become neutral, not only did the photons cease scattering, all they do is redshift subject to the expanding spacetime they exist in, diluting as the Universe expands while losing energy as their wavelength continues to redshift. Image credit: E. Siegel.

If this framework were correct — if the Big Bang had it right — there should be this souvenir from the Universe's youth, radiation that:

1) Comes from all directions in the sky.
2) Is practically identical in temperature everywhere.
3) Follows a blackbody spectrum, and
4) Peaks at just a few degrees above absolute zero, which corresponds to microwave wavelengths.

Although Gamow and his coworkers, Alpher and Hermann, had first calculated the properties of this primeval fireball, it was not until the early 1960s that the attributes of this radiation were worked out in much greater detail. A team of scientists at Princeton, led by Bob Dicke, Jim Peebles, David Wilkinson and Peter Roll, not only worked out those extensive details, but planned to build a Dicke Radiometer that would be capable of detecting this leftover glow from the Big Bang: the **cosmic microwave background** (CMB) radiation. But just 30 miles away, another team — working on an entirely different project — was having all sorts of problems, and they didn't know why.

* * *

Over in Holmdel, New Jersey, Robert Wilson and Arno Penzias were working for Bell Labs, and were using a new horn-shaped antenna that was incredibly sensitive to long wavelengths of light. They were attempting to detect radio waves bounced off of balloon-borne satellites launched by the U.S. Navy, but needed to make sure what they were trying to detect was not contaminated by background sources of this same type of low-energy radiation. Background sources included radio broadcasts coming from transmission towers and/or bouncing off the atmosphere, as well as radar sources. The antenna itself would emit radiation as well, so to mitigate that they cooled it down with liquid helium, which — at just **four K** above absolute zero — should have suppressed any thermal noise (Fig. 6.5).

Figure 6.5 The Holmdel Horn Antenna, built in 1959, was a reflector antenna designed to detect radio waves as part of NASA's ECHO satellite program. Image credit: NASA, taken 1962.

After taking their first sets of data, Penzias and Wilson were perplexed: even after accounting for radar and radio, and even after cooling the antenna down to these ultra-low temperatures, they were still seeing an intense background noise they could not account for. Even more puzzling were the following two facts about this background noise:

1) It was approximately *two orders of magnitude*, or a factor of 100, stronger than the background they were expecting.
2) It appeared no matter where they looked in the sky, in all directions and with equal magnitude.

Other sources of background noise would vary depending on factors such as where you pointed the antenna, on whether there were clouds overhead, on air temperature, and other terrestrial effects. But *none* of these factors seemed to affect this new source of noise. Quite quickly, they were able to rule out the three most conceivable sources of this noise: the Earth, the Sun and the galaxy.

So what could it be? Their first assumption was totally reasonable: something must be wrong with the antenna itself. So they ran all the diagnostics they could, checked out the individual components, made sure all the electrical connections were good, and vetted that everything was working up to a very rigorous set of standards without flaw. After finding no faults and repeating their observations, they again arrived at the same set of results. What else could it have been? As a further inspection would have it, there were pigeons living *inside* the cusp of the horn, having built a nest there. Could pigeon droppings have been responsible for this spurious signal? They removed the nest and cleaned the horn — making them the only known Nobel laureates who cleaned up shit as *part of their Nobel-winning research* — but still this unexplained noise remained. After an exhaustive search, Penzias and Wilson both concluded that the only explanation could conceivably be a low-energy source well outside of our own galaxy, although neither had any idea of what sort of source that would be. They had no choice but to seek out help from other scientists to explain what they were seeing.

Dicke, Peebles, Roll and Wilkinson had not yet published the work that would provide the framework Penzias and Wilson were seeking. But on

1. Assuming that the Universe expanded from a sufficiently highly contracted phase, the early Universe would have been opaque to radiation. As a result the radiation field would have achieved thermal equilibrium with the matter—the Universe would have been filled with black-body radiation. This fireball radiation suffers the cosmological redshift, so that it is very much cooled by the expansion of the Universe, but it retains its thermal, black-body character.

Figure 6.6 While the paper itself was about how this radiation would affect the early formation of galaxies, many of the radiation's details were worked out for the first time here, including what its detectable signatures would look like on instruments such as the Holmdel Horn Antenna. Image credit: P. J. E. Peebles, (1965). *Astrophysical Journal*, 142, 1317.

his own, Jim Peebles had submitted a separate theoretical paper, complete with detailed calculations for the prediction of what this background radiation should look like, and what type of signature it should leave behind: a blackbody spectrum that would be detectable from all directions at certain microwave and radio frequencies (Fig. 6.6).

Penzias and Wilson did not know about this work, but one of the scientists they told about their mysterious, unexplained noise — the radio astronomer Bernard Burke — had seen a preprint of Peebles' work, and connected the dots: what they were interpreting as a low-energy radio source in all directions on the sky was actually the leftover glow from the highest energy event ever to occur in our Universe, the Big Bang itself!

The characteristics of the Princeton group's prediction fit what Penzias and Wilson had seen better than any other explanation. As they came to realize this, Penzias called Dicke in a frenzy to ask for a preprint of the Peebles paper, a request which Dicke immediately obliged. After reading and working through the paper, Penzias and Wilson realized that not only was their prediction a better fit than other explanations, it was practically an exact match! Penzias made a second call, this time inviting Dicke and the rest of the Princeton group down to Bell Labs to see the same thing that he and Wilson had seen, and to observe the background noise originating from the Big Bang (Fig. 6.7).

For the first time, the Big Bang was no longer just one possible theory concerning the origin of our Universe; it was the *only* theory that predicted this leftover, low-energy radiation that would be uniform in all directions on the sky. For the first time, we had learned how our Universe began all those billions of years ago.

* * *

Figure 6.7 This is a simulated image of what the entire night sky would have looked like to Penzias and Wilson using the horn antenna. The green color represents a constant signal — present everywhere you look — and exists in all directions. The lone exception is a small plane of noise, shown in white, that corresponds to the location of our Milky Way galaxy, which has its own foreground emission in microwave wavelengths. The elliptical shape is due to the type of projection used in showing maps of the sky: a Mollweide projection. Inset, the Earth is shown in a Mollweide projection for comparison. Image credit: NASA/WMAP Science Team; Wikimedia Commons user Strebe under c.c.-by-s.a.-3.0 (inset).

Not everyone was convinced by this discovery, however. The Steady-State camp, upon learning of the existence of a uniform background of low-temperature radiation, came up with an alternative explanation: perhaps this was not radiation left over from the Big Bang, but rather was very old starlight, emitted from stars and galaxies strewn across the Universe. Since the Universe is expanding, that light would be redshifted, and since the Universe was constantly generating new matter, that redshifted light would interact with the newly created atoms, scattering and being re-emitted in all directions (Fig. 6.8).

In other words, they contended, finding a uniform background of low-temperature radiation does not necessarily mean that the Universe began from a hot, dense, expanding state a finite amount of time ago. Instead, it was possible that the low-temperature radiation is simply the light from stars arriving after billions of years, having been scattered by the matter present in the expanding Universe.

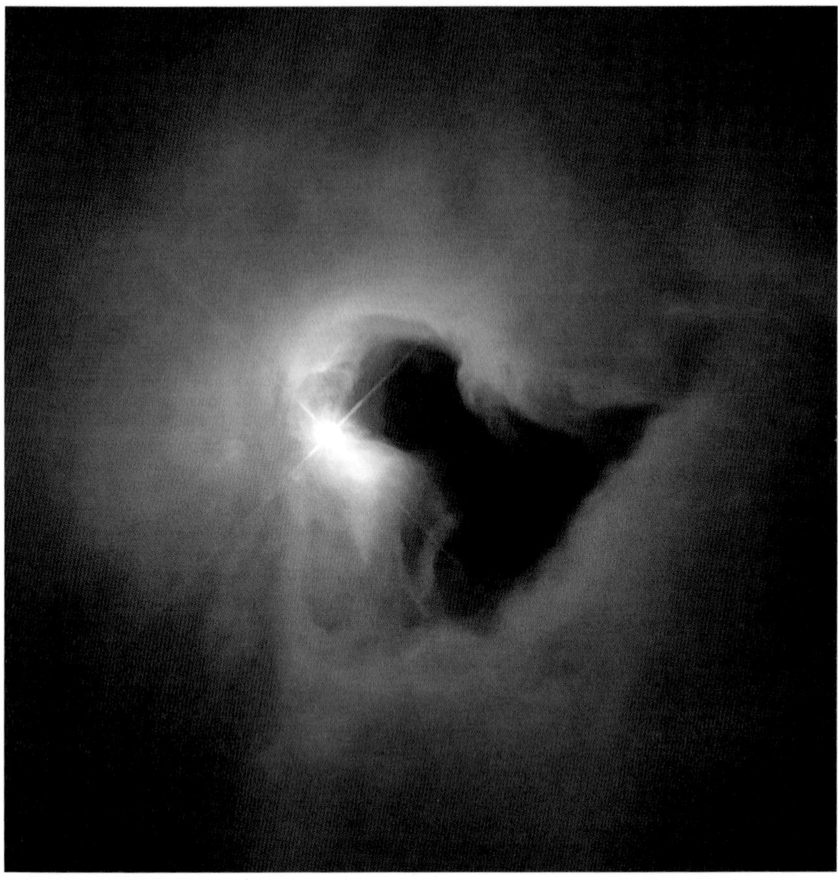

Figure 6.8 Just as NGC 1999, shown here, consists of a region of light-absorbing dust but also has regions that reflect light, the proponents of the Steady-State model claimed that a uniform background of radiation could have been caused by starlight that was absorbed and re-emitted by gas and dust throughout the Universe. Image credit: NASA and The Hubble Heritage Team (STScI).

While the discovery of the CMB was enough to convince the majority of scientists that the Big Bang had it right, this minority opinion was one that needed to be taken seriously. In science, when two explanations can account for the same phenomenon, it is not simply enough to choose the simpler one or the one that you feel better about. Instead, you need to look a little deeper, and find where the explanations differ from one another in their predictions for something you can then go and look for, either observationally or experimentally.

In this case, the Big Bang was very specific in its predictions: if the Universe originated from a hot, dense state, the leftover radiation background should possess a particular type of spectrum: that of a perfect blackbody. In theory, if you heat up a perfectly dark material — that is, something that absorbs 100% of the radiation around it and reflects 0% — to an arbitrary temperature, it will begin emitting radiation. If you have ever seen flowing lava (or a stove's heating element) glow red, that is due to the fact that it is heated up to a temperature of at least 525°C (977°F), so hot that it begins to emit visible light as part of its radiation. The degree to which something like the early Universe was a perfect blackbody ought to have been incredible, as the early Universe should have been nearly perfectly homogeneous, or uniform in density. This translates into a prediction for the blackbody spectrum of the CMB that corresponds to deviations being no larger than one part in a thousand.

Stars, however, are also pretty good blackbodies. The Sun, for example, is to a good approximation a blackbody of temperature 5,777 K, and is the one star in the Universe we have studied best. However, the starlight that we see does not come from a single surface, since the Sun is not a solid body, but rather a huge ball of plasma. The outermost portion of the Sun's photosphere (the part that emits the light we see) is significantly cooler than this temperature, while layers that are closer to the Sun's interior are significantly warmer. In other words, even if we ignore the fact that the Sun has (and all stars have) significant absorption lines, its light should not appear as a single blackbody, but rather as a sum of many blackbodies of different temperatures. When summed together, we see only an approximate blackbody (Fig. 6.9).

We should, in principle, be able to tell these two models apart by measuring what the intensity of the CMB radiation is at many different frequencies. Penzias and Wilson, in 1964, only were able to measure a single frequency, extrapolating a temperature for the CMB of 3.5 K. But over the course of the 1960s, 1970s and 1980s, many other measurements came in, establishing that the temperature was closer to 3 K, and later refined even further to 2.7 K. Most importantly, it was confirmed that the spectrum of this radiation, to the limits of our equipment, was indeed as consistent with a perfect blackbody as we could measure.

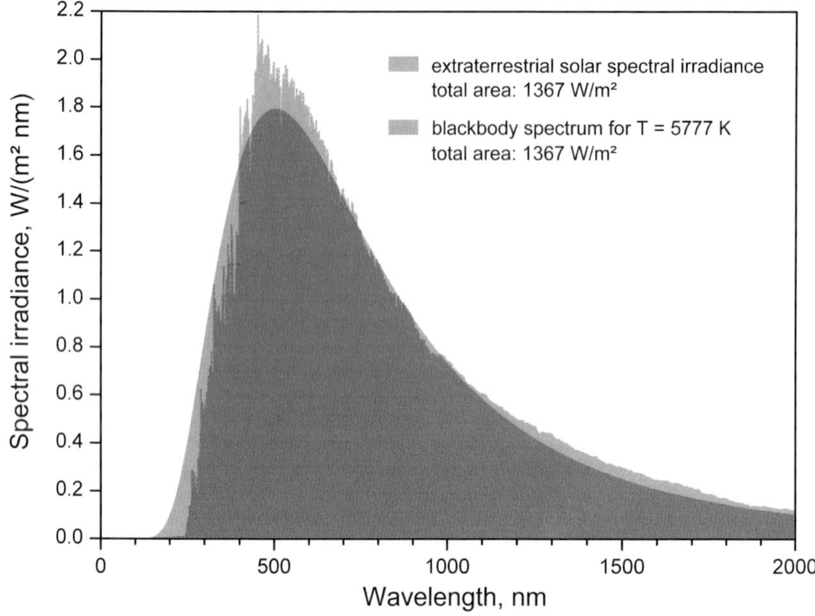

Figure 6.9 The Sun's spectrum can be approximated well by a blackbody at 5,777 K (in grey), but is much better fit (to the orange curve) by summing up a number of blackbody spectra many hundreds of degrees in either direction. While the blackbody approximation is quite good (with a match of around 97%), it's easily differentiated from a true blackbody of a single temperature. Image credit: Wikimedia Commons user Sch, under c.c.-by-s.a-3.0.

But departures from a perfect blackbody would be difficult to detect. At its peak frequency, the CMB was measured to have a maximum intensity of 400 MegaJanskys per steradian, which is a measurement of the amount of flux per unit area on the sky. If the "starlight" explanation held, deviations from a perfect blackbody would only be on the order of 10 MegaJanskys per steradian, a difficult measurement to make. In addition, there should be, compared to a perfect blackbody, a greater flux at large frequencies and a smaller flux at lower ones. The ultimate test would come when, in 1989, the Cosmic Background Explorer (COBE) satellite was launched, with the specific intent of measuring the temperature, spectrum and — if possible — the deviations from a perfect blackbody in the CMB. After years of careful study, the COBE team released their results in 1992, finding that the CMB was, in fact, a perfect

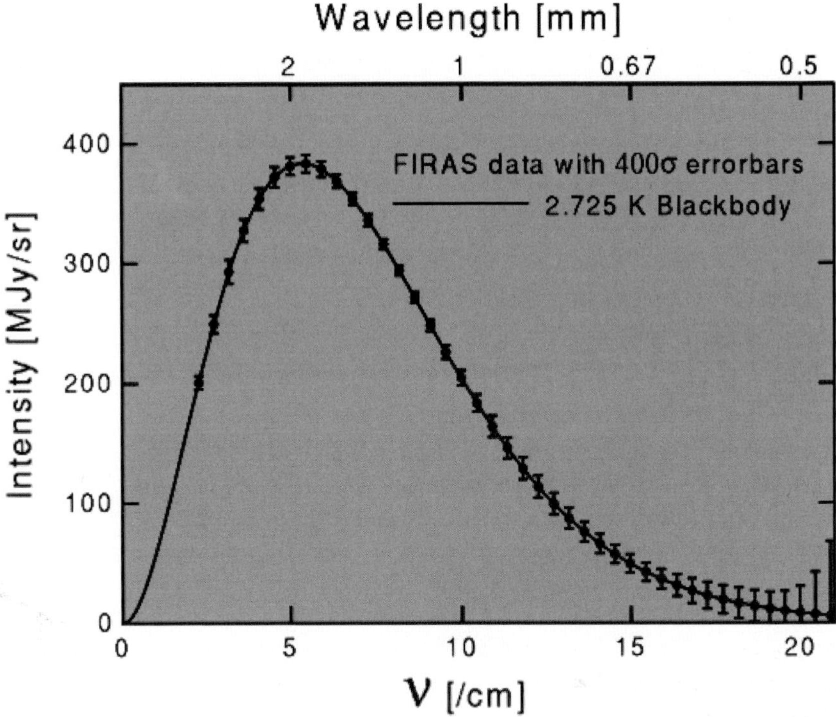

Figure 6.10 The match between the predictions of the Big Bang (along the curve) and the actual CMB data (points, with the measurement error bars exaggerated by a factor of 400) was outstanding once the adequate measurements were made. The Steady-State model disagrees with the observations, having been definitively ruled out by the properties of the CMB. Image credit: COBE/FIRAS instrument, via NASA/JPL-Caltech.

blackbody at *all* frequencies and of equal energies in *all* directions, to a remarkable precision of 0.01 MegaJanskys per steradian. Moreover, we had measured the temperature precisely: 2.725 K, with an uncertainty of only around 470 *micro*K (Fig. 6.10).

At last, any reasonable doubters had nowhere to turn. The conclusion that the Big Bang model had it right was now inescapable, as the Steady-State model's predictions failed to match the key observations.

* * *

The discovery and precision measurement of the CMB, first predicted by Gamow back in the 1940s and first measured by Penzias and Wilson in

1964, was no doubt one of the greatest achievements of 20th century physics. As our scientific capabilities increased, so did the precision to which we could measure not only the CMB in all directions and at a myriad of frequencies, but the precision to which we could measure the *other* two pillars of the Big Bang — the Hubble expansion of the Universe and the primordial abundance of the light elements — also improved dramatically.

When Hubble first published his results on the expanding Universe in 1929, he used but one technique (the measurement of Cepheid variable stars) to measure the redshifts and distances of 22 galaxies, out to approximately 7,000,000 light years. Today, we have measured precise distances to hundreds of thousands of galaxies through a variety of well-understood techniques, including Cepheids, the Tully–Fisher relation, the Faber–Jackson relation, surface brightness fluctuations and more, out to a distance of more than 2,000,000,000 light years! The longest-ranging accurate technique is precision measurements of the light curves of type Ia supernovae, and those have allowed us to probe a maximal distance (so far) of 10 billion light years, with the current record-holder, SN UDS10Wil, observed in 2013 (Fig. 6.11).

Hubble's original measurement of the rate of expansion was around 600 km/s/Mpc, with an uncertainty of around 100%. Keep in mind that this is an unusual type of rate: it is a speed *per unit distance*, which means that the apparent speed a galaxy recedes from us depends on three things:

1) The expansion rate of the Universe.
2) The galaxy-in-question's distance from us.
3) And the galaxy's peculiar velocity, or how quickly it is moving relative to our line-of-sight from nearby gravitational effects.

The value of the expansion rate has been refined many times as our observations have improved, and the full suite of measurements now available gives us an expansion rate of 68 km/s/Mpc, with a combined uncertainty from all effects of only ±4 km/s/Mpc. In all directions and at all distances, the Hubble expansion of the Universe holds, with the history of the Universe's expansion dependent only on what forms of matter and energy are actually present.

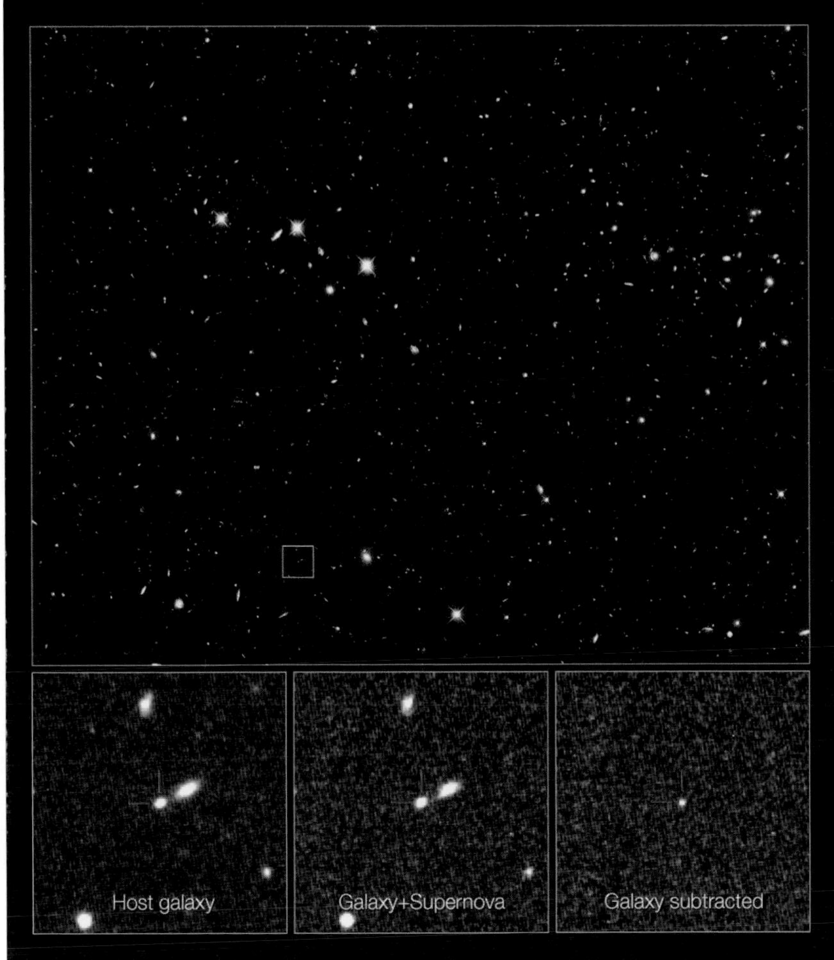

Figure 6.11 The supernova shown here, SN UDS10Wil, emitted its light back when the Universe was only a *third* its current size. Our ability to measure the redshift–distance relation out to distances more than 1,000 times as far as Hubble's original measurements enable us to make much more precise measurements of the Universe's expansion rate, both today as well as in the distant past. Image credit: NASA, ESA, A. Riess (STScI and JHU) and D. Jones and S. Rodney (JHU).

Simultaneously, as our observing techniques and technology improved, so did our ability to measure the abundances of the light elements. This is not a straightforward task to tackle, since what you want to measure is how much hydrogen, deuterium, helium-3, helium-4 and lithium-7 were present in the Universe before any stars formed. Yet, the only way we can measure the distant Universe is by receiving light from distant sources. With the exception of the CMB, *all* of that light originates from either stars themselves or the light-emitting processes associated with the births and deaths of stars. But there is hope.

Even though our earliest measurements of the CMB showed us a Universe that was perfectly uniform in temperature and density, we know for certain that it could not have been so. If it were perfectly smooth and uniform, where no region of space was even the slightest bit hotter, colder or more-or-less dense than any other, then no region of space would have preferentially attracted more matter to it than any other. No region would have grown more massive over time than any other due to gravity. And we never would have had regions of space grow to be rich in stars, galaxies or clusters of galaxies, while also developing great voids in the regions between them. We must have had a Universe that had initial density imperfections at some level, even if that level was less than 1%, 0.1% or even 0.01%. The Universe could not have begun in an arbitrarily uniform state.

If those initial fluctuations were present, then the regions with the largest overdensities would be the first to gravitationally collapse to form stars, while the ones with only modest overdensities would take far longer to do so. This leads to a Universe where the first stars and proto-galaxies could form quite early on, as early as 50 to 150 million years after the Big Bang. But also present would be regions of space where neutral, molecular gas clouds formed, but took many hundreds of millions or even, perhaps, billions of years to form stars for the first time. Considering that when we look at greater and greater distances in the Universe, we are also looking farther and farther back in time, there ought to be some regions of space where either completely pristine or very nearly pristine gas not only exists, but where it exists *with a luminous source behind it*. Just as we can perform emission spectroscopy on luminous sources, breaking up their light spectra into different wavelengths to see what elements are present, we can also perform absorption spectroscopy to see what elements are present in an intervening molecular cloud in between our line-of-sight and a luminous source.

There is a set of systems that lend themselves to this type of observation, and they are the most luminous single sources of electromagnetic radiation in the Universe: **quasars**. Originally formulated as an acronym for **Q**uasi-**S**tellar **R**adio **S**ource (QSRS), a quasar is now known to be a supermassive black hole — usually found at the centers of galaxies — that is actively feeding on some sort of matter. Quasars emit a tremendous amount of electromagnetic radiation not only in the radio frequency, as originally thought, but also often in the visible and even the X-ray. As the quasar's light passes through the expanding Universe, it redshifts, and as the light then passes through an intervening molecular gas cloud (or multiple clouds), some of that light is absorbed at just the right frequencies. Finally, the light continues to redshift until it reaches our eyes, giving us a window into what elements are present and in what relative abundances in each intervening gas cloud (Fig. 6.12).

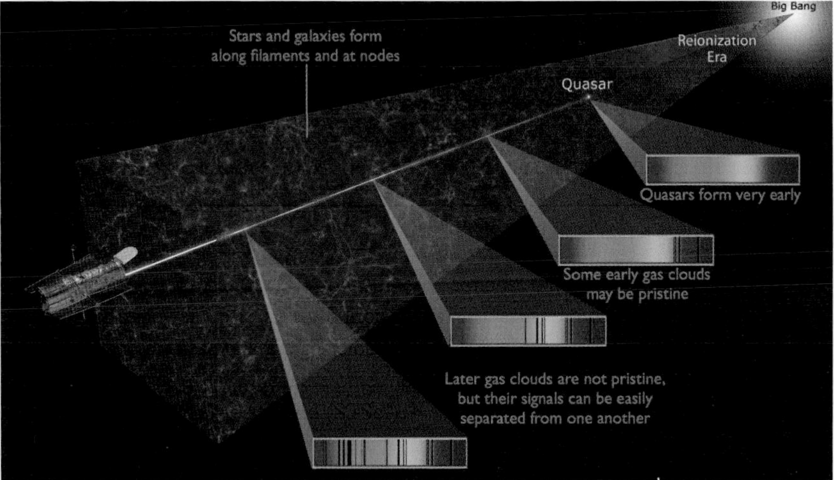

Figure 6.12 While some quasars form very early in the Universe, there are plenty of regions where neutral gas takes a long time to collapse into star clusters and galaxies, remaining pristine for up to a few billion years. If we have a fortuitous alignment of a quasar with one of these pristine gas clouds, we can detect the absorption signatures from that gas. This is true even if there are other non-pristine clouds also along the line-of-sight. Image credit: NASA/ESA, A. Feild (STScI).

Because each atomic nucleus has a unique combination of charge and mass, each type of neutral atom has its own unique absorption spectrum. Even different isotopes of the same element — hydrogen and deuterium, helium-3 and helium-4 — have slightly different spectral signatures from one another. And even though completely pristine samples are hard to find, by measuring progressively more and more pristine samples, we can extrapolate how much of each element must have been present before any stars formed. When all is said and done, we arrive at a universal, coherent picture: the Universe started off with (by mass) 76% hydrogen, 24% helium-4, about 0.0022% deuterium (hydrogen-2), about 0.0011% helium-3 and approximately two-or-three parts in 10^{10} lithium-7. This picture was confirmed spectacularly in 2011 when the very first truly pristine gas clouds were discovered — clouds with no carbon, oxygen or any elements made from stars *at all* — which gave the precise hydrogen, helium and deuterium abundances that we expected from all our other observations (Fig. 6.13).

* * *

In addition to all these successes of the Big Bang, the 1992 data release of the COBE satellite also brought with it one more revolution: the first definitive detection of *imperfections* in the radiation background dating back to the Big Bang. Although COBE could only measure angular scales down to a resolution of around 7°, it found that although all regions of sky exhibited the same 2.73 K temperature, some regions were ever so slightly warmer or cooler than average. By measuring the temperatures in two different directions simultaneously, COBE was able to pick up subtle temperature differences to far greater precision than the absolute temperatures it could measure. The magnitudes of these fluctuations were only on the order of 100 μK (or *micro*kelvin), but these hot spots and cold spots told us something critical about the Universe: it *did not* start off perfectly uniform after all!

This isn't because the radiation itself was non-uniform; it's the one thing that was truly perfect in all directions and locations. No one part of the Universe was intrinsically any hotter or colder than any other at the moment of the Big Bang, but some regions did begin as slightly more-or-less dense than the average. It could not have been by very much,

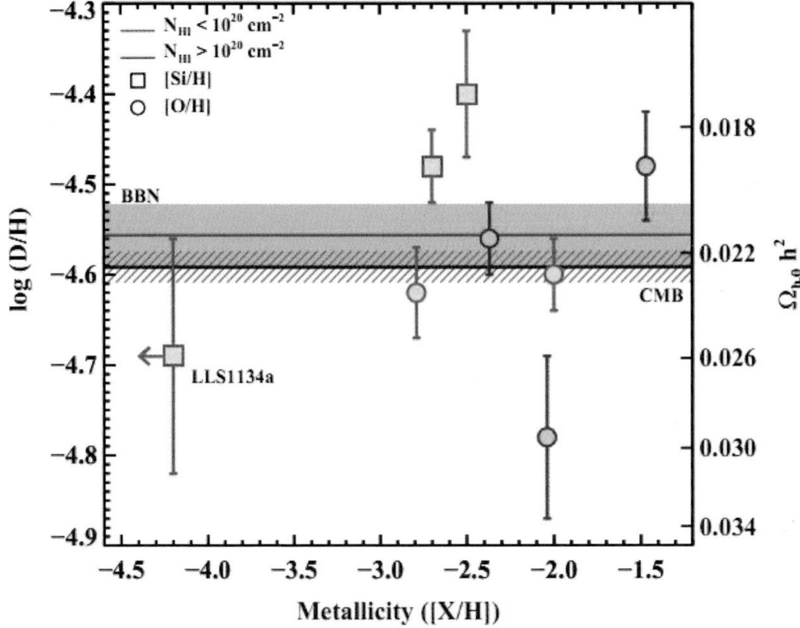

Figure 6.13 The most pristine gas ever was first discovered in 2011, along the line-of-sight of a known quasar. Not only were we able to find that the hydrogen-to-helium abundance (by mass) ratio was 3 to 1, exactly as expected, but we were even able to measure the deuterium abundance, obtaining a ratio that was in perfect agreement (within the allowable errors) with what was predicted. Image credit: Michele Fumagalli, John M. O'Meara and J. Xavier Prochaska, (2011). *Science*, 334 (6060), 1245.

otherwise we would see significantly asymmetric structures on the largest scales in the Universe. But fluctuations on the scale of around 0.003% from the average density are not only perfectly consistent with the large-scale structure we see in the Universe, but they were the cause of temperature fluctuations that the COBE satellite observed.

How could density imperfections take a uniform bath of radiation and make some spots hotter and other spots colder? Imagine that the density of the young Universe is like the surface of an ocean. You might have waves on the surface that are typically a few centimeters (or, on occasion, even tens of centimeters) from top to bottom, but the ocean extends downwards for many kilometers at its deepest. The highest peaks of the waves are regions where there is just slightly *more* water than average, while the lowest troughs are regions where there is slightly less water than a typical region would have. If we averaged the ocean's depth over a large number of waves and troughs, we would get the same standard value everywhere. The density of the young Universe is very much like the ocean's surface in this regard, with some small regions having a slightly greater than average density and others having a slightly lower than average density. On far greater scales, the largest regions — when their density is averaged overall — should display the same densities as any other region of comparable size (Fig. 6.14).

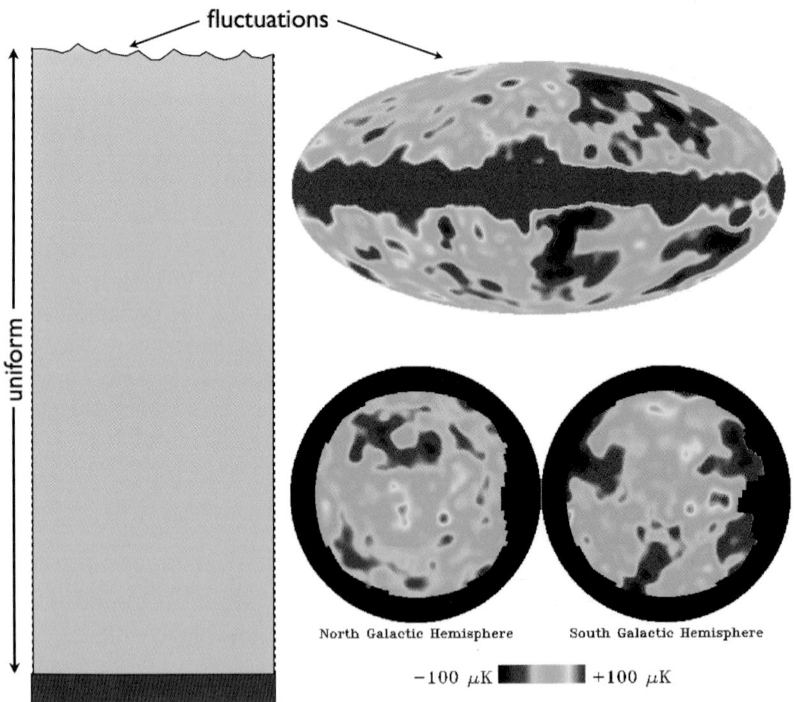

Figure 6.14 If we look at the surface of the ocean, we can see sizable fluctuations on the surface, but they're minuscule compared to the actual depths involved. Similarly, if we look at the temperature fluctuations in the CMB, they appear large relative to one another, but are only on the scale of approximately 100 *micro*K, compared to the average CMB temperature of 2.725 K, corresponding to fluctuations of just 0.003% the average value. The red swath across the middle of the CMB is the galactic plane, which COBE's technology was unable to subtract. Image credit: E. Siegel; COBE/DMR/NASA/Caltech/LBL.

How do those slight density differences show themselves to us? It is not like we can directly measure the density of the Universe in different directions when stars have not yet formed, after all. But remember that you have this uniform background of radiation that is arriving at our eyes after originating from this ocean-like surface, with peaks and troughs, which correspond to overdense and underdense regions. At the instant it leaves the surface of last scattering, the radiation really *is* the same everywhere: the same blackbody spectrum, the same number density of photons, the same exact temperature. (Technically, there are very, very slight imperfections in the radiation itself, particularly on smaller scales, but that is a discussion for a more in-depth treatment of this subject.) But although the radiation is uniform, the last surface that the radiation interacts with before it streams freely to our eyes *is not*. Instead, that radiation is coming from a location that is as diverse in density as the surface of the ocean is in height at any given moment: some regions have slightly more (or less) matter than average clumped together, while a few rare regions have significantly more (or less) matter than average.

Every photon, when the atoms in the Universe at last become neutral, will finally stop scattering off of the previously ionized plasma, which it spent the first 380,000 years of the Universe's history interacting with at an alarming frequency. After each photon scatters off of its last ion, it undertakes an epic journey, traveling in a straight line and having its wavelength stretch as the Universe expands. But there is one huge task every photon needs to accomplish before it can stream across the Universe, eventually arriving at our eyes: it needs to climb out of the gravitational potential well that the matter creates at the surface of its last scattering. There are three ways this can proceed:

- For a photon in a region of space that has the *average* density of the Universe, the photon will climb out of an average-sized well, losing a typical amount of energy due to gravitational redshift in the process. As it streams across the Universe, it has an energy consistent with the average blackbody spectrum of all the photons in the Universe.
- For a photon in a region of space that has an *above average* density, the photon will climb out of a deeper-than-average well, losing a greater-than-typical amount of energy due to gravitational redshift. As it streams across the Universe, it has an energy that is below average compared to the average blackbody spectrum for all the photons in the Universe.

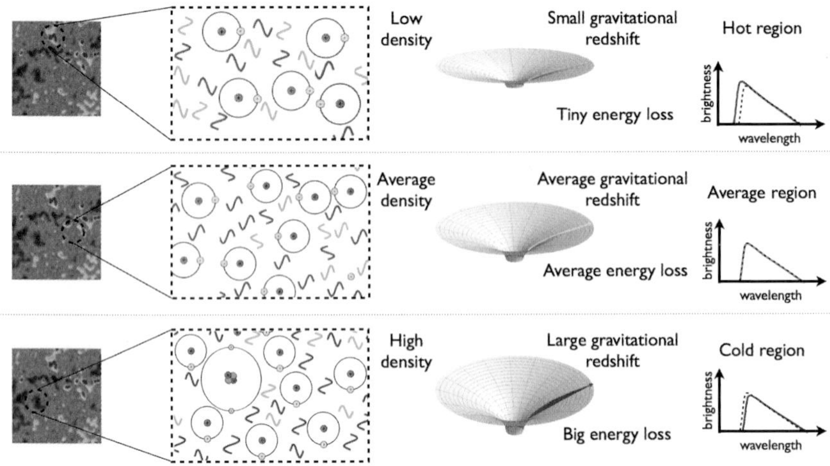

Figure 6.15 The temperature differences seen in the CMB are due to density differences at the surface of last scattering, with cold spots corresponding to higher densities, deeper gravitational potential wells and hence greater gravitational redshifts, while hot spots correspond to lower densities, shallower gravitational potential wells and hence smaller gravitational redshifts. Overdense and underdense regions lead to cooler and warmer temperatures, respectively, compared to photons emerging from regions of average density. Image credit: E. Siegel.

- For a photon in a region of space that has a *below average* density, the photon needs only to climb out of a shallower-than-average well, losing less energy than photons typically do owing to gravitational redshift. As it streams across the Universe, it possesses as an energy that is greater than average compared to the average blackbody spectrum for photons in the Universe (Fig. 6.15).

So when we observe hotter-than-average temperature fluctuations in the CMB, that tells us we are looking at a region of space, when the Universe was 380,000 years old, that had a *lower* than average density. Similarly, when we see a colder-than-average temperature fluctuation, we are looking at a region that a long time ago had a *higher* than average density. The higher density regions are more likely to preferentially attract more and more matter over time, leading to the formation of structures

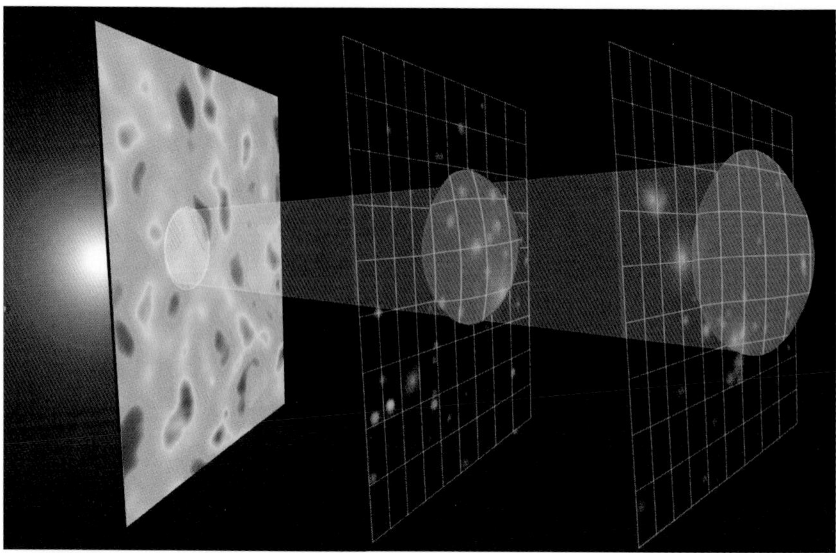

Figure 6.16 The cold spots (shown in blue) in the CMB are not inherently colder, but rather represent regions where there is a greater gravitational pull due to a greater density of matter, while the hot spots (in red) are only hotter because the radiation in that region lives in a shallower gravitational well. Regions in green indicate an average density, and hence the photons emerging from them have an average temperature. Over time, the overdense regions will be much more likely to grow into stars, galaxies and clusters, while the underdense regions will be less likely to do so. A look back at the CMB gives a window into the initial seeds that grow into today's large-scale structures in the Universe. Image credit: E.M. Huff, the SDSS-III team and the South Pole Telescope team; graphic by Zosia Rostomian.

like stars, galaxies and galaxy clusters, while the lower density regions are less likely to do the exact same thing (Fig. 6.16).

The COBE satellite was the first all-sky map of these temperature fluctuations, and subsequent satellites — including WMAP and Planck — have measured these hot and cold spots in many different frequency ranges and down to angular scales of less than half-a-degree. At long last, we have a coherent picture of not only how uniform the temperature and density of the early Universe was, but also what the seeds of structure looked like, and how they grew, over millions and then billions of years, into the stars, galaxies, clusters and great cosmic voids filling our Universe today. With the expanding Universe, the abundances of the light elements, and the CMB all pointing to the same, singular conclusion, the Big Bang has today been overwhelmingly accepted by the scientific community (Fig. 6.17).

* * *

I wish I could tell you that every scientist who worked on an alternative theory to the Big Bang was eventually convinced by the overwhelming suite of observational data that has come to support it. Just as Kepler, who was once wedded to the idea that planets moved in perfect circles whose sizes were prescribed by the five perfect polyhedral solids, was able to throw that idea out and replace it with the notion that planets moved in ellipses around the Sun *based on the incontrovertible data*, you would hope that all scientists would be able to change their conclusions when the evidence became incontrovertible. I wish I could tell you that the Steady-State supporters, the Tired-Light proponents and the Plasma Cosmology theorists (among others) were willing to change their conclusions on the basis of the overwhelming observational evidence we have accumulated. But even though science marches forward, weeding out old, invalid theories and leaving only the most robust ones to continue onwards, not all scientists are capable of the same evolution.

Geoffrey Burbidge, one of the leading proponents of the Steady-State model (and one of the "B"s in the famous B^2FH paper that correctly worked out the nuclear physics inside stars), never accepted the Big Bang, insisting that many of the most distant objects (such as quasars) were actually nearby until the 2000s. Fred Hoyle, another Steady-State proponent (and the "H" in B^2FH) would give talks proclaiming that we mysteriously lived in a fog of radiation, which is what he called the CMB. Despite the failure of scattered starlight to produce a blackbody spectrum, Hoyle never abandoned his discredited hypothesis, with many of his former students and collaborators still adhering to it. Halton Arp, a pioneer in the study of interacting galaxies, continued to claim that quasar redshifts were quantized, despite the eventual accumulation of hundreds of thousands of these objects that showed no such effects at all. Hannes Alfvén, a revolutionary scientist in the astrophysical field of magnetohydrodynamics (or how plasmas flowed under extreme

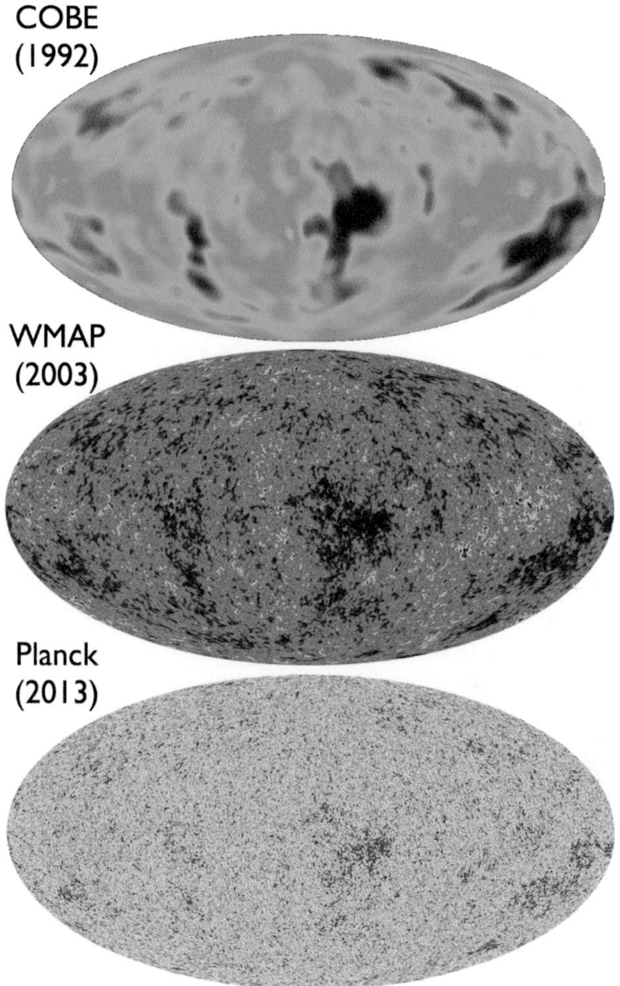

Figure 6.17 The years given correspond to the first major data release of each mission, and show the increased resolution as time progresses. The COBE data has been retro-fitted with the galactic sky subtraction measurements from later missions, although was only sensitive to angular scales of 7° and larger, measuring three different frequency bands of radiation. WMAP was able to measure resolutions down to 0.3° in five different frequency bands, with Planck measuring all the way down to just 5 arcminutes (0.08°) in nine different frequency bands in total. Images credit: NASA/COBE/DMR; NASA/WMAP science team; ESA and the Planck collaboration.

conditions, such as in the Sun), correctly stated that many astrophysical calculations in the 1960s did not properly account for electromagnetic phenomena and effects, which may have been especially important in the early Universe, which experienced a fully ionized, plasma phase for a time. But although such effects were important for *some* phenomena, such as magnetic fields associated with galaxies and black hole jets, they could not explain the larger-scale gravitational phenomena, nor the CMB.

These outstanding scientists, and many others, were never convinced by the overwhelming evidence in favor of the Big Bang, and were never able to reject their own pet ideas despite the overwhelming evidence to the contrary. To this very day, there is no other model that is both consistent with General Relativity and explains the Hubble expansion of the Universe, the abundances of the light elements and the existence and properties of the CMB; the Big Bang is the only one. Despite the lifelong objection of many such prominent scientists, fewer and fewer people working in the field took their ideas seriously, and today the Big Bang is disputed little more than the fact of heliocentrism. After all, as Max Planck famously said,

> "A new scientific truth does not triumph by convincing its opponents and making them see the light, but rather because its opponents eventually die, and a new generation grows up that is familiar with it."

With the Big Bang firmly on solid observational as well as theoretical ground, its limits could now be probed, paving the way for an even deeper understanding of what our Universe is made of, where it came from and where it would head in the future.

Chapter 7

What Does It Matter: Why There Is More Matter Than Antimatter In The Universe

The greatest successes of the Big Bang, of its predictions for what we would see in the Universe *today*, came from extrapolating the known laws of physics using this model extraordinarily far back into the distant past. A cooling, expanding Universe that has billions of years to form stars, galaxies and clusters of galaxies on the largest scales also has the extraordinary property that as we look farther away at greater distances in the Universe, we also see back in time. When we look at galaxies and clusters that are farther away, we not only see that they appear to be speeding away from us more quickly due to the Hubble expansion of the Universe, we also see them when the Universe was around for less time, and hence when these objects were less evolved. This should mean a huge variety of things, a great number of which have been tested observationally (Fig. 7.1).

If the Universe is not only hotter and denser but also *younger* as we look to greater and greater distances, there should be a great number of things we see as we observe what was present at earlier and earlier times. We should see that galaxy clustering becomes sparser at great distances, as fewer large mergers and less gravitational collapse had time to have occurred in the distant past. We should see that the temperature of the Universe — the 2.725 K cosmic microwave background (CMB) that we observe today — rises to higher temperatures the farther away we look. We should see that there are fewer heavy elements in stars and galaxies at

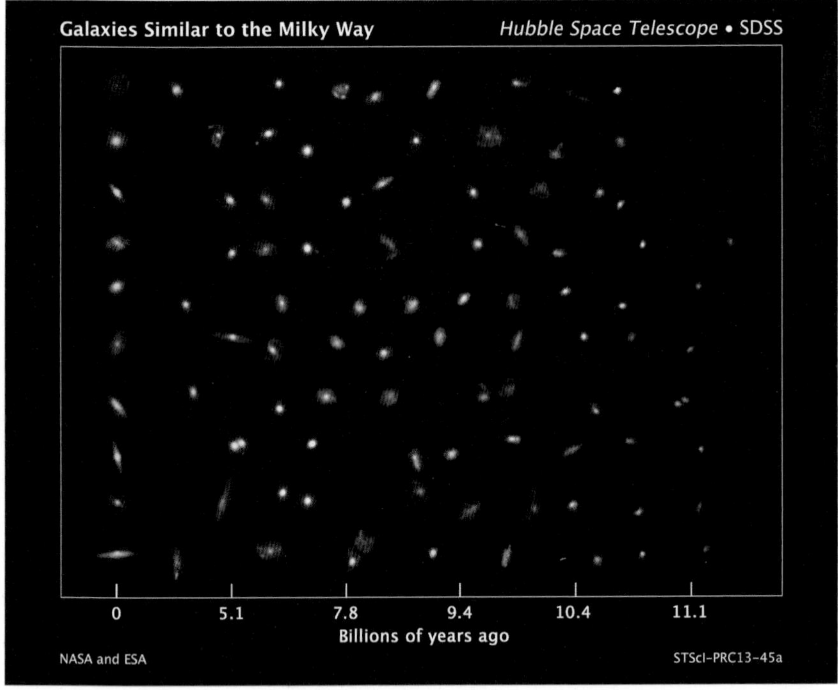

Figure 7.1 As we look farther back in time, we find that galaxies analogous to our Milky Way were smaller, more actively forming stars, lower in heavy element content and less structurally evolved than comparable galaxies today. Image credit: NASA, ESA, P. van Dokkum (Yale University), S. Patel (Leiden University) and the 3D-HST Team.

great distances, since the Universe had less time to form many generations of stars. And, thinking along those same lines, if we look back far enough, we ought to find completely pristine clouds of intergalactic gas that have never yet been "polluted" with heavy elements by the formation-and-eventual-death of even a single star in the Universe.

All of these are inescapable consequences of the Universe's history arising from the Big Bang: from the fact that the Universe was hotter, denser and expanding more rapidly in the past as compared to today. When you confront these predictions with observations — none of which were available at the time that the Big Bang theory was first formulated — you find that all of the predictions mentioned earlier (in addition to many others) have been confirmed observationally to a very high degree of precision.

In order for the Universe to come to be as it is today, there were a number of important events that needed to happen. Going back in time step-by-step, in reverse chronological order, they are:

- Rocky planets with the ingredients for life needed to form, requiring many prior generations of stars to have lived, burned through their fuel and died, recycling the heavy elements created deep within back into the Universe.
- To form those stars, gravitation needed to preferentially attract huge amounts of matter from very large volumes of space into cold, dense molecular clouds capable of collapsing in order to trigger star formation.
- To form those cold molecular clouds, the atoms that would come to make them up needed to lose the energy they once had in the early stages of the Universe, requiring time for their temperatures and kinetic energies (and densities) to drop sufficiently.
- For those atoms to exist in the first place, the Universe needed to transition out of the phase where it consisted of a hot, dense, ionized plasma, where photons were energetic enough to prevent neutral atoms from stably forming out of the nuclei and electrons for very long.
- And to form the first stable, complex atomic nuclei, the Universe needed to cool from a time where the photons were so hot that even the simplest atomic nucleus that was more than just a simple proton or neutron on its own, deuterium, would be blasted apart before it could undergo any further steps in a nuclear chain reaction.

In the context of the Big Bang, each and every one of these steps *naturally* occurs, leading to a Universe that develops nuclei, atoms, molecular clouds, and at last stars. Eventually, after many generations of them have lived and died, new stars with rocky planets around them, full of the ingredients for life, came to be, with one such world giving rise to us.

But this story assumes something that we have not yet considered: that the Universe *began* with protons and neutrons. This is an assumption that is unlikely to be true. To see why, all we have to do is extrapolate the Big Bang back even further: to the very highest energies humanity has ever probed, and even beyond! (Fig. 7.2)

* * *

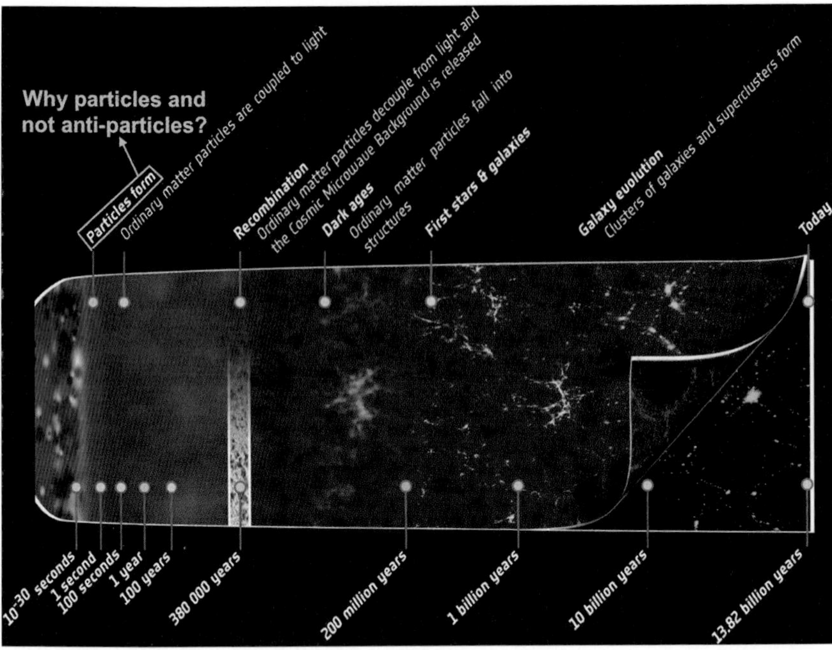

Figure 7.2 The Big Bang allows us to extrapolate very far back into the past. It provides us with a cosmic story that includes the origin of complex molecules, heavy elements, stars and planets, galaxies, the formation of neutral atoms and the first atomic nuclei from a hot, dense past. But there is a missing piece to this story: why do we have a Universe that is filled with matter particles at all? Image credit: ESA and the Planck Collaboration, modified by E. Siegel.

Whenever any two particles collide, which happens with increasing frequency the farther back in time you look, there are a number of possible ways in which they can interact with one another. They can collide elastically, bouncing off of one another and turning all of that initial energy from before the collision into the kinetic energy of the outgoing particles. They can collide inelastically, perhaps causing one of the particles to explode apart or perhaps causing both of them to stick together. Or, if the energies are high enough, they can spontaneously cause the creation of new particles: a combination of matter and antimatter in equal amounts. This is something that happens without any provocation so long as there is enough energy, and the amount of energy required to do so is given simply by Einstein's most famous equation: $E = mc^2$.

Figure 7.3 If we go back to early enough times, it will be too energetic to form even single protons and neutrons, as they will be dissociated into their constituent quarks and gluons. At the high temperatures achieved in the very young Universe, not only can particles and photons be spontaneously created, given enough energy, but also antiparticles and unstable particles as well, resulting in a primordial particle-and-antiparticle soup. Image credit: Brookhaven National Laboratory.

If we go back to earlier and earlier times in the Universe, both the mean kinetic energies of all the massive particles and the mean energies of photons continue to increase, and therefore so does the amount of energy available to create new particles (Fig. 7.3).

But we cannot simply create whatever particles we like; there are conservation rules we need to obey. In particular, we need to create particles in pairs, with only equal parts matter and antimatter being permissible, in the following fashion:

- Every particle in the Universe has a specific set of properties that uniquely describe it: a combination of rest mass, electric charge, baryon number, lepton number, lepton family number and spin, among others.
- For every particle, there is a counterpart to it: an *antiparticle*, with the same mass and spin, but the *opposite* electric charge, opposite baryon number, opposite lepton number and opposite lepton family number.
- Some particles, such as chargeless bosons, happen to be their own antiparticles.

- And finally, when antiparticles and particles collide, they annihilate into two photons, each with an energy equal to the rest mass of the particle–antiparticle pair, again given by $E = mc^2$.

So if we continue to go farther back in time, to when the Universe was approximately *one second* old (after the Big Bang), we reach energies high enough to begin spontaneously creating electron–positron pairs. At even earlier times, we can spontaneously create particle–antiparticle pairs of even heavier particles like muons, pions, protons and neutrons. If we go farther back still, we can produce all the known particles (and antiparticles) in the Standard Model, including quarks, leptons, gluons, heavy bosons and even the Higgs! (Fig. 7.4)

But there is a problem here. If the Universe started off as an ultra-high energy sea of photons (which are their own antiparticles), along with copious amounts of matter and antimatter, which can only be created *in equal*

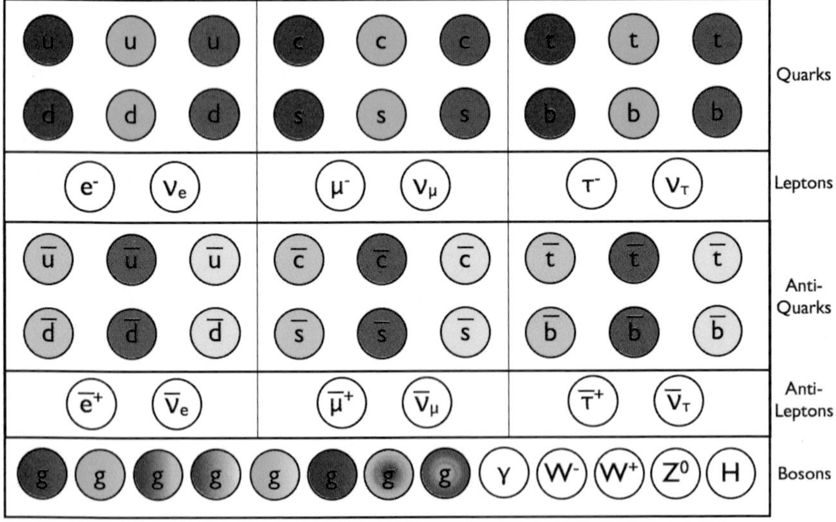

Figure 7.4 These are all the known fundamental particles and antiparticles in the Universe. Particles that experience the strong nuclear force are shown colored; all the quarks and the electron, muon, tau and the W-bosons are electrically charged; the eight gluons and the photon (symbolized by γ) are massless while all the others (even neutrinos) have mass. At high enough energies, all these particles and antiparticles — even the unstable ones — should have existed in roughly equal abundance in the hot, early Universe. Image credit: E. Siegel.

amounts, then why, when the Universe expanded and cooled, were we left with a Universe full of matter *and not* antimatter? At least, exclusively matter-filled is what the Universe appears to be in our local neighborhood. Is this truly the case? Let us find out.

* * *

Let us start by considering what would happen if every reaction in the Universe *were* completely symmetric between matter and antimatter. After all, that is what we would expect within the Big Bang framework and with the presently known laws of physics; would that situation lead to a Universe that looked anything like ours? All we would need to do is consider beginning with the Universe in an arbitrarily hot, dense state, full of radiation and equal amounts of matter and antimatter, and then have it expand and cool according to the laws of General Relativity. If that were *our* Universe, what would happen to it? What would we be left with today?

Imagine a primordial soup of particles, all so energetic that they move ultra-relativistically, meaning not only that the massless ones are moving at the speed of light, but that the massive ones are moving extraordinarily close to that asymptotically unreachable speed: at 99% the speed of light or more. Particles collide with one another, sometimes merely exchanging energy, sometimes creating new particle–antiparticle pairs, and sometimes — when they collide with their own antiparticle — simply annihilating into two photons. Similarly, two photons frequently collide as well, sometimes producing particle–antiparticle pairs, and sometimes not (Fig. 7.5).

So long as the energies are high enough, these two processes of spontaneous particle–antiparticle *creation* and particle–antiparticle *annihilation* will occur at the same rates, giving the Universe some equilibrium number of particles, antiparticles and photons at any given time. But as the Universe expands and (more importantly) cools, that equilibrium changes. Because of the expansion, the collision rate, and hence both the creation and the annihilation rates, drop. But because the Universe cools as well, the particles lose energy, too. You do not need any extra energy to annihilate matter with antimatter, but you *do* in order to create new particles. As a rule-of-thumb, when the average kinetic energy of a particle, antiparticle or photon drops below the equivalent rest mass

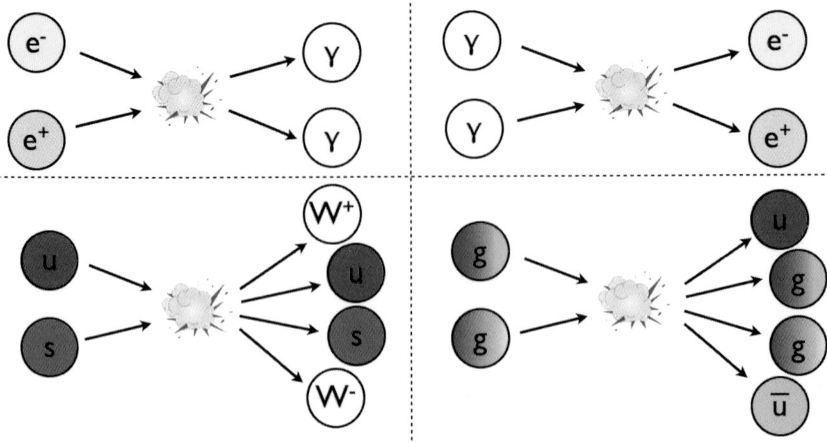

Figure 7.5 When the energies are high enough, any particle collisions (including photon collisions) have the potential to spontaneously produce new matter–antimatter pairs, and matter–antimatter pairs can annihilate back into photons just as easily. Image credit: E. Siegel.

energy (again, via Einstein's $E = mc^2$) that you would need to create that particle, the creation rate plummets towards zero. For the reactions shown in Fig. 7.5, the lower left reaction would cease first, followed by the one on the lower right as the Universe continued to cool, and then followed by the spontaneous electron–positron creation at the upper right. As less energy becomes available, it gets more difficult to create new particles, but *annihilations*, like the upper left reaction in Fig. 7.5, can still proceed without any barriers. Eventually, the annihilation is so efficient that all you are left with is a tiny fraction of both particles and antiparticles, in equal numbers, too sparse to find one another and annihilate, living among a sea of photons.

If this completely symmetric scenario were representative of our Universe, we would first see all the unstable particle–antiparticle pairs annihilate into photons, with the remnants decaying into stable particles like electrons, positrons and neutrinos (and antineutrinos). The unstable quarks would all decay into up and down (and antiup and antidown) quarks, which would condense into protons, neutrons, antiprotons and antineutrons in equal numbers. As the temperature continued to cool, spontaneous particle–antiparticle creation would cease for massive particles, while annihilation would continue unimpeded. The proton–antiproton pairs (as well as the neutron–antineutron pairs) would annihilate away until they could not find one another anymore, and then the electron–positron pairs would do the same thing. Eventually, the neutrons and antineutrons would decay (since they are only stable when bound as part of a larger nucleus), leaving us with only protons, electrons, antiprotons and positrons, in addition to photons, neutrinos and antineutrinos. We would be left with a Universe that was filled mostly with radiation, with only trace amounts of ionized hydrogen, antihydrogen, and nothing more (Fig. 7.6).

This is very clearly *not* representative of our Universe! Sure, we have many more photons than matter particles — a ratio that is a little over a billion-to-one — but that ratio would be worse than 10^{20}-to-one if there were no matter–antimatter asymmetry! You might hold out hope that perhaps there was just some cosmic event that segregated matter from antimatter in the distant past, so that they wouldn't annihilate with one another. But if that were the case, we should see unmistakable signatures coming from the regions where matter-dominated space bordered antimatter-dominated space. Because matter and antimatter annihilate with one another, and the great cosmic web of structure appears to be unbroken, we would expect to see signatures of particle–antiparticle annihilation where stars, galaxies and intergalactic gas collide.

Yet our Universe is observationally incompatible with that. We have surveyed too much of the Universe too deeply for that to be a possibility; the stars, galaxies and gas present in our Universe are entirely made up of matter and *not* antimatter. Somewhere, somehow, at some early time, this present understanding of the Universe must be incomplete. Either the Universe was *born* with some fundamental asymmetry of matter over antimatter, where there were more quarks than antiquarks and more leptons than antileptons, or the Universe had to somehow *create* an asymmetry where there was none initially (Fig. 7.7).

In science, we always try and avoid the assumption of what we call **finely-tuned initial conditions**, where the Universe needed to be in a very particular, contrived state to produce what we see today. Instead, we look for explanations based on *dynamics*, which is the great hope that there are

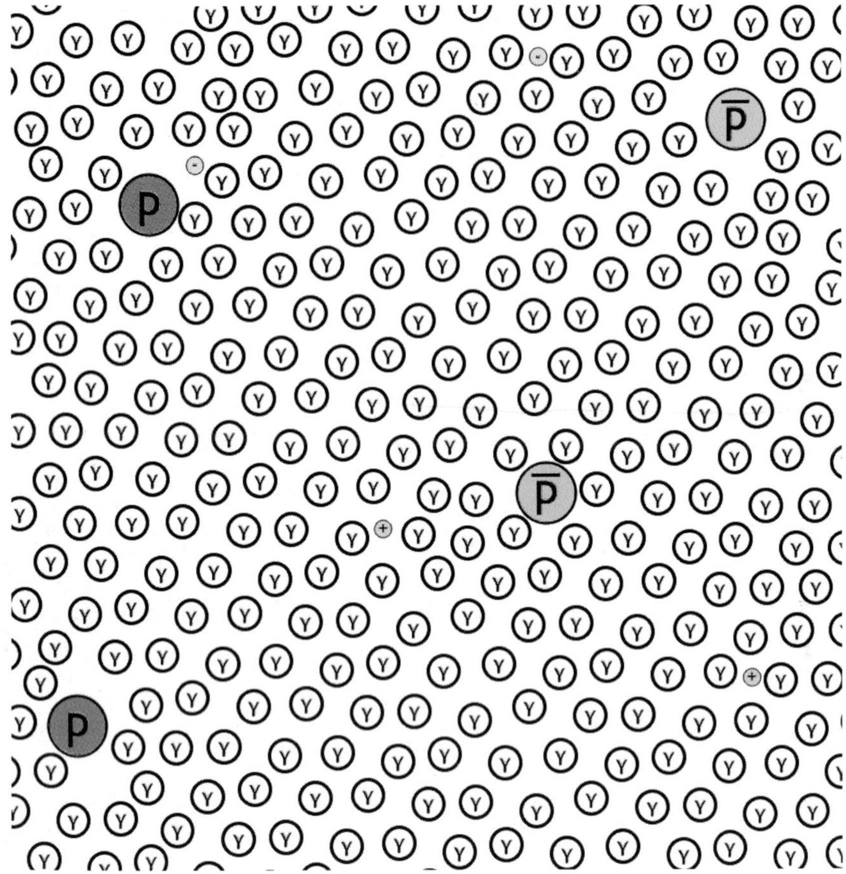

Figure 7.6 A Universe without a fundamental asymmetry between matter and antimatter would not produce stars, galaxies or any of the large-scale structure we observe today. If matter and antimatter were totally symmetric, all that would exist would be a sparse smattering of protons, antiprotons, electrons and positrons amidst a sea of radiation, approximately 10^{18} times less concentrated than this illustration shows. Image credit: E. Siegel.

some underlying physical laws and mechanisms that give rise to the Universe we observe. This is the greatest power of theoretical physics, but also its greatest challenge. You see, in order to become universally accepted, a physical theory needs to not only account for the previously unexplained phenomena already observed, but it needs to make *new predictions* that can be verified and validated as well. We are about to take

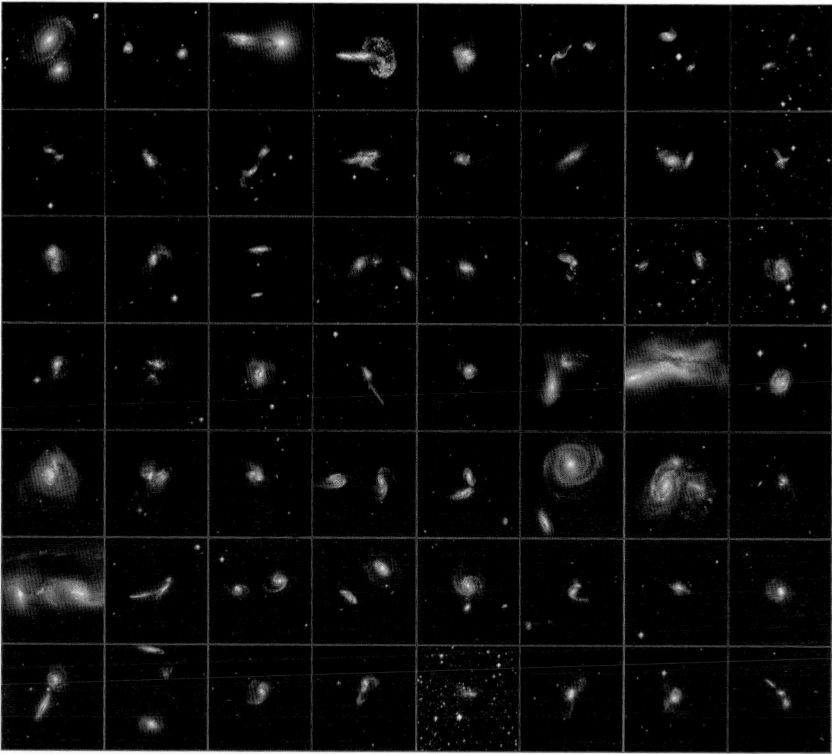

Figure 7.7 All surveyed regions of the observable Universe consist of galaxies in the vicinity of other matter. If there were any interface region between matter and antimatter, it would reveal itself via an incredible outburst of gamma rays. The lack of such radiation combined with such strong evidence for the interaction of the material in various regions of space with adjacent ones — such as the sampling of interacting galaxies shown here — demonstrates to us that our Universe is everywhere made of matter and *not* antimatter. Image credit: HubbleSite, NASA and the Space Telescope Science Institute.

our first step into presently unknown territory: where we believe that something happened in the Universe's past to give rise to the matter–antimatter asymmetry we observe today, but where we have not yet accumulated sufficient evidence to know exactly what that "something" was.

* * *

Even though we do not know exactly *how* the Universe came to have a matter–antimatter asymmetry, there are a number of things we have been able to work out concerning its existence. Observationally, we can be certain of a few things:

1) There *really is* more matter than antimatter in the Universe today.
2) All the stars, galaxies and clusters of galaxies that we have discovered are made up of matter and *not* antimatter.
3) At the energies we have reached in the Universe in our laboratories here on Earth, we have neither created more baryons (e.g., protons and neutrons) than antibaryons (e.g., antiprotons and antineutrons) nor created more leptons (e.g., electrons and neutrinos) than antileptons (e.g., positrons and antineutrinos) in any observed reaction.
4) And, to the best of our measurements, the Universe contains one baryon (and one lepton) per **1.6 billion** photons.

Although that baryon-to-photon ratio might seem like a tremendously small number, remember how big the Universe is! If you were to smear out all the atoms in the Universe and make all of space a uniform density, you would wind up with about one hydrogen atom for every four *cubic meters* of space. For comparison, there are around 411 photons left over from the Big Bang per cubic *centimeter* of space. Although the difference between these two numbers is huge, both numbers themselves are independently large and important! (Fig. 7.8)

So how could this creation of more matter than antimatter have happened? If we wanted to begin in a high-energy state where matter and antimatter existed in equal amounts and wind up in a low-energy state with slightly more matter than antimatter, something must have happened as the Universe expanded and cooled to create that asymmetry. It was a Soviet physicist, **Andrei Sakharov**, who in 1967 became the first to work out that there are only *three things* that need to be true in general for this to happen. These three properties that the Universe must have, today known as **the Sakharov conditions**, are as follows (and do not worry; we will explain them):

1) The Universe must be out of thermal equilibrium.
2) Two of the three fundamental symmetries in the Universe — charge conjugation (C) and charge conjugation *plus* parity (CP) — must be violated.

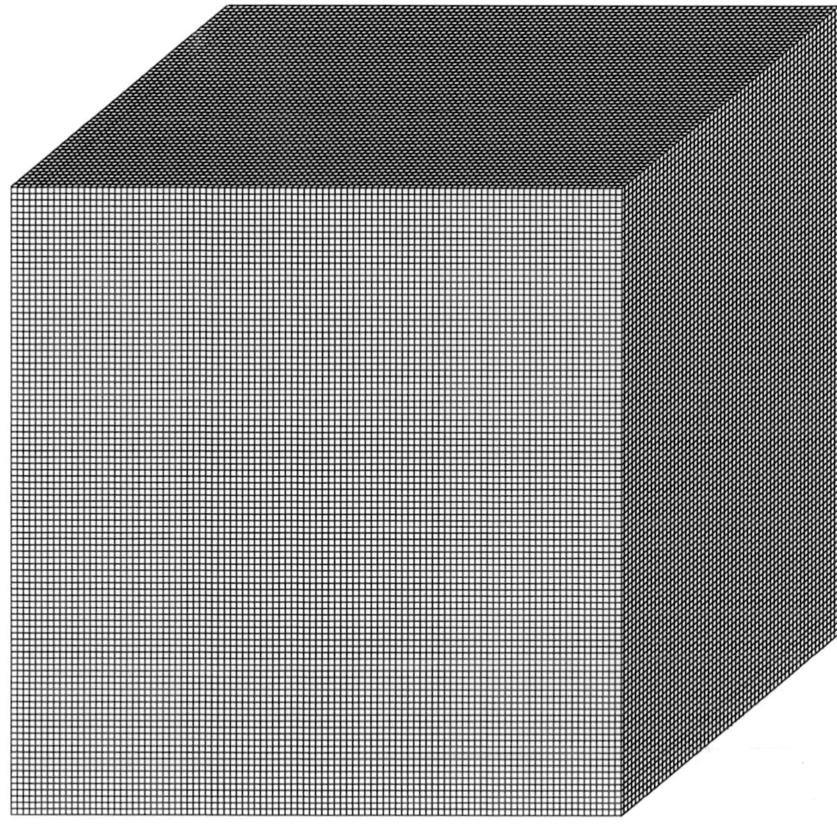

Figure 7.8 The large cube shown here represents one cubic meter of space, where each meter contains 100 centimeters. As illustrated, a cubic meter of space contains 100 cm × 100 cm × 100 cm, or one million cubic centimeters. In our Universe, each tiny cube shown here — representing a cubic centimeter — contains 411 photons within it, while to contain a single hydrogen atom, it would take the volume contained in four of the cubic meters shown here! Image credit: E. Siegel.

3) And finally, there must be interactions that violate the conservation of baryon number.

Before we talk about what these mean individually, we should emphasize that all three of these things need to occur *together*, otherwise you would not generate a difference in the amount of matter vs. antimatter present in the Universe.

Thermal equilibrium is the easiest one to understand: it is the same idea that, if you turn on a heater in one part of a cold room, eventually the entire room becomes the same temperature. In the case of the room with a heater in it, the reason why the temperature equilibrates is because the heater gives energy to the molecules closest to it, which in turn speed up, collide with the molecules that are further away and give energy to them, which then in turn collide with molecules even further away and exchange energy, and so on. Given enough time (and given a room where no heat is exchanged through the walls, ceiling or floor), the room will eventually reach a state of thermal equilibrium, where every part and component of it has the same, stable temperature.

But the Universe is not a stable, static system: it is both *expanding* and *cooling*! A region of the Universe that is a slightly higher temperature than a region just a short distance away might not be able to exchange heat (or any type of information) with that region for hundreds, thousands or even millions of years thanks to the Universe's expansion. On top of that, the fact that the Universe is non-uniform and cooling means that different regions of space will have different relative abundances of particle–antiparticle pairs at any given time, dependent on how much energy is available for particle–antiparticle creation as the Universe cools down. In short, this condition is the easiest one to satisfy: the expanding Universe is perhaps the ultimate out-of-thermal-equilibrium system! (Fig. 7.9)

What about the violation of two of the three fundamental symmetries at once? This one is a little harder to understand, so I want you to picture a particle of matter. Imagine this particle as a little sphere and that it spins counterclockwise around its North Pole as we move forward in time. There are three types of fundamental symmetries that we can apply to this particle:

- **C**-symmetry, known as charge conjugation. This is the same as replacing our particle with an antiparticle: it has the same mass and spin, but certain other properties — electric charge, color charge, baryon number, lepton number and lepton family number — have the exact *opposite* value.
- **P**-symmetry, known as parity, or mirror-symmetry. This is the same as reflecting our particle in a mirror: all its properties remain the

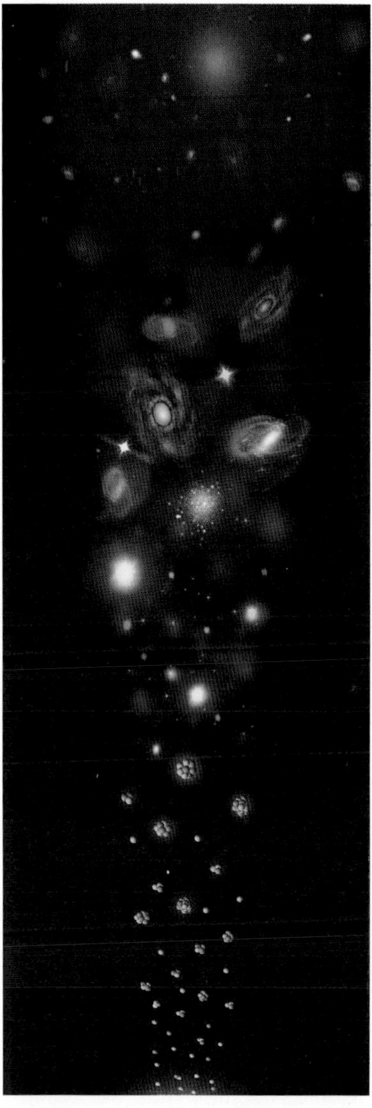

Figure 7.9 This illustration shows a number of different stages that the Universe goes through as it expands and cools. Different regions of space are separated by large enough distances that it takes up to many billions of years for them to exchange information (e.g., photons, temperatures, etc.) with one another, while each individual region finds itself in a meta-stable state every time it cools through a critical transition. Owing to the Universe's rapid expansion and cooling, the Universe is not in thermal equilibrium on large scales. Image credit: NASA/CXC/M. Weiss.

same *except* spin (and, for multi-particle systems, orbital angular momentum), which has the exact opposite value. For example, a particle spinning clockwise would have a reflection spinning counterclockwise in a mirror.

- And **T**-symmetry, or time-reversal symmetry. Instead of a particle moving forward through time as it interacts with the Universe around it, it would move *backwards* through time when time reversal symmetry is applied.

Most interactions in the Universe are completely symmetric under each one of these three transformations: if a particle and an antiparticle exhibit the same physical behavior as one another under certain conditions, they *conserve* **C**-symmetry. If a particle and its mirror reflection behave the same way, they conserve **P**-symmetry. And if particles behave the same way whether you move them forwards or backwards in time, they obey **T**-symmetry as well.

In addition to the individual symmetries, we can look at symmetries in tandem: if a particle behaves the same way that its antiparticle does when it is reflected in a mirror, that is an example of **CP**-symmetry. There are cases where it is possible to violate **C**-symmetry and **P**-symmetry individually but to still conserve **CP**-symmetry. And finally, there is a theorem that says that all physical systems in the Universe *must* conserve **CPT**-symmetry; if we ever find a system where a particle moving forward in time behaves differently from its antiparticle, reflected in a mirror moving backwards in time, we will have overthrown our present understanding of how the Universe works!

With this understanding of symmetries in mind, what do we need to do in order to meet the second Sakharov condition? We need for some particles that existed in the early Universe to behave *differently* than their antiparticles (**C**-violation). Not only that, but we also need these particles to behave differently from how their antiparticles would behave if we reflected them in a mirror (**CP**-violation). To envision this (see Fig. 7.10), imagine again our particle of matter — a sphere — rotating counterclockwise around its North Pole. Imagine further, now, that this is an unstable particle, and when this particle decays, it always spits out an electron along its North Pole. If we applied **C**-symmetry, we would expect the

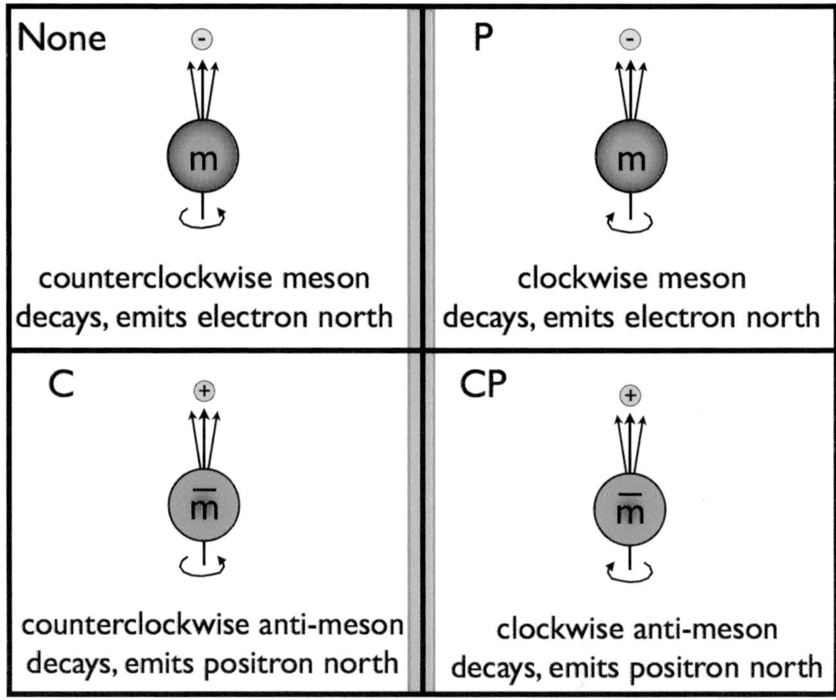

Figure 7.10 A normal meson spins counterclockwise about its North Pole and then decays with an electron being emitted along the direction of the North Pole. Applying **C**-symmetry replaces the particles with antiparticles, which means we should have an antimeson spinning counterclockwise about its North Pole decay by emitting a positron in the North direction. If it does not, this violates **C**-symmetry. We can also apply **P**-symmetry, or mirror symmetry, to both of these cases, which ought to result in a particle spinning clockwise about its North Pole, but still emitting a particle (either an electron or positron) in the North direction. If the particles and antiparticles do not behave exactly the same — with equal probabilities — in all these situations, then **C**, **P** and/or **CP** can all be violated. In the real world, mesons are seen to violate both **C** and **CP** by having their antiparticles decay in different ratios from their particle counterparts. Image credit: E. Siegel.

antiparticle to rotate counterclockwise around its North Pole and to always spit out a positron along its North Pole when it decays. If this does not happen 100% of the time, **C**-symmetry is violated. But what if we applied **CP**-symmetry? We would expect the antiparticle to rotate *clockwise* around its North Pole and then to always spit out a positron along its North Pole when it decays. If it does not do that — and if it does not do that *100% of the time* — we have **CP**-violation on our hands as well. So far, we have discovered three classes of particles that definitely exhibit both **C**- and **CP**-violation: neutral mesons (quark–antiquark pairs) that contain either strange, bottom or (as of 2012) charm quarks. Since these particles exist and decay in the early, expanding and cooling Universe while it is out of equilibrium, our Universe satisfies *both* of the first two Sakharov conditions (Fig. 7.10).

But what about the third one? When we talk about "baryon number," we mean the total number of baryons (e.g., protons, neutrons, etc.) minus the total number of antibaryons (e.g., antiprotons, antineutrons, etc.), where each proton has a baryon number of one (and each antibaryon has a baryon number of negative one). A baryon number violating interaction needs to occur in order to create a matter–antimatter asymmetry, and (because the Universe seems to have the same number of electrons as protons) there ought to be a corresponding lepton number violating interaction as well, so an electron is created for every proton created. Of all the particles and interactions known, we have never violated either baryon number or lepton number by themselves in the lab, nor do we know how to create the conditions to do so. But according to the Standard Model of elementary particles and their interactions, it *should be* possible to violate both baryon number (B) conservation and lepton number (L) conservation, so long as the combination of baryon number *minus* lepton number (B – L = 0) is conserved. This last of the Sakharov conditions is the source of our greatest uncertainty concerning the matter–antimatter asymmetry, and is the only condition that has yet to be confirmed experimentally.

* * *

If we took everything known to exist in the Universe — both from a particle perspective and also from the conditions the Universe starts off with — and went back to the earliest times while applying the laws of physics, what would we wind up with today? Remember that as far as we are concerned, the Universe starts from a hot, energetic, dense state with equal amounts of matter and antimatter. That Universe then expands and cools, with energies dropping, collisions becoming less frequent and matter–antimatter creation becoming more infrequent compared to annihilations as we drop below certain energy thresholds. With that in mind, we have the three Sakharov conditions to simultaneously meet if we want the Universe to wind up with more matter than antimatter. (Or, I suppose, more antimatter than matter would have worked just as well!) Remember what they are:

1) The Universe must be out of thermal equilibrium.
2) The fundamental **C** and **CP**-symmetries must be violated.
3) We must violate the conservation of baryon number.

Based on our current laws of physics, with all the known particles and interactions, we actually *would* generate an asymmetry between matter and antimatter, despite our limitations!

The way this would happen would actually be pretty straightforward. Imagine we produce all the particles and their antiparticle counterparts in equal numbers, including the heavy, unstable quarks such as the strange, charm and bottom quarks. They spontaneously form baryons (groups of three quarks), antibaryons (groups of three antiquarks) and mesons (quark–antiquark pairs) when the Universe is hot enough. Over time, as the Universe cools, these classes of particles all decay, with the matter and antimatter versions decaying along slightly different pathways: a consequence of **CP**-violation. Every time we produce an extra three quarks over antiquarks (i.e., enough to make a baryon), we *also* produce an extra lepton over an antilepton, so that for every extra proton we wind up with in the Universe (made up of three quarks), we also wind up with an extra electron. In other words, the Universe has a mechanism for making atoms! (Fig. 7.11)

But don't get too excited just yet. We have to ask ourselves the biggest and most important question in the midst of all this: can we generate *enough* atoms this way to explain what we actually observe in the Universe? The answer, to the best of our knowledge, looks to be *not quite.* Although we can generate significant amounts of matter over antimatter via the mechanisms we currently know, it looks like we fall about a factor of *10 million* short of what we need to produce what we observe in the Universe today. This might change as we discover new types of exotic mesons and baryons (e.g., heavier-mass bound states of quarks, antiquarks and quark–antiquark pairs) that may have even greater amounts of **CP**-violation, but the change would have to be much more significant than anticipated to account for such a massive difference.

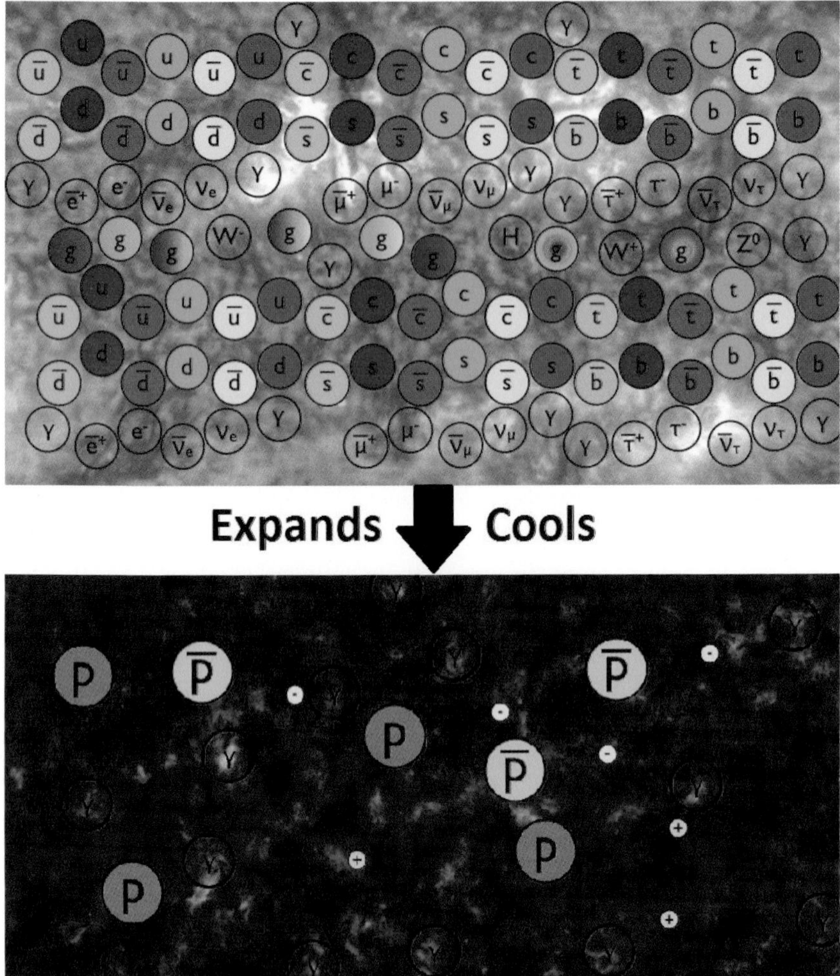

Figure 7.11 Starting with just the particles and antiparticles allowed by the Standard Model at high energies and allowing our Universe to expand and cool under the known laws of physics, we will generate a slight matter–antimatter asymmetry. The Universe will wind up with a small excess of matter over antimatter, and specifically of baryons over antibaryons. But based on what is known, we do not get enough extra matter to account for the Universe we see today. Image credit: E. Siegel.

Instead, if we want to account for the observed matter–antimatter asymmetry, it likely requires some new physics (and most likely, some new particles as well) beyond what is currently known. And to look for that, we have to turn to one of the cutting-edge fields of theoretical physics that is right at the frontiers of knowledge: **baryogenesis**.

* * *

Baryogenesis is just what it sounds like: the creation (or genesis) of baryons, and in particular of baryons *over* antibaryons. If we want to increase the number of baryons the Universe generates as it passes through an early, hot dense phase and evolves, we have to do one of two things. We can either increase the amount of **CP**-violation, or we can increase the amount (and ease) of violating the conservation of baryon number. Interestingly enough, there are many theoretical pathways through which this can occur, but they all have one thing in common: they all point towards there being more to the Universe, at a fundamental level, than we presently are aware of! If you look at the particles and interactions that govern the Universe, you find that there is no way to increase sufficiently either **CP**-violation or baryon number violating interactions without adding something extra and new to what is known. But what is this new, extra thing that must be added? This is one scientific mystery we have not yet solved.

There are four main pathways that can lead to baryogenesis, all requiring a more advanced level of detail. It might seem counterintuitive that the secret to the matter in the vastness of the Universe — which permeates all of space on the largest scales — ought to reside in particles and interactions at the *smallest* scales, but that is exactly what the Universe tells us about itself! The very early Universe, from a certain point of view, is actually the *ultimate* particle accelerator, breaking everything in the Universe up into its most fundamental constituents, even the most unstable, shortest-lived particles. To understand where the matter–antimatter asymmetry came from, we have to speculate as to what happened at earlier times, and hence at higher energies, than we have ever recreated in a laboratory here on Earth. Theoretically, a number of possibilities can account for the observed matter–antimatter asymmetry. Although they differ significantly in their details, each one could be either partially or even wholly responsible for our Universe's dominance of matter over antimatter. (For those who do not want that level of detail, feel free to skip to the final section in this chapter.)

One pathway is to have new physics at the electroweak scale, which is particularly exciting because this is exactly the energy scale that the Large Hadron Collider (LHC) will probe over the coming decade. There is an energy scale *below* which we have four fundamental forces in the Universe: the strong nuclear force, responsible for holding protons and neutrons together; the gravitational force, responsible for everything from why we are bound to Earth to the orbits of the stars around the galaxy; the electromagnetic force, responsible for electricity, magnetism, and how our physical bodies bind together; and the weak nuclear force, responsible for radioactive decays. We currently live *below* this energy scale, and hence see four forces wherever we look. But *above* a certain energy scale — one the Universe achieved when it was less than a nanosecond old — there are only three fundamental forces, as the electromagnetic and weak forces unify into a single force: the electroweak force! If enough new particles or interactions exist at the electroweak scale, then the transition through that scale, when the Universe cools from having three fundamental forces to four, could significantly increase the violation of baryon number, leading to the creation of all the matter (and *not* antimatter) we observe today (Fig. 7.12).

A second pathway would occur at very high energies, and is known as leptogenesis. We have already mentioned that in the Standard Model, you can violate baryon number so long as you violate lepton number by an equal amount. It might be difficult, as our best understanding shows, to create a large baryon asymmetry in the early Universe, but it might be much easier to create a large lepton asymmetry and then to convert a portion of that into a baryon asymmetry. This is the basic idea behind leptogenesis. Right now, the Standard Model contains both neutrinos and antineutrinos as examples of leptons, but they possess what appear to be two very odd properties. First, neutrinos and antineutrinos all have an intrinsic angular momentum to them, colloquially known as the property of "spin," which you can envision by pointing your thumb in the direction of their motion and curling your fingers in the direction that they spin around. You would expect that you would have some neutrinos that were "left-handed," meaning that if you

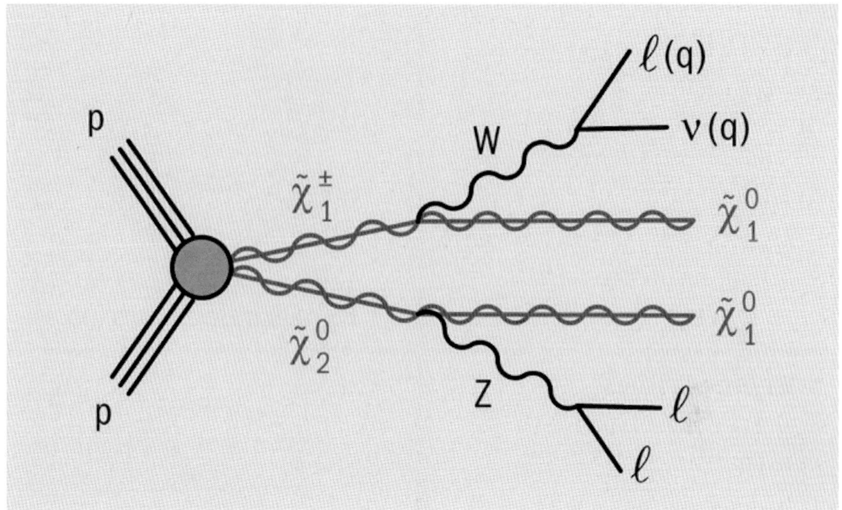

Figure 7.12 New physics at the electroweak scale, whether via new interactions or particles (like the various χ particles shown in the diagram here), could enhance the baryon asymmetry, *possibly* accounting for all the normal matter in our Universe today. Image credit: CERN Courier/ATLAS Collaboration 2014/International Journal of High Energy Physics.

pointed your thumb upwards, the particles would spin clockwise, and some that were "right-handed," where if you pointed your thumb upwards, they would spin counterclockwise. Yet when we observe these particles, we find that all neutrinos are left-handed, while all antineutrinos are right-handed! And second, we find that neutrinos and antineutrinos both have very, very small *but non-zero* masses. They are more than a million times lighter than the electron (the next lightest known particle), and there is no good explanation for why this would be so. At least, the known laws of physics offer no satisfactory explanation. But one solution to all these conundrums would be to postulate new, very heavy right-handed neutrinos (and left-handed antineutrinos), that were created in great numbers in the early Universe and then decayed, creating a lepton asymmetry when they did. That lepton asymmetry could then get converted into a baryon asymmetry, creating the observed ratio of matter and antimatter we see today (Fig. 7.13).

The third pathway requires a very specific extension to the Standard Model of elementary particles known as supersymmetry (or SUSY, for short), and is known as Affleck–Dine baryogenesis, after the two physicists

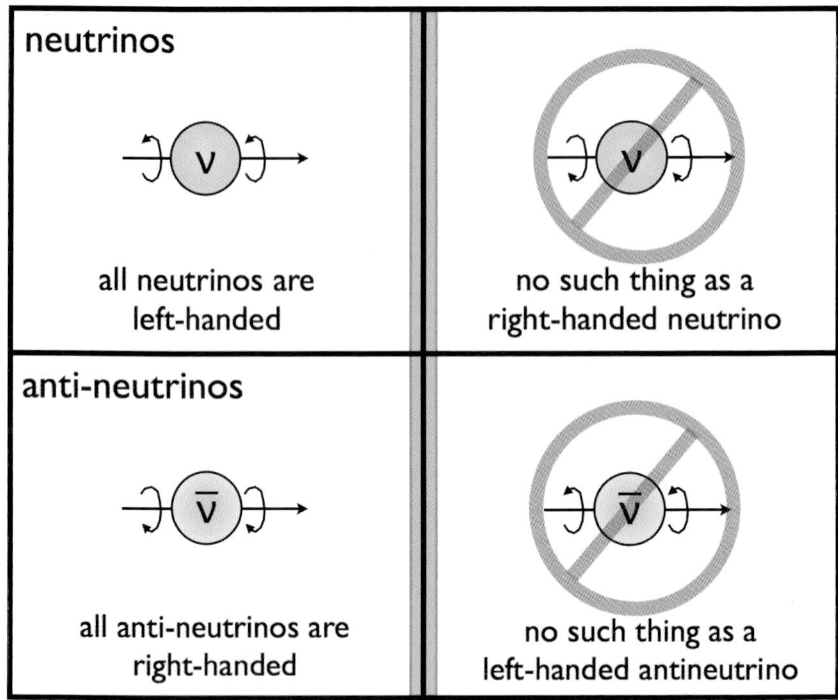

Figure 7.13 So far, every neutrino we have ever observed has been left-handed, meaning that if you point your left thumb in the direction of its motion, it spins in the direction your fingers curl around its axis. Similarly, every antineutrino is right-handed, meaning that you point your right thumb in its direction of motion and it spins on its axis the way your right hand's fingers curl. So far, we have never found a left-handed antineutrino or a right-handed neutrino, but it is conceivable that versions of neutrinos and antineutrinos that have this "handedness" exist, are very heavy, and could give rise to a lepton asymmetry via leptogenesis in the early Universe. That lepton asymmetry could then give rise to a baryon asymmetry. Image credit: E. Siegel.

(Ian Affleck and Michael Dine) who discovered this mechanism. In supersymmetry, all of the particles of the Standard Model have unstable "superpartner" particles that correspond one-to-one to the normal ones, which have higher masses, carry baryon and lepton numbers, and also decay. Unlike the fields associated with quark and lepton particles, the superpartner particles are predicted to be scalar fields, which have the advantage that the highly energetic early Universe can easily put them in an excited state that leads to significant asymmetries. If SUSY is real, there is expected to be greatly increased amounts of baryon number (and, for that matter, lepton number) violation at the electroweak scale. When those superpartners decay, that asymmetry gets transferred into the quarks and leptons present in our Universe today. *If* SUSY turns out to be real.

And finally, the fourth pathway postulates that just as the electromagnetic and weak forces unify at high energies in the early Universe, there are even higher energies at which the strong force unifies with the electroweak force! We call this class of models Grand Unified Theories (GUTs), and they not only bring along new particles with them, they allow for both new opportunities for **CP**-violation and also for direct violation of both baryon and lepton number (Fig. 7.14).

In the interest of demonstrating how baryogenesis might have actually occurred in our Universe, let us take a look at the specifics of this last possibility.

* * *

Imagine the Universe as it was when it was hotter and denser than anything we have considered so far. Not minutes old, not seconds old, not even microseconds or nanoseconds old, but rather when it was some 10^{-34} seconds old! We have to go to these supremely early times in order to realistically portray things as they might have been when there were only *two* fundamental forces in the Universe: gravitation and the unified strong-electromagnetic-weak force. In this scenario, there are at least two

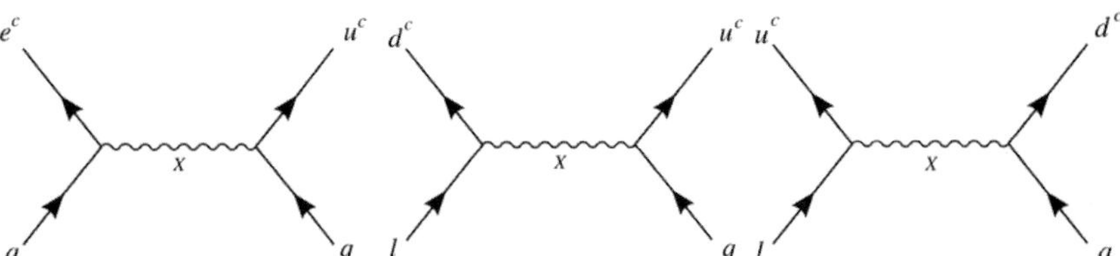

Figure 7.14 In Grand Unified Theories, additional super-heavy particles exist, such as X and Y bosons, that couple to both quarks and leptons. These particles, *if* they exist, are expected to violate both lepton number and baryon number, and might represent the most likely pathway towards creating the matter in our Universe. Images credit: Wikimedia Commons user GreenRoot.

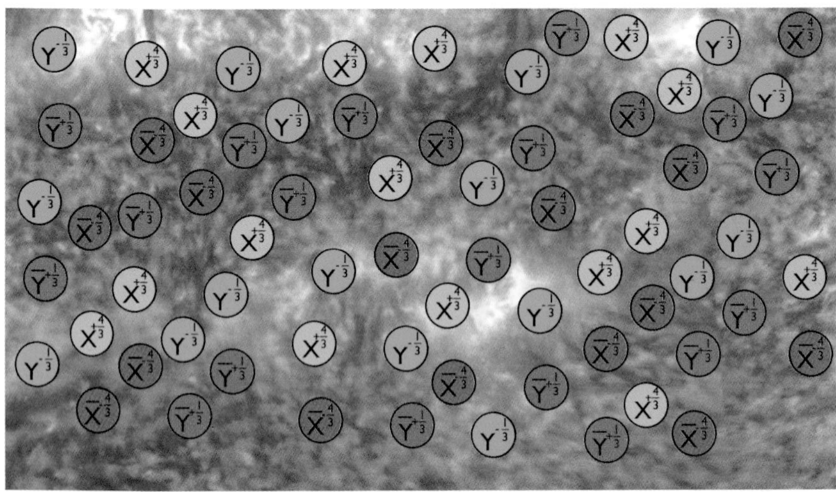

Figure 7.15 In addition to the other particles in the Universe, *if* the idea of a Grand Unified Theory is relevant to our Universe, there will be additional super-heavy bosons, **X** and **Y** particles, along with their antiparticles, shown with their appropriate charges amidst the hot sea of other particles in the early Universe. Image credit: E. Siegel.

new types of unified particles that would exist: the **X**, which has a charge of $+\frac{4}{3}$ and the **Y**, which has a charge of $-\frac{1}{3}$. Of course, these are the *matter* versions of these particles; there would also be antimatter versions, too: the anti-**X** with a charge of $-\frac{4}{3}$, and the anti-**Y** with charge $+\frac{1}{3}$. Just like everything else, these particles would be created *in equal numbers* to their antiparticles, so that the Universe would start off completely symmetric between matter and antimatter (Fig. 7.15).

But the Universe will expand and cool and all of these particles — **X**, anti-**X**, **Y** and anti-**Y** — will decay away if they fail to annihilate rapidly enough. While these particles and antiparticles must conserve the full combination of **C**, **P** and **T** symmetries together, however, they are expected to violate the combination of **CP**. What this means, when we look at the details, are incredible.

Imagine that the **X**, since it has an unusual charge of $+\frac{4}{3}$, has two ways it can decay: into two up quarks, *or* into an antidown quark and a positron. If this were the case, then the anti-**X** would have to be allowed to decay into either two antiup quarks, or into a down quark and an electron. Note that these decays violate baryon number (B) and lepton number (L)

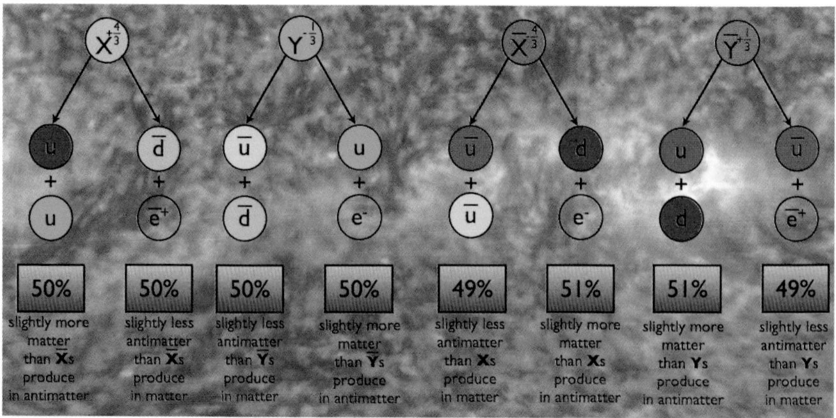

Figure 7.16 If we allow **X** and **Y** particles to decay into the quarks and lepton combinations shown, their antiparticle counterparts will decay into the respective antiparticle combinations. But if **CP** is violated, the decay pathways — or the percentage of particles decaying one way versus another — can be different for the **X** and **Y** particles compared to the anti-**X** and anti-**Y** particles, resulting in a net production of baryons over antibaryons and leptons over antileptons. Image credit: E. Siegel.

individually, but that the total baryon *minus* lepton number (B − L) is conserved at $+\frac{2}{3}$ for the **X** and at $-\frac{2}{3}$ for the anti-**X** (Fig. 7.16).

So far, everything is symmetric. But one remarkable thing that **CP**-violation allows us to have is that the individual decay *fractions* can be different for the **X** and anti-**X**, so long as the total decay rates for the particle and antiparticle are the same. So we could have the **X** decay into two up quarks 50% of the time and into an antidown quark and a positron 50% of the time, but the anti-**X** decay into two antiup quarks only 49% of the time, while it can decay into a down quark and an electron 51% of the time. That would mean, for every 50 **X** and anti-**X** pairs that we created, we would get a total of 151 quarks, 51 leptons, 148 antiquarks and 50 antileptons. The quark–antiquark pairs and the lepton–antilepton pairs would annihilate away, leaving us with three quarks and one lepton *left over*, or the equivalent of one extra baryon and one extra lepton. This pathway would allow us to create a significant asymmetry of matter over antimatter!

Similarly, if we allow the **Y** particle to decay into either an antiup quark and an antidown quark, or an electron and an up quark, and the anti-**Y** to decay into either an up quark and a down quark or a positron and an antiup quark, we can get a similar asymmetry through that same **CP**-violating process. As long as the total decay rate is the same, **CP**-violation allows the individual pathways to have different ratios for particles as compared to antiparticles, giving rise in the cases of both the **X**/anti-**X** and **Y**/anti-**Y** pairs to a net baryon (and lepton) number. However, the energies to probe the existence and properties of these hypothetical particles are well beyond what we can access, not only from terrestrial colliders but even from the highest energy cosmic rays that the Universe itself produces. Nevertheless, this is a possibility for baryogenesis worth keeping in mind, as it may yet prove to be the very reason our existence is possible (Fig. 7.17).

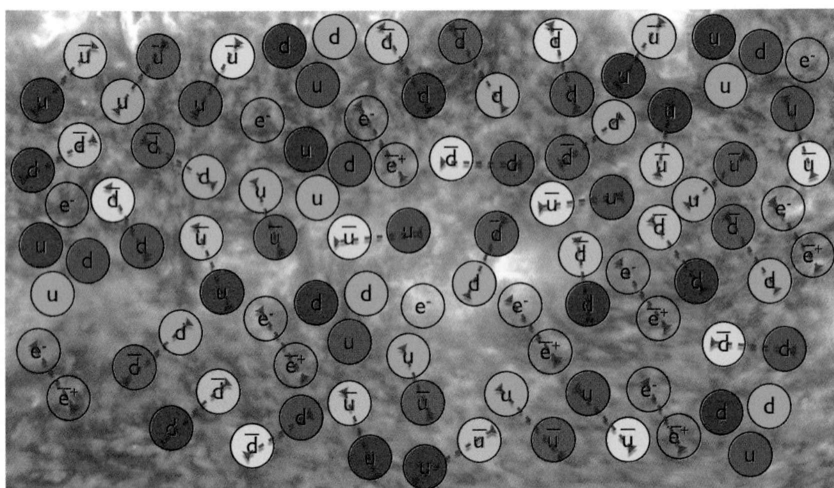

Figure 7.17 If the particles decayed away according to the mechanism described in Fig. 7.16, we would be left with an excess of quarks over antiquarks (and leptons over antileptons) after all the unstable, superheavy particles decayed away. After the excess particle–antiparticle pairs annihilated away (matched up with dotted red lines), we would be left with an excess of up-and-down quarks, which compose protons and neutrons in combinations of up–up–down and up–down–down, respectively, and electrons, which will match the protons in number. Via the mechanism outlined, only one in approximately 100,000 of the created **X**, **Y**, anti-**X** and anti-**Y** particles would need to decay in this **CP**-violating fashion to produce the observed matter–antimatter asymmetry; the rest could safely annihilate away in the early stages of the Universe (to photons, for example) and we would obtain enough matter in the Universe to match what we observe today. Image credit: E. Siegel.

Whether GUT baryogenesis turns out to be responsible for the matter–antimatter asymmetry in our Universe or not is still an open question. In fact, any of the possibilities we mentioned could, in the end, be the mechanism responsible for the matter we observe in the Universe. For that matter, there are other possibilities as well, that we have not considered here. In the end, no matter when it happened in our Universe, or by what mechanism, it will be experiments, not the theory we most favor, that enables us to determine why we live in a Universe that is primarily made up of matter and *not* antimatter.

* * *

The topic of why we live in a Universe exclusively composed of matter-based stars and galaxies is one of the most tantalizing questions open to both theorists and experimentalists alike. It is very rare that we find ourselves in a scientific situation where we know *that* something happened, we know multiple pathways as to *how* it could have happened, but we do not yet know exactly *what* occurred in our Universe. This is part of the joy and also the frustration of probing the Universe at the limits of our knowledge: the vast majority of scientists working on this exact problem — and *possibly* all of those scientists — will have gone down the wrong path.

There is no shame in that! Part of the freedom that science gives you is the freedom to guess wrong; so long as you let the evidence guide the direction you go in once it arrives, you are bound to get it right in the end. It was once reasonable to think that the Universe may have consisted of large regions of antimatter rather than matter; we let the evidence guide the way, and learned that is not the case at all as a result.

In the same vein, the next decade of particle physics results should teach us whether the Universe's matter–antimatter asymmetry owes its origins to physics arising at the electroweak scale or not and — if we are lucky — it may shed light on the existence of supersymmetry as well, either pointing to, disfavoring or conceivably ruling out up to two of the four most commonly considered scenarios for baryogenesis.

There is also no shame in admitting that science does not have all the answers at any given time. In fact, when we *do* discover the origin for the matter–antimatter asymmetry, it will likely raise more questions about the nature of baryogenesis, and indicate new puzzles and implications for other unexplained phenomena in the Universe. Part of the beauty and power of science is that even when we do not know everything about a particular topic, we can point to the things that are known and have them guide us. And another part of that beauty and power is that science never ends, as each new discovery brings with it a whole new set of things to investigate, details to understand, and a new aspect of the Universe to both appreciate and become aware of. If nature is kind to us, one of the greatest unsolved mysteries about the Universe may be resolved within our lifetimes!

Chapter 8

Before The Big Bang: How The Entire Universe Began

If you ask the scientific question of where the Universe came from, "the Big Bang" is likely to be the answer you get from almost anyone you can ask. But it is actually a relatively new idea, scientifically speaking; just a few decades ago, the Big Bang would have been hotly contested, and a few decades before that, it hadn't even been considered. As far as we came to confirm it, and for all of the Big Bang's successes, there are a number of points where it does not provide a satisfactory answer to the question of where our Universe and everything in it comes from. Remember that the whole concept was born from two simple facts: (1) Einstein's description of gravitation as a changeable spacetime fabric whose curvature is determined by the matter and energy in it, and (2) the observed relationship between the measured distance of galaxies beyond our own and their redshift. When taken together, these two pieces of evidence — both verified to incredible degrees of accuracy through multiple lines of observation — lead us to conclude that we live in a Universe whose spacetime fabric is expanding over time.

When space expands, the energy of everything in it drops. Radiation, a form of energy defined by its wavelength, sees its wavelength stretch, causing it to fall to lower energies. Matter, whose kinetic energy is determined by its speed, sees its velocity (and hence its energy) drop. And to make matters worse, even as the energies of the individual particles in the Universe are dropping due to the expansion, the number densities — or the number of particles per unit of volume — are also dropping rapidly, again a consequence of space's expansion. But this picture opens up some tremendous possibilities for the Universe's *past.* A Universe that is expanding, cooling, becoming less dense and lower in energy as time moves forward was therefore hotter, denser and higher in energy long ago.

The Big Bang started off as an idea that took this possibility and ran with it. If you extrapolated backwards, you would find that the farther and farther into the Universe you looked, distance-wise, the faster galaxies would continue to appear to recede, and you would measure ever greater redshifts to their spectral lines. You would find that galaxies would appear less evolved in the past, and would be more closely packed together on average, since the great distances separating them would have been less in the past. And as you continued to go back, you would eventually find that galaxies and star clusters would first be lower in mass and sparser, and then cease to exist entirely. These luminous objects only came to exist through the clustering of matter via gravitational collapse over time. The farther away in space you looked, the farther *back in* time you would also be looking, and hence, to an earlier stage of the Universe, when everything was much younger. If you went back even farther than that, you would come to a time where the Universe was energetic enough that neutral atoms themselves could not exist, then to a time where atomic nuclei could not exist, and then to a time when energies were so high that matter and antimatter pairs spontaneously sprang into existence.

In principle, there is a step even farther back than that, one that is perilous to take but also seems inevitable. If we go to higher and higher densities, higher and higher temperatures and energies, and pack all the components of the Universe into an increasingly smaller space, eventually we cross a threshold where the physics of spacetime breaks down: a singularity. This would, conceivably, correspond to the origin of space and time itself, and a true moment of birth for everything we know of in the Universe. There would be nothing outside of it, as questions of "where" make no sense if we take away space; there would be nothing before it either, as questions of "when" are equally nonsensical if we take away time. General Relativity has nothing more than that to say about a singularity, save that the laws that govern the Universe return results that are known to be non-physical at those arbitrarily high energies, densities and temperatures. Going back arbitrarily far seems to pose a problem (Fig. 8.1).

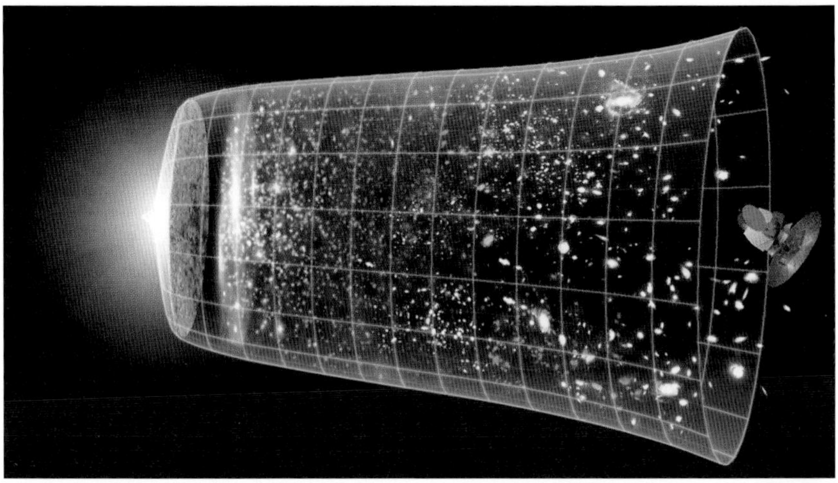

Figure 8.1 If we extrapolate all the way back to earlier and earlier times, to higher and higher energies and increasing densities, we would eventually run into a singularity: a place where the laws of physics break down. Image credit: NASA/WMAP Science Team.

Yet, despite that problem, and despite how crazy of an idea the Big Bang might seem — and it *did* appear far-fetched to a majority of physicists up until the 1960s — it has actually made a huge number of successful predictions, which correspond to very particular times and events in the Universe's past. It predicted the existence of the cosmic microwave background (CMB) radiation, a uniform glow just a few degrees above absolute zero, corresponding to the radiation that was left over from the primeval state of high densities and temperatures, having cooled over billions of years. The radiation itself was last scattered off of a sea of ionized electrons (and some nuclei, but mostly electrons) that ceased to exist when the Universe formed neutral atoms, back when the Universe was only some 380,000 years old. It predicted that the Universe would have started off with about 75% hydrogen/25% helium-4 by mass (92% hydrogen/8% helium-4 by number), with a fraction of a percent of deuterium and helium-3, and a tiny sliver of lithium-7 as well, formed when the Universe cooled to a temperature that allowed the formation of nuclei without them immediately being blasted apart by collisions with sufficiently energetic particles. In 2011, pristine gas clouds were detected for the first time, consisting of these early elements and nothing else. It represented a sample of gas that had never undergone even a single generation of star formation, whose composition dated back to when the Universe was less than four *minutes* old.

There are predictions arising from the Big Bang that were made subsequent to this, such as the formation of the first stars and the reionization of the intergalactic medium (supported by quasar absorption lines), the hierarchical formation of galaxies through mergers (supported by ultra deep-field observations), and the measurement of the CMB temperature at great distances, which is in fact seen to increase (as measured through the Sunyaev–Zel'dovich effect) exactly in line with the predictions of the expanding Universe and the Big Bang. There are a whole host of cosmic predictions that the Big Bang makes that have observable consequences, and each one has been verified once we reached the technical capacity to detect it (Fig. 8.2).

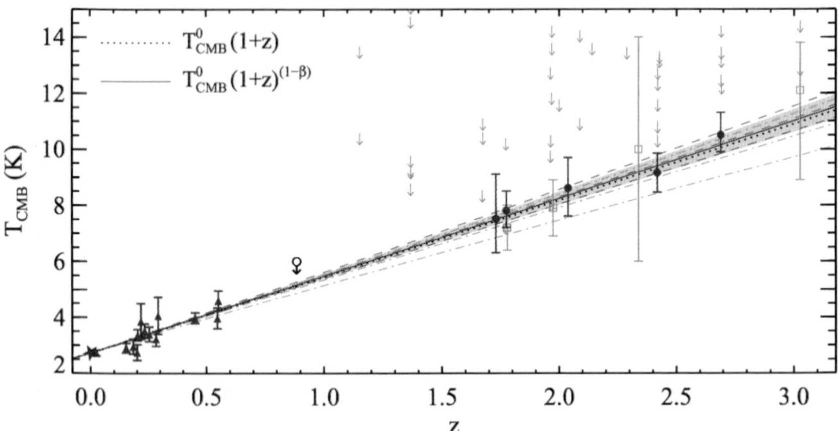

Figure 8.2 By measuring the light coming from distant galaxies and how that light is affected by the CMB, which should have been *hotter* in the past since the Universe was smaller in the past, we can infer the temperature of the CMB at earlier times. We find that the observations are in fantastic agreement with the predictions of the Big Bang, which is shown as the dotted line indicated. Any modifications that could be consistently made are extremely small in magnitude, as shown by the various other lines. Image credit: P. Noterdaeme, P. Petitjean, R. Srianand, C. Ledoux and S. López, (2011). *Astronomy & Astrophysics*, 526, L7.

However, our technological capabilities, although great, are not infinite. In addition to the confirmed consequences mentioned above, there are earlier predictions that are still yet to be confirmed, such as a cosmic neutrino background dating back to when the Universe was just one *second* old, the origin of the matter–antimatter asymmetry, which could date back to any point when the Universe was between 10^{-36} to 10^{-12} seconds old, and — if we accept the prediction of a singularity — a time when the laws of physics broke down, back when the Universe was only 10^{-43} seconds old. But some of these early-time extrapolations may turn out to not necessarily be so robust. For all of its successes, and for all of its verified predictions, there are a number of very substantial puzzles that arise from the idea of the Big Bang. And many of these problems seem to get worse the farther back in time we are willing to look.

* * *

When the CMB was first detected, it was a surprise to Penzias and Wilson, who never suspected the existence of such a signal. But to cosmologists familiar with the possible mechanisms that could have brought about our present Universe, it was a tantalizing hint of our cosmic origins. When this radiation was discovered to be uniform on the sky — the same in all directions — and to follow a precise blackbody curve, all the alternatives to the Big Bang fell away as scientifically unreasonable. And yet, the very properties that so strikingly supported the Big Bang also raised a harrowing question that seemed to defy conventional expectations: *why* was the temperature of the CMB the same in all directions?

Think about it for a moment. Look towards the eastern horizon, and focus in on a narrow region, no bigger than your pinky finger's nail held at arm's length. Imagine seeing through the atmosphere, past the stars in our galaxy, out beyond all the known galaxies and finally — after a journey of many billions of light years — your eyes arrive at the surface of last scattering, where the transition from an ionized to a neutral Universe occurred. You are seeing the very photons that were emitted when the Universe was just 380,000 years old, and they display a spectrum indicative of a very precise temperature. You should not be *surprised* that you get a single temperature for this small region; the particles that are forming neutral atoms all shared a very similar cosmic history. Because they are all so close to one another, they have had the opportunity to collide with one another, to transmit photons between one another, and in general to *exchange information* with each other. So it is no surprise that this small region has a uniform temperature.

But now look towards the west, at a similarly sized region that is also many billions of light years away. Look towards the north, or the south, or up at the zenith, and again imagine a similarly sized region. You would expect each one of these regions to have a temperature that was mostly uniform throughout, but you *would not* expect all of these different regions to have the same temperatures as one another! That would be incredibly puzzling, since these regions are separated by distances that are greater than any signal, even light, could have traveled in the time since the Universe was born (Fig. 8.3).

Yet somehow, this is exactly what we observe: that regions of the Universe that could have had no way to exchange information with one another, that could not have possibly had the opportunity to thermalize and reach the same temperature through interactions with each other, actually are the same exact temperature! You would have expected one disconnected region to typically have half or double the temperature of another, and yet the biggest temperature differences are only a few parts in 100,000. We call this puzzle **the horizon problem**, since regions that are outside one another's causal horizons — that have not had the chance to interact or exchange information — somehow, inexplicably, have the same exact properties as each other.

The Hubble expansion of the Universe, similarly, raises its own puzzle that is just as troubling, but one that requires a little bit more background. You will remember that we think of space and time as a four-dimensional "fabric" in the context of General Relativity, and it is the presence of matter and energy that determines how that fabric is shaped. But it isn't merely that matter and energy curve the fabric of space; the space itself is expanding, and it's the matter and energy present within that space that can change the expansion rate as time goes on. The expansion rate of the fabric of space must have started off with some initial value, with the matter and radiation's presence, density and type determining how that expansion rate changed in the past, and how it will change, even into the far future, as time goes on. These two cosmic forces — the

Figure 8.3 No matter where we look in the sky, we see the same 2.725 K temperature for the CMB, even though these regions are separated by distances so great they preclude any information from being exchanged between them. Yet somehow, the Universe is the same average temperature everywhere, a puzzle known as the horizon problem. Image credit: E. Siegel.

initial expansion on one hand, and the matter-and-energy-induced gravitational pull on the other — battle to determine the Universe's fate.

Imagine these two great cosmic presences, and the struggle that takes place between them as each one fights to dominate the Universe. The initial expansion rate must have been tremendous, working to drive all the matter and radiation present in a given region of space apart and into oblivion. On the other hand, there is all the matter and energy in the Universe, and the tremendous gravitational force it brings along with it, working to slow the expansion rate — and, if it can, reverse it — in spectacular fashion as the Universe goes on. In the context of this picture, there are three possibilities you can imagine for the Universe:

1) **Gravitation wins:** the initial expansion starts off incredibly fast, but the density of matter and energy is incredibly large as well, so much so that it not only slows the expansion rate down, but eventually stops it completely! The Universe reaches a maximum size, and with the gravitation from all the matter and energy still present, causes the Universe to begin contracting, making everything denser and hotter once again, until finally everything collapses into a fiery *Big Crunch*.
2) **Expansion wins:** the initial expansion still starts off incredibly fast, and although the density of matter and energy is large enough to slow the expansion rate, it is not quite enough to overcome how quickly things were expanding initially. Gravity might make the Universe expand slower and slower over time, but the expansion rate never drops to zero, much less reverses, and the Universe (and all the matter and radiation in it) keeps expanding away into an infinite abyss. A scenario like this has many names, including a *Heat Death*, a *Big Chill*, or — my personal favorite — a *Big Freeze*, as the Universe keeps expanding and cooling forever and ever.
3) **The Critical Case:** do you remember the story of Goldilocks, who tries the bears' porridges (one is too hot, one is too cold, one is just right), chairs (the first two are two big, the third is just right), and beds (the first is too hard, the second is too soft, and the third is just right)? You can imagine a Universe on the border that separates these two different scenarios, where only *one more proton* would cause the Universe to recollapse in a Big Crunch, but that proton is not present. Instead, the initial expansion rate and the matter and energy density are so precariously balanced that the Universe's expansion rate asymptotes towards zero, but never recollapses. This is known as either a *critical Universe* or, sometimes, a *Goldilocks Universe*.

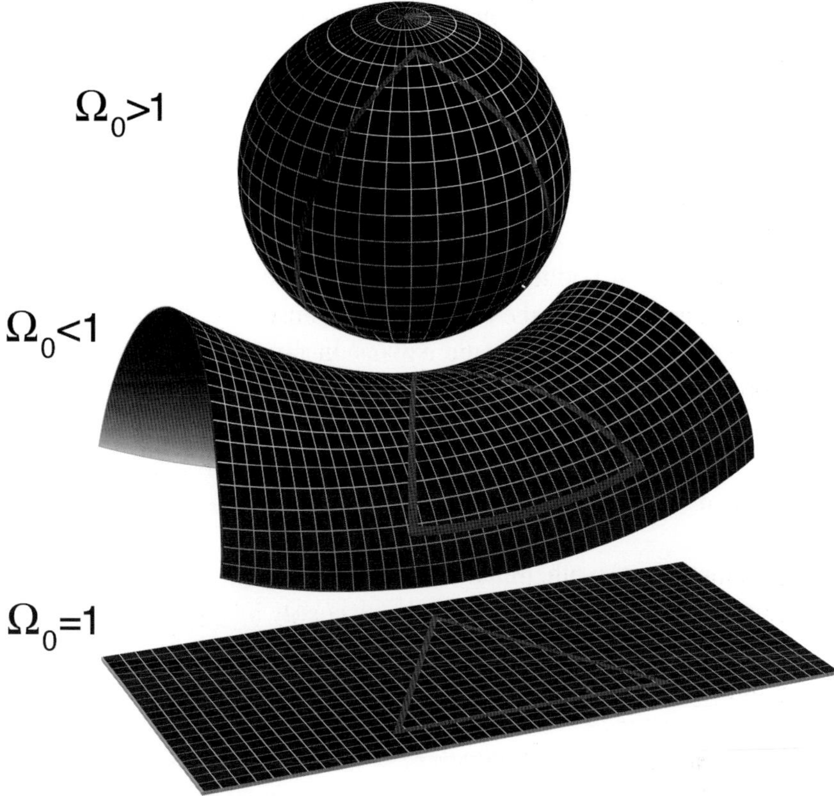

Figure 8.4 Different spatial curvatures — positive curvature (top), negative curvature (center), and a flat, zero-curvature Universe (bottom) — correspond to different fates of the Universe (recollapsing, expanding forever, and a critical Universe, respectively) but also lead to different geometries, which can be measured. Image credit: NASA/WMAP Science Team.

These three cases also correspond to three different *shapes* for the fabric of spacetime. The first case, where gravitation wins and the Universe recollapses in a Big Crunch, corresponds to a closed Universe with positive curvature, similar to the surface of a sphere. The second case, where the expansion wins and continues forever, culminating in a Big Freeze, corresponds to an open Universe with negative curvature, similar to the surface of a saddle, which curves downwards along a horse's flanks but upwards along a horse's spine. And the final case, the critical or "Goldilocks" case, corresponds to a flat Universe, with absolutely zero curvature (Fig. 8.4).

If there is an intrinsic curvature to the fabric of space itself, that is physically interesting because it is something that, in principle, can be measured! Imagine you send out two receivers/transmitters into space a long distance away from you. You measure the angle between the two transmitters, and have each one measure the angle between the other one and you. Add those three angles up, and see what you get. At first glance, you might expect to *always* get 180°, since that is what the three angles of a triangle always add up to be. But that is *only* if space is flat! As an example of how this could be different when space is curved, consider three people on the surface of the curved Earth: one at the North Pole, one in Quito, Ecuador and one in Macapá, Brazil. The Ecuadorian and the Brazilian would each say the other two sites were separated by an angle of 90°, while Santa Claus (at the North Pole) would say that Quito and Macapá were separated by 21°, for a positively-curved total of 201°, not 180°!

In practice, we use other methods to measure the spatial curvature, but the principle is the same. What we find is that — to the best we can measure — the Universe is truly flat. But this in itself is a puzzle! There is one particular density that corresponds to the Goldilocks case, and it needed to be incredibly finely tuned. If the Universe were just one part in 10^{25} less dense, that is, 0.00000000000000000000001% less dense, it would be more than *double* its current size by now. And if it were just one part in 10^{25} *more* dense, it would have recollapsed billions of years ago! So for some unexplained reason, the Universe is arbitrarily, perfectly flat. This puzzle is known as **the flatness problem**, and is another condition that is simply not addressed by the Big Bang (Fig. 8.5).

And finally, we know that there needs to be new physics — and most likely, new particles — at high energies to account for the observed asymmetry between matter and antimatter. There are some pretty general predictions that come out of all the models that could account for these, and one of those predictions is the existence of ultra-massive particles that should be relatively abundant in our Universe at early times. In practically all of these models, at least one of the new particles created should be *stable*, meaning it should exist in detectable quantities today. If the Universe passed through a phase where it reached the extremely high energies necessary to create them, which is a very general prediction of the Big Bang going back to a hot and dense state, these particles should

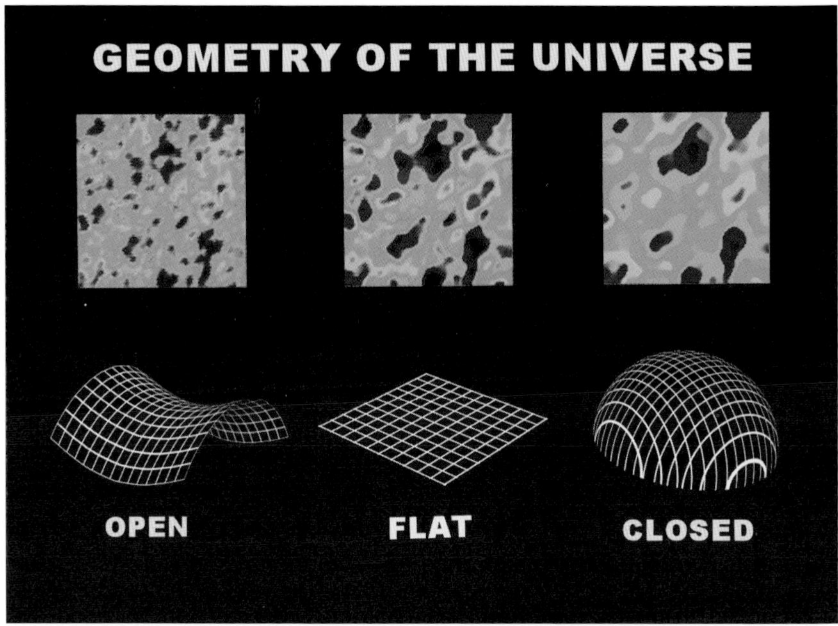

Figure 8.5 Different amounts of spatial curvature would lead to different signatures for the fluctuations in the CMB. From measurements of the geometries of these fluctuations, we have determined that the Universe is incredibly flat, with a curvature that is constrained to be less than 0.4% the size of the observable Universe. Image credit: NASA/WMAP Science Team.

have been created, and the stable ones should persist until the present day. In particular, one of these stable particles should be observable as a super-heavy magnetic monopole.

Just like electric charges come in positive and negative varieties, magnetic poles come in North and South varieties. But *unlike* electric charges, magnetic poles are only found in pairs, never in isolation. However, at ultra high energies — in particular, if the electroweak force unifies with the strong force — there should exist not only electric charges, but fundamental magnetic ones as well! Instead of solely North–South pairs, or a dipole, there should have been isolated magnetic monopoles at extremely high energies. Even if the "North" monopole and the "South" monopole are antiparticles to one another, enough should have escaped annihilation to still exist in sufficient numbers to be detected by a dedicated experiment here on Earth (Fig. 8.6).

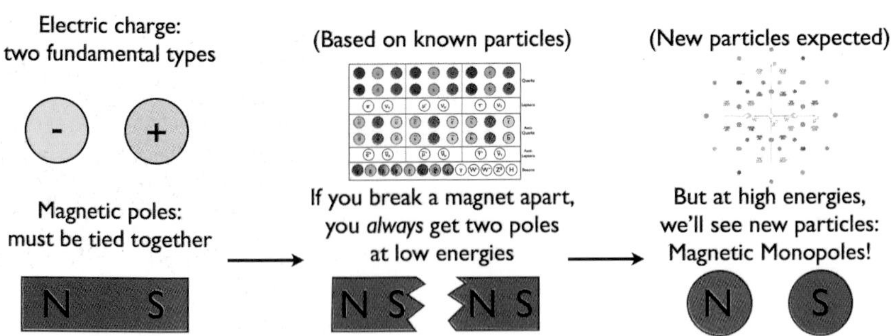

Figure 8.6 At energies we have been able to experimentally attain on Earth, we find only conventional particles, which include electric charges (or electric monopoles) and magnetic dipoles, but no magnetic monopoles. However, at higher energies, like the ones reached in the early Universe, we should have produced new types of particles, including magnetic monopoles: individual particles with a fundamental magnetic "charge," equivalent to a "North" or "South" pole on its own. These particles, although they were expected to be left over from the Big Bang, have failed to show themselves despite intensive, dedicated searches. Image credit: E. Siegel.

These experiments began in earnest in the 1970s, to no avail. Things got very exciting for a time in 1982, when a single candidate monopole was detected by Stanford physicist Blas Cabrera. This generated a flurry of interest in the field, and much larger, more sensitive detectors were built to try and not only replicate but to improve upon this finding. Unfortunately, no other monopole candidates were ever detected over more than three decades of searches, leaving that one positive result as an experimental outlier, unable to ever be verified. This unfortunate lack of magnetic monopoles is known as **the monopole problem**, and is another result that flies in the face of the expectations of the Big Bang (Fig. 8.7).

* * *

It is tempting to look past these three problems as not problems at all, but rather as simply the conditions that the Universe started off with. After all, there is no reason that the Universe *could not* have started with the same initial temperature conditions in regions that had never been connected to one another. There is no reason that the fabric of spacetime *could not* have simply been perfectly flat without any perceptible imperfections. And there is nothing forcing magnetic monopoles to exist in the numbers

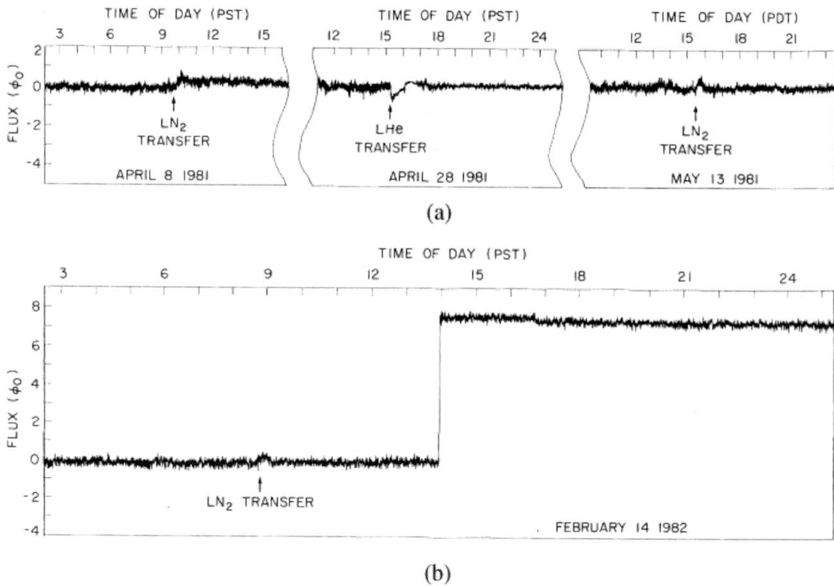

Figure 8.7 This was the candidate monopole event that set the scientific world on fire in 1982, where a clear signal emerged that looked exactly like what a magnetic monopole was expected to look like. But no other monopoles were ever found. For the most part, the search has been abandoned at present. Image credit: Blas Cabrera, (1982). *Physical Review Letters*, 48(20) 1378–1381.

we predict; perhaps there is some additional new physics that is hiding them from our view. Or perhaps our ideas about magnetic monopoles are completely off-base, and even at arbitrarily high energies, perhaps they never existed. The Big Bang has enough successes that it is easy enough to sweep these problems under the rug, and not be bothered by them.

But that would be a terrible, non-scientific attitude to take! If we simply said to ourselves, "these three issues — the **horizon problem**, the **flatness problem**, and the **monopole problem** — must simply have been initial conditions that the Universe started off with," our scientific progress on this front would have abruptly ended. We never would have been able to investigate the origins of the Universe back to times earlier than this hot, dense, rapidly expanding state. As soon as we convince ourselves that something is a question that science cannot answer, and hence we do not investigate it scientifically, it becomes a self-fulfilling prophecy.

So we have to try! There is a tremendous opportunity for scientific advancement here, even in these cases where we cannot either directly observe what the Universe was doing at these incredibly early times or recreate those conditions in a laboratory setting. There are still scientific avenues of investigation that are open to us, if we are willing to take a more theoretical approach. Rather than simply say, "these were the initial conditions," we could ask ourselves, instead, whether there are any *dynamical* occurrences that could create these initial conditions instead.

In other words, we can make an assumption that there was a physical state that set up these conditions, and ask ourselves what such a state would look like. When we run up against a wall in science, where we reach the limit of what our observations and experiments can teach us, that is the point where we should consider the full extent of the theoretical possibilities available. Some of these possibilities will be preposterous, and will make physical predictions that clash dramatically with other things we have already observed. Others will show some promise as offering possible after-the-fact explanations, but will wind up being dead-ends that fail to make any new, potentially observable predictions. But the best of these ideas will do the same three remarkable things that General Relativity wound up doing when it was proposed:

1) It will reproduce all the successes of the model it was designed to replace.
2) It will explain a suite of observational signatures that the old model was unable to account for.
3) And perhaps most importantly, it will make *new predictions* that have not yet been tested, but that are not only testable in principle, but in practice as well.

If we can find a dynamical process that could have set up the initial conditions for the Big Bang *without* disturbing any of its major successes, and if that process also winds up making new predictions that can be tested observationally, we are in business!

* * *

Starting in the late 1970s, these were exactly the types of solutions that many of the world's top theoretical physicists were considering. It would not do to simply throw away the Big Bang, as its successes were too many, and there were no alternatives that were able to duplicate all three of the major ones: the Hubble expansion, the abundances of the light elements, and the uniform, blackbody CMB. The Universe certainly experienced a very hot, dense, expanding state early on, and evolved into the cosmic web of stars and galaxies we see today. But the earliest stages of all seemed to be where the greatest problems arose; perhaps, if there were some sort of phase preceding the hot, dense state identified with the Big Bang, these problems could be solved. In December of 1979, that is exactly what a young theorist named Alan Guth was thinking.

When spacetime is filled with matter and radiation, its expansion works in a very particular way: the expansion rate *drops* accordingly as the density of matter and energy drops. But it is the very expansion of the Universe that causes the density to drop, so the slowing of the expansion is inevitable. Not all expanding spacetimes, however, have their expansion rates drop as time goes forward. One classic example is de Sitter spacetime, where instead of matter or radiation, the expansion rate is determined by energy *inherent to space itself*! This is a remarkable change from a Universe filled with matter or radiation, where the amount of energy-per-unit-volume dilutes as the Universe expands. But if there is a large amount of energy inherent to space itself, then even as the Universe expands, the energy density does not decrease at all, as every cubic centimeter of space still has the same amount of energy to it. Because changes in the expansion rate and changes in energy density are related, this also means that the expansion rate would not drop at all as the Universe expands, which leads to a much more spectacular type of expansion than we are used to: *exponential* expansion! (Fig. 8.8)

Exponential expansion is a far different case than everything else we have considered so far. If you have a Universe that is primarily filled with matter, for example, because the matter density *drops* as the volume of the Universe grows, the expansion rate slows down by a certain amount. If the Universe starts out as a certain size after a certain time,

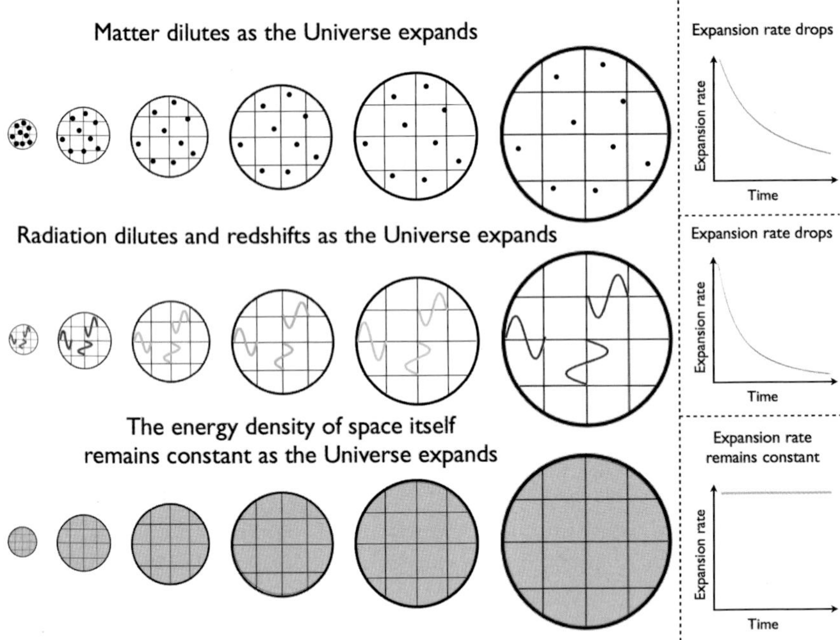

Figure 8.8 In a Universe dominated by matter or radiation, the energy density drops as the Universe expands. But in a Universe dominated by the energy inherent to space itself, the energy density remains *constant*, and hence does not drop over time. Since the energy density and the expansion rate of the Universe are tied together, a matter-dominated or radiation-dominated Universe sees its expansion rate plummet over time. Spacetime still expands, but does so at an ever-decreasing rate. But in a Universe dominated by energy intrinsic to space itself, the expansion rate does not drop at all, and hence the Universe expands in a very different fashion: *exponentially*. Image credit: E. Siegel.

then when twice as much time has passed, it will be 59% larger in all directions; when four times as much time has passed, it will be 152% larger in all directions; when ten times as much time has passed, it will be 364% larger in all directions. In other words, the Universe gets bigger, but the *rate* at which it gets bigger slows down. This slowdown is even more severe if your Universe is primarily filled with radiation, because not only does the radiation density drop, but the energy of each individual quantum of radiation drops as its wavelength stretches along with the expanding Universe. If the Universe again starts out as a certain size at a certain time, but is filled with radiation, when twice as much time has passed, it will be just 41% larger in all directions; when four times as much time has passed, it will be 100% larger in all directions; and when ten times as much time has passed, it will be 216% larger in all directions. But if we allow the Universe to expand *exponentially*, with energy intrinsic to space itself, then the expansion rate remains constant, and sizes grow faster than you might be comfortable with. In the exponential case, if the Universe is dominated by energy intrinsic to space itself, and we still allow it to start off as a certain size at a certain time, then when twice as much time has passed, it will be 172% larger in all directions; when four times as much time has passed, it will be 1,909% larger in all directions; and when ten times as much time has passed, it will be 810,208% larger (or more than *eight thousand* times the original size) in all directions. In other words, exponential expansion allows the Universe to grow larger in less time than in all the other scenarios! (Fig. 8.9)

Guth's big realization was to imagine a phase where the Universe expanded exponentially prior to entering a hot, dense, matter-and-radiation-filled state: a phase of the Universe *before* the Big Bang. He gave this early stage a name: **cosmic inflation**. Consider how such a state would affect the three major problems impacting the Big Bang:

1) **The horizon problem**: without this early phase, there is no reason to expect disconnected regions in different parts of the sky to have the same properties. If they were once connected together — if they once shared the same properties — but were then *driven apart* by a phase of exponential expansion, that could explain how different areas of the observable Universe have the same temperatures today. They only appear disconnected now because we were not around to see the early, exponential phase. But if such a phase existed, it means that regions that are presently tens of billions of light years away from one another could have been arbitrarily close together during

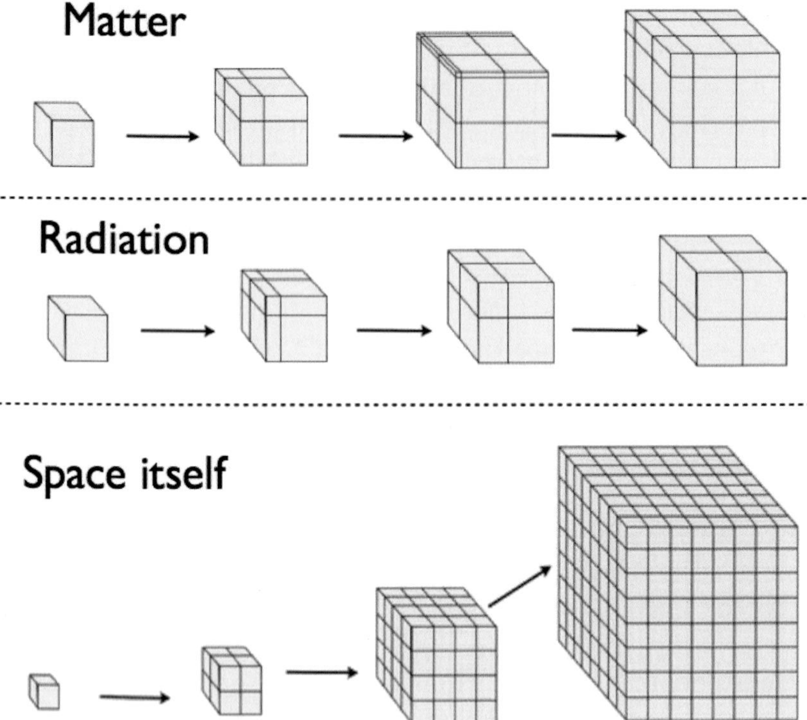

Figure 8.9 These three images illustrate how a region of space expands from a certain size after double, triple, and quadruple the initial amount of time, assuming they all started with the same expansion rate, if these regions were dominated by matter (top), radiation (middle), or energy intrinsic to space itself (bottom). While the expansion rate slows dramatically in the top two cases, the fact that it remains a constant in the bottom case leads to an exponential expansion, where space begins to *inflate*. Image credit: E. Siegel.

this inflationary phase, and were stretched across the Universe as a result.

2) **The flatness problem**: without any causal reason, the Universe would be expected to have either a large positive or negative curvature. The idea that two unconnected things — the expansion rate and the matter/energy density — would be so perfectly balanced without a cause sounds a bit too much like magic. But if you have an inflationary phase, it takes any type of Universe,

whether initially closed, open or flat, and stretches it so that the portion visible to us becomes indistinguishable from a flat Universe. The Universe may truly, on some extraordinarily large scales, actually be positively or negatively curved, but we would perceive it to be flat, the same way that someone unable to venture beyond a single city block would be unable to tell whether the Earth were truly curved or not.

3) **The monopole problem**: if the Universe truly went back to an arbitrarily hot, dense state, there should be all sorts of high-energy relics that populate our Universe today. Yet we have searched for them and have not been able to find them. In an inflationary Universe, there is some energy *cutoff* to how hot the Universe actually got in the past. The reason for this is that regardless of what came before inflation, that phase happened at a specific energy, and when that inflationary phase ended, the energy got transferred into matter and radiation. But the maximum temperature that the hot Big Bang (occurring *after* the end of inflation) reached couldn't have exceeded or even reached the energy density of space itself during inflation, because once you start creating matter and radiation, the energy density *drops* as the Universe expands! (Fig. 8.10)

Now, it is worth pointing out that Guth's initial idea did not do *all* the things we require of a new scientific theory to supplant the pre-existing one. Despite all of these successes, his first model for inflation could not reproduce a Universe that was of a uniform density at all locations and of similar properties in all directions in space; it failed to reproduce *one* of the observations that the pre-existing model was able to account for. But the potential for success that inflation brought with it was so tremendous that many scientists went to work on it immediately, and that particular problem was solved just a single year later, independently, by Andrei Linde and by the team of Andy Albrecht and Paul Steinhardt. By the early 1980s, we had a working, viable model of inflation.

How does one visualize inflation, and how this inflationary phase comes to an end? I like to imagine a very flat surface, suspended high off the ground, made up of a tremendous number of rectangular blocks. These blocks are not locked together, but are held in place by some

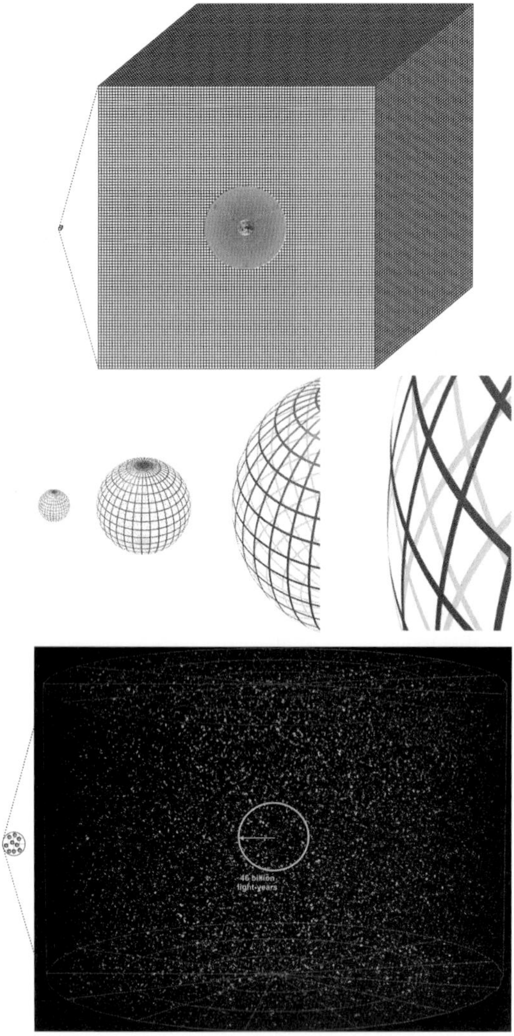

Figure 8.10 In inflation, the horizon problem is solved because even though distant regions of space are still separated by tremendous distances, they were once causally connected in the distant past (top), prior to the end of inflation. The flatness problem is solved by the stretching of space, because no matter its initial shape, inflation stretched it to be so large that the part we can see today is indistinguishable from flat (middle). And the monopole problem is solved because the particles that existed prior to inflation were pushed so far apart that they would not be visible in our Universe today (bottom), so long as the temperature post-inflation never got hot enough to create new ones. Image credit: E. Siegel, using an image from Wikimedia Commons users Frédéric MICHEL and Azcolvin429.

unseen force pushing in around their edges. And at the same time, there is a massive ball — maybe a bowling ball — rolling over the blocks. So long as the blocks stay in place and the ball rolls over them, the Universe inflates. Every additional block that the ball rolls over gives the Universe enough time to more than double in size. By time the ball rolls over 64 blocks, enough inflation has occurred to take something the size of the smallest possible particle in our Universe today and stretch it to the size of the entire visible Universe. It is true that the ball may have wound up rolling over many more blocks than that; we can only place a lower limit on *at least* how many blocks it ran over. But at some point, the rolling ball either encountered a weak spot in the blocks, or simply rolled for long enough that its cumulative effects caused just one single block to give way. When that happens, there is a cascading chain reaction around the ball, and all the blocks in your vicinity fall away, plummeting towards the ground. When the ground is finally reached by both the ball and the blocks, that signifies the end of inflation and the beginning of a Universe — one that is the same everywhere you look — that is filled with matter, antimatter and radiation, whose energy is determined simply by what height the blocks fell from. (So long as the blocks did not fall from a height that was too great, they would be not be able to produce those high energy relics like magnetic monopoles.) What we are left with at the end of inflation is a Universe that can be described by a hot, dense, expanding *but cooling* phase: the very thing we identify with the Big Bang (Fig. 8.11).

* * *

By 1982, there was a new game in town for the origin of the Universe, as it could reproduce all the successes of the Big Bang *and* potentially explain some of the problems that the Big Bang could not address. By adding an early inflationary period to the Universe — a period where the Universe was not filled with matter and/or radiation, but rather with energy inherent to space itself — we were able to explain why the hot Big Bang would have started with the initial conditions that it did. Not only can we reproduce a hot, dense, expanding and cooling Universe that is set up to produce a matter–antimatter asymmetry, the light elements via Big Bang nucleosynthesis, neutral atoms and the CMB, and then matter that

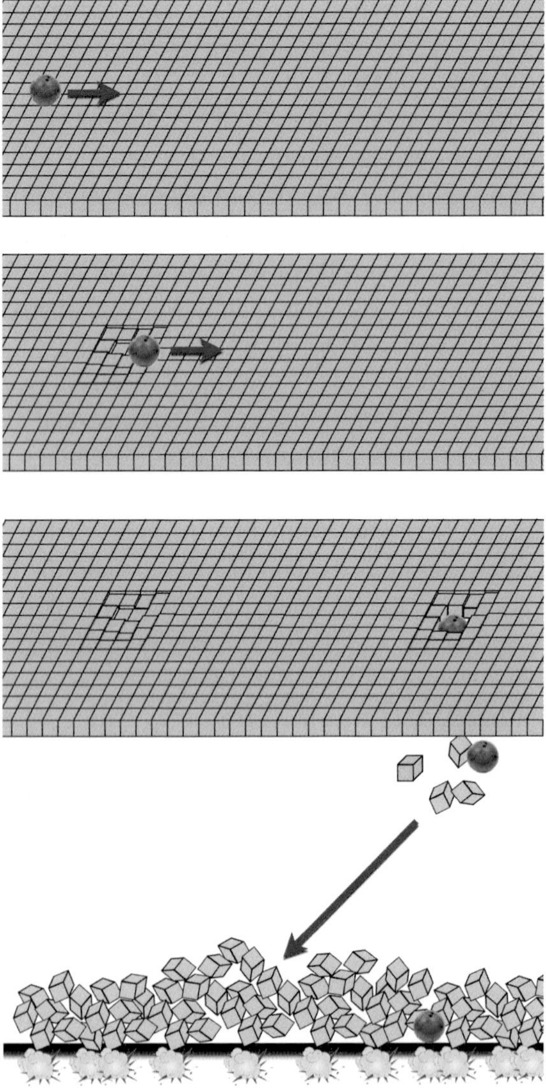

Figure 8.11 During inflation, so long as the ball rolls along the surface-of-interlocking-blocks that represents energy intrinsic to space itself, the expansion rate continues on at an exponential pace. But eventually, there will be a transition — as the ball falls through a spot in one of the blocks — that causes all of the blocks to fall practically simultaneously. When the ball (and all the blocks) fall down to their minimum value, the energy intrinsic to space itself gets converted into matter, antimatter and energy, causing the rate of expansion to begin dropping and marking the start of the hot Big Bang. Image credit: E. Siegel.

gravitationally collapses into stars, galaxies and clusters of galaxies, but we can solve the three problems it was designed to address:

1) Why did the Universe start off with the same temperature everywhere?
2) Why does the Universe appear indistinguishable from spatially flat?
3) Why are there no ultra-high energy relics?

In this new story, there is a period of cosmic inflation that precedes the Big Bang, and the hot, dense, matter-and-radiation filled state only comes about after inflation ends. Even then it does not get *arbitrarily* hot, but is truncated at the energy scale where inflation comes to an end. In our falling-block picture earlier, it is only the magnitude of the "bang" from the blocks hitting the bottom that re-energize (in a process known as *cosmic reheating*) the Universe.

That is a very nice story so far. We have now met the first two criteria for a new theory: we have reproduced all the predictions encompassed by the old one, and have also explained some other phenomena we could not previously: the uniform temperature everywhere, the flatness of space, and the lack of ultra-high-energy relics. But unless we can coax some *new predictions* out of inflation — unless we can go out and search for something we had neither seen nor predicted before — it will not turn out to be a very physically interesting theory! Thankfully, inflation gives us two possible *observable* signatures, and they come from one of the most surprising sources: quantum physics! You might think this would be an unlikely source for anything that has to do with astronomy or astrophysics, since quantum physics is generally only important on subatomic scales. But remember what is happening during inflation: the expansion rate is incredibly rapid, the energy density intrinsic to space itself is incredibly high, and most importantly, scales that start off corresponding to subatomic distances can very quickly — on timescales corresponding to some 10^{-32} seconds — be stretched across the Universe onto cosmic scales.

In your conventional experience, you are used to being able to make arbitrarily good measurements, where the only limitation is set by the quality of your equipment. In addition, those measurements you make in no way physically change whatever it is you are measuring. Your height does not change, for example, when you go to measure it with a tape measure, and if you used a more precise device, such as a laser, the only change you would expect is that instead of measuring your height to be 5′9″ (or 176 cm), you might measure it to be 5′9.185″ (or 175.73 cm). But in the quantum world, measuring a quantity such as distance not only has the potential to change that quantity itself, but that act of measurement can change *other* quantities as well, such as momentum, or how quickly that massive body is moving! This is due to a quantum principle first formulated by Werner Heisenberg — the uncertainty principle — which states that there are certain *pairs* of quantities that have an inherent uncertainty between them. Position and momentum are one such pair: the more precisely you measure one, the greater the uncertainty in the other becomes, and it is never possible to have an uncertainty of *zero* in either one (Fig. 8.12). But another pair of inherently uncertain quantities is

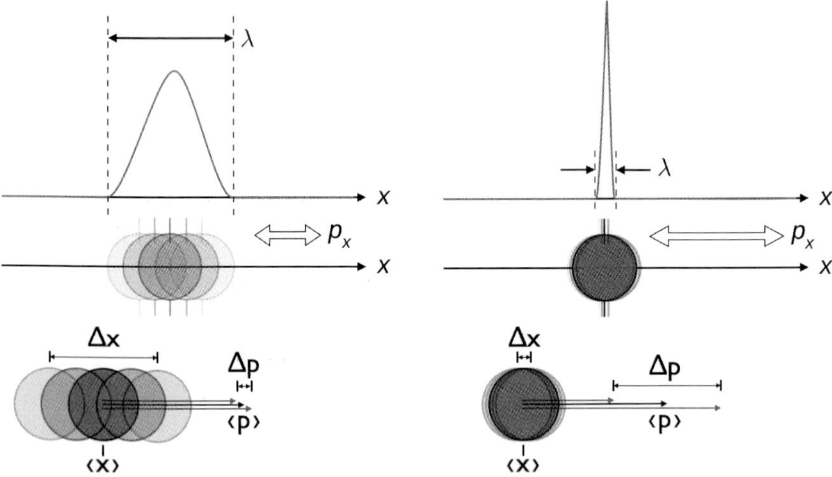

Figure 8.12 In our common experience, we are used to being able to measure quantities like *position* or *momentum* as arbitrarily well as we like. But there is an inherent relation between those two quantities, and in particular between their intrinsic uncertainties. The more accurately you measure a particle's position, the less accurately you're capable of measuring its momentum, and vice versa. It is also impossible to know either quantity exactly, as there is always a minimum inherent uncertainty between the product of the two. Although it is not shown here, that same uncertainty relationship exists between two other quantities: energy and time. Image credit: E. Siegel, based on work from Wikimedia Commons user Maschen.

energy and time: the shorter a timescale something happens on, the greater the uncertainty in its energy.

When we talk about cosmic inflation, the fabric of the Universe is expanding so quickly — on such short timescales — that there are inherent uncertainties in energies that are very large. Because all the energy in the Universe at this time is energy inherent to space itself, that means that the energy fluctuations lead to different regions of space having slightly different amounts of energy inherent to them. These fluctuations are still small compared to the overall energy scale, but because of the short timescales involved, the energy fluctuations are substantial. Remember how we talked about visualizing inflation as a bowling ball rolling along a flat surface? Now imagine adding imperfections to that surface. It is a lot like the surface of the ocean in the middle of the Pacific: even though the total depth is huge, perhaps many miles (or kilometers), and the surface looks perfectly flat from a distance, in reality it has imperfections that are typically a few inches (or centimeters) in size. These become tiny imperfections in the fabric of space, and translate into imperfections in the *energy* of space itself. As inflation stretches the fabric of space itself to grow exponentially larger, those fluctuations get stretched across the Universe as well, impacting all scales, from the subatomic all the way up to the scale of the observable Universe itself (Fig. 8.13).

When inflation ends, those fluctuations in space remain present across the entirety of our observable Universe. What inflation predicts is that these quantum imperfections will show up in two different ways, with two different impacts on our Universe:

1) When inflation ends, those fluctuations in energy are on all sorts of different scales, *almost* perfectly equally. The fluctuations on the smallest scales turn out to be *slightly* smaller in magnitude than the fluctuations on the largest scales, and these translate into tiny imperfections in the matter-and-energy content in the Universe once the Big Bang phase begins. As far as today's observables are concerned, we would expect to see an almost scale-invariant spectrum of density fluctuations, with a very particular pattern of imperfections that show up in the CMB. In detail, there should be a very slight favoring of larger scales as compared to smaller ones. These

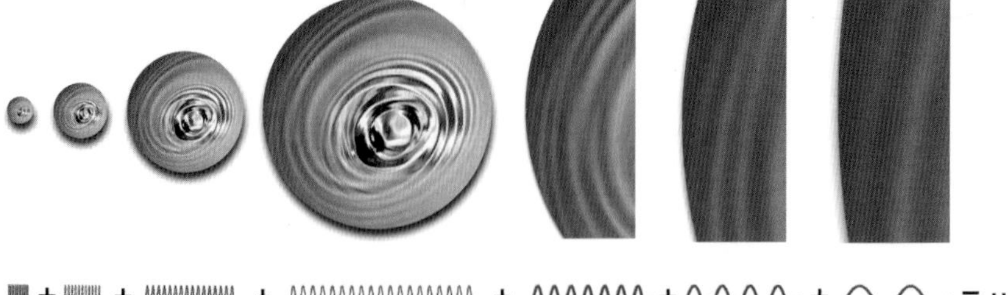

Figure 8.13 The quantum fluctuations in the energy inherent to space itself — fluctuations that are fairly large due to the incredibly short timescales inherent to the quantum world — are normally too small, scale-wise, to have a measurable impact on our Universe. But during inflation, those fluctuations are stretched along with the fabric of space itself, leading to imperfections in the energy of space itself that are large in both magnitude and scale (top). Since quantum fluctuations are always newly occurring on the smallest scales and, during inflation, space is exponentially expanding without any pauses, we will have to add up the fluctuations on all different scales to accurately describe the energy imperfections of space (bottom). Image credit: E. Siegel, with assistance from a Wikimedia Commons photo by Roger McLassus.

imperfections should be no greater than a few parts in ten thousand, with the differences between small scales and large ones only a few percent of that tiny difference.

2) There should also be fluctuations in the gravitational field itself, which should create gravitational waves — again, on all scales — that impact today's Universe. We're only in the infancy of gravitational wave astronomy, still in the stage of testing the very first prototype gravitational wave detectors. (As of the time of this book's writing, we have not yet directly detected a single gravitational wave from *any* source.) However, gravitational waves can affect the polarization signature of the light from the CMB, something that has been observed by both the WMAP and Planck satellites. Although we know the theoretical shape of this signature, there is a wide range of how big the magnitude of this polarization signature should be, dependent on which model of inflation is correct. The more optimistic models could lead to a detection sometime this decade, while the more pessimistic models would not expect to see one with any planned observatory or experiment.

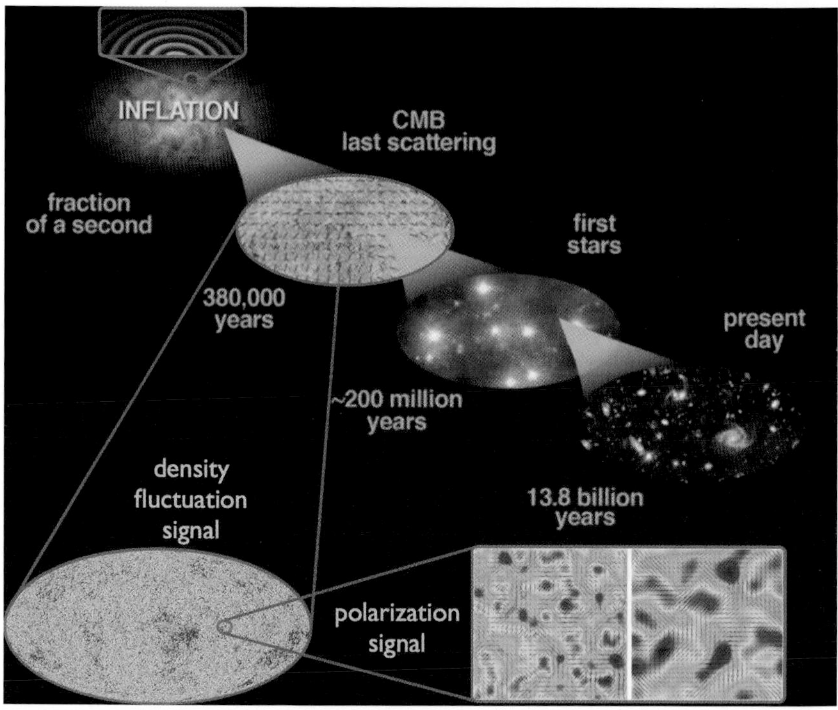

Figure 8.14 Fluctuations in spacetime itself at the quantum scale get stretched across the Universe during inflation, giving rise to imperfections in both density and gravitational waves. These imperfections should show up imprinted in the fluctuations in the CMB, in the form of temperature fluctuations due to differing density from one region to the next (lower left) and in the polarization signal from those temperature fluctuations (lower right). The density fluctuations originally generated by inflation will continue to grow, under the influence of gravity as time goes on, into the stars and galaxies of the present day. Image credit: E. Siegel, with images derived from ESA/Planck and the DoE/NASA/NSF interagency task force on CMB research.

Relying on the second signature alone would be difficult, since our prospects for directly detecting gravitational waves are not so strong. But measuring fluctuations in the CMB was exactly what astrophysicists in the 1980s were planning to do (Fig. 8.14).

* * *

Measurements of the CMB's fluctuations, unfortunately, cannot be taken the same way Penzias and Wilson took them: from a large telescope on the ground. As it turns out, Earth's atmosphere — excellent at letting visible light through — makes a terrible window when it comes to infrared and microwave frequencies. If we want to avoid atmospheric effects, and in particular, if we want to look for temperature differences that might be as small as a few parts in 10,000 (or even smaller), we should really be going to space to make these observations. In 1989, the COBE (COsmic Background Explorer) satellite was launched, the same satellite whose FIRAS (Far InfraRed Absolute Spectrophotometer) instrument measured the blackbody shape of the CMB. But there was another instrument on board that satellite: the DMR, which stood for Differential Microwave Radiometer. The fluctuations it was looking for were too small to measure absolutely; that type of precision was (and still is) beyond our technology's capability to directly measure it. Instead, the trick that the DMR instrument used was to examine temperatures in *two* different locations of the sky at once and measure the *difference*. That way, it could record results that were far more sensitive than an absolute temperature measurement could see (Fig. 8.15).

The first results from COBE were released in the early 1990s, and definitively showed, first off, that there *were* fluctuations in the CMB! They were not the biggest fluctuations allowed by inflation, but were rather a little bit smaller: at about three parts in 100,000 (or the 0.003% level). Although the COBE satellite did not have the greatest resolution — it could only measure down to scales of 7° at the smallest, about the angular size of four fingers held together at arm's length — it was able to measure whether these fluctuations had been born with the same magnitude on all the scales it could measure: *and they were*! For the first time, we had observational evidence that addressed a new prediction that cosmic inflation had made, and inflation passed that first test.

Since the time of COBE, two superior generations of satellites have been launched, made measurements of the imperfections in the CMB and returned their data: NASA's Wilkinson Microwave Anisotropy Probe (WMAP) in the 2000s, and the European Space Agency's Planck satellite in the 2010s. WMAP was able to measure temperature fluctuations down to angular scales of less than half-a-degree, and Planck was able to get all

Figure 8.15 The 31.5 GHz radiometer. Note the two horn antennae on top of the DMR instrument, pointed in different directions. This setup is what allowed the COBE satellite to measure temperature differences, even though the absolute temperature is so consistent between any two arbitrary regions of the sky. Image credit: NASA/COBE/DMR team/LBL.

the way down to 0.07° resolution! (See Fig. 6.17 for an illustration of the differences in angular scale.) What these two next-generation satellites have been able to teach us has been that:

- The density fluctuations that the Universe started off with are consistent with a mostly scale-invariant spectrum across a wide range of scales, from the largest in the Universe down to the smallest measurable.
- The process that gives rise to a hot, dense state at the end of inflation — known as reheating — actually places an upper limit on how hot the Universe got in the distant past, consistent with inflation's hoped-for solution to the monopole problem.

- There are patterns in the CMB where temperature fluctuations are larger in magnitude on some scales than others, which tells us how matter and radiation interacted during the first 380,000 years of the Universe.
- If we take all of this data and the known laws of physics (in detail) into account, the density fluctuations actually favor, *very slightly*, large scales as compared to smaller ones. This gives us information as to what type of inflationary model best describes our Universe. (In detail, there is a quantity called the scalar spectral index, n_s, that should equal 1 exactly for a perfectly scale-invariant spectrum. It is measured to be about $n_s = 0.96$ to 0.97, with a slight preference for large scales over small ones, which is precisely what the models of inflation proposed by Linde and Albrecht & Steinhardt predict. (Fig. 8.16)

This information is enough to confirm the first of our two new predictions that inflation made! The data from these satellites definitively tells us that the Universe did *not* get arbitrarily hot and dense in its distant past. If we follow the best evidence that we have, it teaches us that the Big Bang *was not* the very first thing that happened in our Universe, nor does it lead us all the way back to a singularity. Instead, there was an inflationary phase — where the Universe's energy was intrinsic to space itself, and the fabric of space itself expanded exponentially quickly — that paved the way for the Big Bang to occur.

* * *

The discovery that there was a phase to the Universe *before* the Big Bang was a huge achievement for science, and in a way that was truly unprecedented. We had built theoretical frameworks before that had encompassed the pre-existing one and also explained other puzzles that had not been explained previously. But never had we taken a theoretical leap into new territory before on the basis that the Universe would have had to be extremely finely-tuned — balanced between expansion and gravitation, balanced with equal temperatures everywhere in disconnected regions, etc. — as the impetus for that advance. Previously, there was always an observation the pre-existing theory gave an incomplete or incorrect prediction for. On the contrary, the Big Bang gave not an incorrect prediction, but *no* prediction for these quantities. We simply

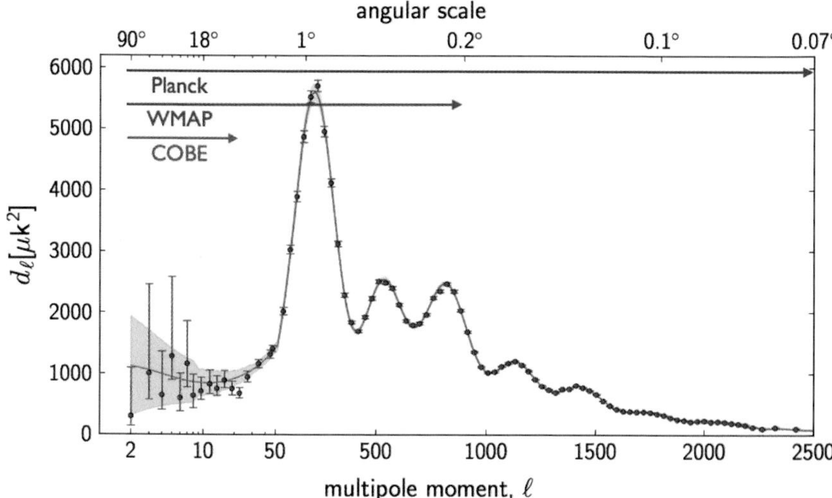

Figure 8.16 The COBE satellite was capable of measuring the "flat" part of the curve, showing the scale-invariant spectrum of fluctuations. Smaller scales were able to yield more information about the history of the Universe, confirming COBE and taking our understanding even further. Image credit: ESA and the Planck Collaboration; P. A. R. Ade *et al.* (2014). *Astronomy & Astrophysics*, 571, A1.

imagined all the possibilities and concluded, "what we see would be unlikely, unless there were something that caused this outcome." That was the impetus for inflation. The most remarkable part of this story is that it turned out to have its new predictions, about the density fluctuations that the Universe was born with, and the fact that the Universe's past should have a maximum temperature, validated by observations (Fig. 8.17).

But there are a number of things we do not yet know about inflation, including what will come of its second great prediction: gravitational waves. In principle, this is something we can observe. When we look at the CMB, we find temperature fluctuations on the order of just a few tens of microK, which is less than 0.01% of the actual CMB temperature. But buried in this signal is another, even more subtle one: the signal of photon polarization. Remember that photons are electromagnetic waves, with oscillating electric and magnetic fields that are perpendicular to one another. Because of this, when photons pass through electrically charged particles that are configured in particular ways, those electric and

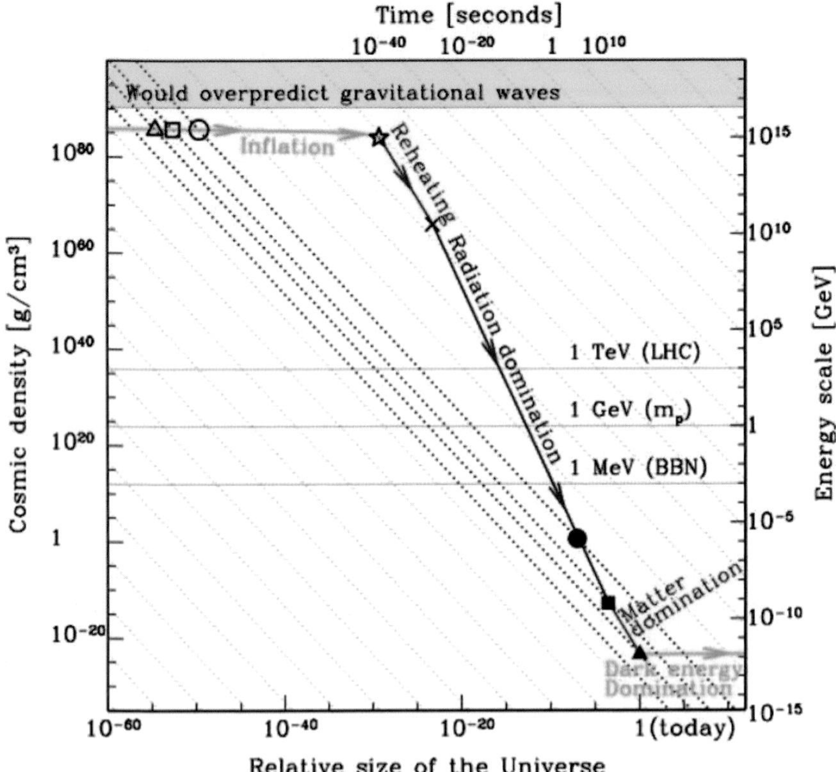

Figure 8.17 The energy density at the end of inflation/the start of the hot Big Bang was more than 20 orders of magnitude (a factor of 10^{20}) larger than it is today. But more importantly, there is an upper limit to how hot our observable Universe could have gotten in the past, an important prediction of inflation that has been borne out by observations of the CMB. Image credit: the DoE/NASA/NSF interagency task force on CMB research.

magnetic fields are affected. They can either have their polarizations altered to be circularly symmetric on a particular region of the sky, which is known as an E-mode polarization, or altered to be sheared and stretched in a very asymmetric way, which is a B-mode polarization pattern. If we can measure both the E-mode and B-mode polarization patterns in the CMB, and we can measure them on a variety of angular scales, from the large all the way down to the very small, we ought to be able to reconstruct what caused these signals (Fig. 8.18).

The difficulty is that *many* things cause these signals: charged particles, distant galaxies, leftover gravitational waves from inflation and also the foreground of our own galactic plane, which emits polarized light of the same wavelengths that the CMB occupies. But if we can successfully account for the origin of all the different components of the E-mode and B-mode polarizations, we should be able to find a signal of gravitational waves that remains — originating from inflation and with a specific pattern across all wavelengths — in the B-modes. And if we can detect those signals, we might be able to discern which of the two major classes of viable inflation models describe our Universe: **new inflation**, which predicts a very small-magnitude level of gravitational waves (and hence, tiny B-modes), or **chaotic inflation**, which predicts large ones (Fig. 8.19).

There are a number of current and planned experiments, observatories and satellites that are designed to measure exactly these B-mode signatures, including the Planck satellite but also including lesser-known efforts like QUIJOTE, ACTPOL, POLARBEAR, SPIDER, SPTPOL, QUBIC, EBEX, ABS and BICEP2. In 2014, BICEP2 made headlines by announcing that they had detected primordial B-modes that supported the chaotic inflation interpretation, and claimed to have detected it at such a significant level that the chances it was a statistical fluke were less than one-in-a-million! Unfortunately, like many grandiose scientific claims, this one turned out to be premature, as the Planck team was able to show that they did not properly account for the foreground emission of the Milky Way galaxy itself. Chaotic inflation is still a viable candidate for the type of inflation that occurred in our Universe, but the BICEP2 data no longer favors it over a model of inflation that produces much smaller gravitational waves. More and better data is needed before we can be certain of exactly how our Universe came to be (Fig. 8.20).

* * *

Regardless of which particular variety of inflation occurred early on in the history of our Universe, we would not be the curious creatures we are if we did not begin to ask ourselves what happened at even *earlier* times! Because of how rapidly the Universe expanded during inflation, and the fact that it's "only" been 13.8 billion years since inflation ended and the

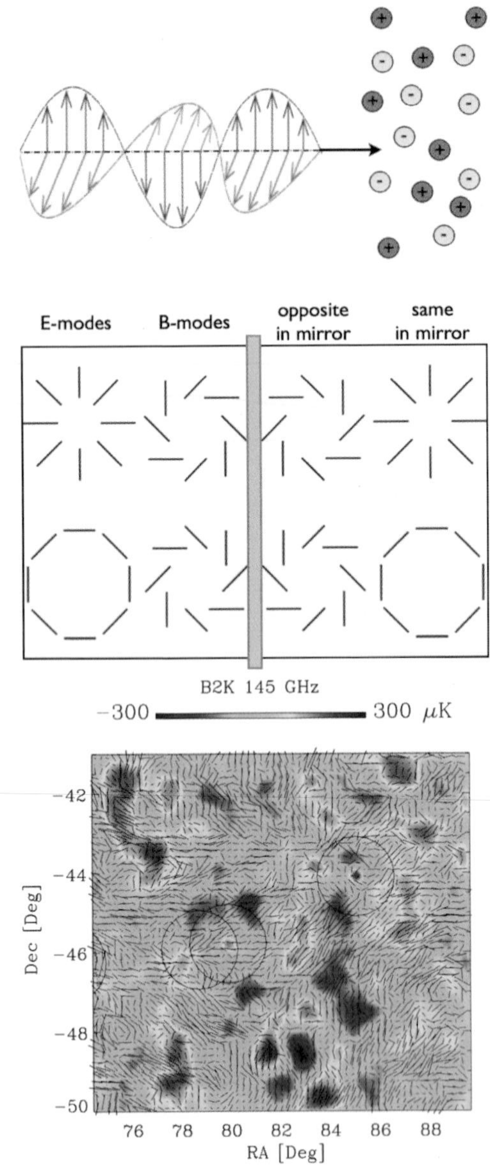

Figure 8.18 When photons pass through charged particles in particular configurations, they can become polarized. E-mode polarizations are symmetric if reflected in a mirror; B-mode polarizations are not. Real radiation (bottom, from the BOOMERANG experiment) displays elements of both E-mode and B-mode polarization signals. Image credit: E. Siegel (top and middle), W. Jones (bottom) (2005). Ph.D. Thesis, California Institute of Technology, AAT 3180590.

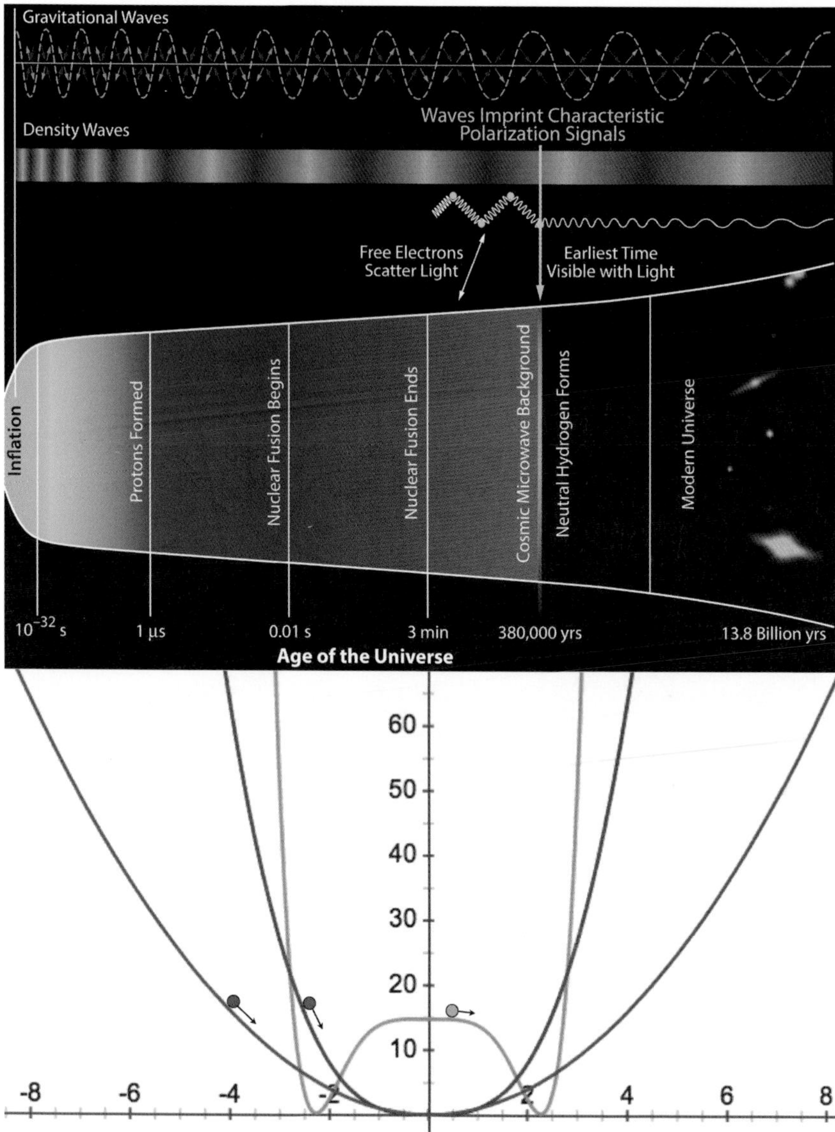

Figure 8.19 All models of inflation predict the creation of density fluctuations and of gravitational waves (top). However, chaotic models (blue and red) predict large ratios of gravitational fluctuations to density fluctuations, while "new inflation" models (orange) predict small ratios (bottom). Images credit: National Science Foundation (NASA, JPL, Keck Foundation, Moore Foundation, related) — Funded BICEP2 Program (top); E. Siegel (bottom).

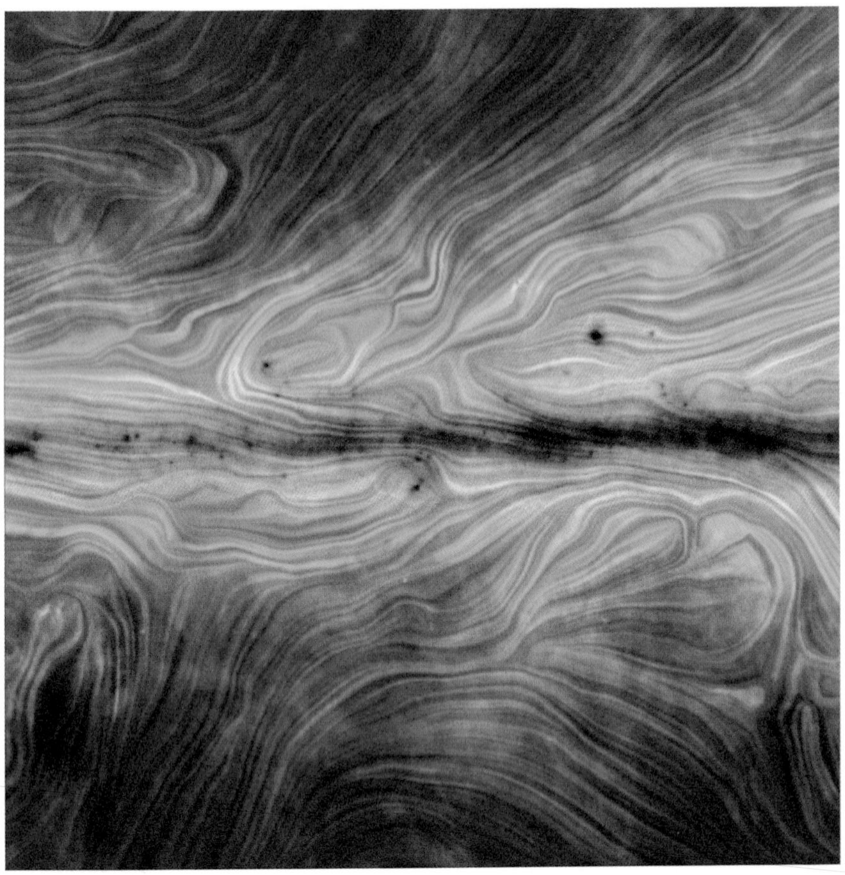

Figure 8.20 In addition to ancient cosmic sources, relatively nearby sources — such as material in our Milky Way — can cause photon polarization. This image shows our galaxy's magnetic field overlaid atop the cosmic density fluctuations, as determined by the Planck satellite in 2014. Image credit: ESA/Planck Collaboration. Acknowledgment: M.-A. Miville-Deschênes, CNRS, Institut d'Astrophysique Spatiale, Université Paris-XI, Orsay, France.

Big Bang occurred, it is only the final 10^{-30} seconds or so of inflation that have any impact on our observable Universe at all. Everything that occurred earlier, because of the exponential nature of inflation, has been pushed to scales that are far larger than our observable Universe, and hence there's no way for us to measure them.

This is very frustrating for us as scientists, since it necessarily means that there is no conceivable way to gain information about what happened *prior* to those final 10^{-30} seconds of inflation. Our period of cosmic inflation could have been an extremely short-lived state, it could have been a very long-lived state, or it could have even been an *eternal* state; we simply do not have the evidence to tell which one is correct. It is tempting to take the same approach that led us to cosmic inflation in the first place, and examine what are the conditions that could have led from a non-inflationary state to an inflationary one. Indeed, there are plenty of theorists who work on exactly this problem! Unfortunately, *none* of their results are conclusive, as no one has yet come up with a surefire observable signature that would remain in our Universe today. As far as we can tell, inflation lasted for at least 10^{-30} seconds or so, but there is no upper bound to how long it lasted, and it could have even been for a truly infinite amount of time!

However, there is a theoretical approach we can take that indicates what lies beyond the portion of our Universe that is observable to us. What we can do is imagine an inflationary Universe, and consider both how that Universe evolves in time and also what needs to happen in order for inflation to end. For simplicity, let us consider the theoretical case of new inflation. We are going to visualize it as a flat plateau that runs down into a valley on either side, that then shoots up with steep walls as you move out of the valley. Inflation occurs when we are towards the middle of the plateau, and inflation ends and converts that energy-inherent-to-empty-space into matter, antimatter and radiation when it rolls down into either one of the valleys.

It is probably your first instinct to imagine this the same way you would imagine a ball rolling down a hill shaped like this. To begin with, you are up at the top, at the flat part, rolling slowly. Things do not appear to change much as you roll along; you are at the same height pretty much the entire time, remaining in an *almost* static state. This is when inflation occurs: the Universe expands exponentially, growing rapidly, getting stretched flatter-and-flatter, with fluctuations stretching across the Universe. After enough time, you finally begin to approach the valley, and suddenly changes not only become perceptible, but quite large. In short order, you're rolling down into the valley, losing your height (and the energy intrinsic to space with it), and converting your energy into particles as you roll back-and-forth into an equilibrium state at the bottom of the valley.

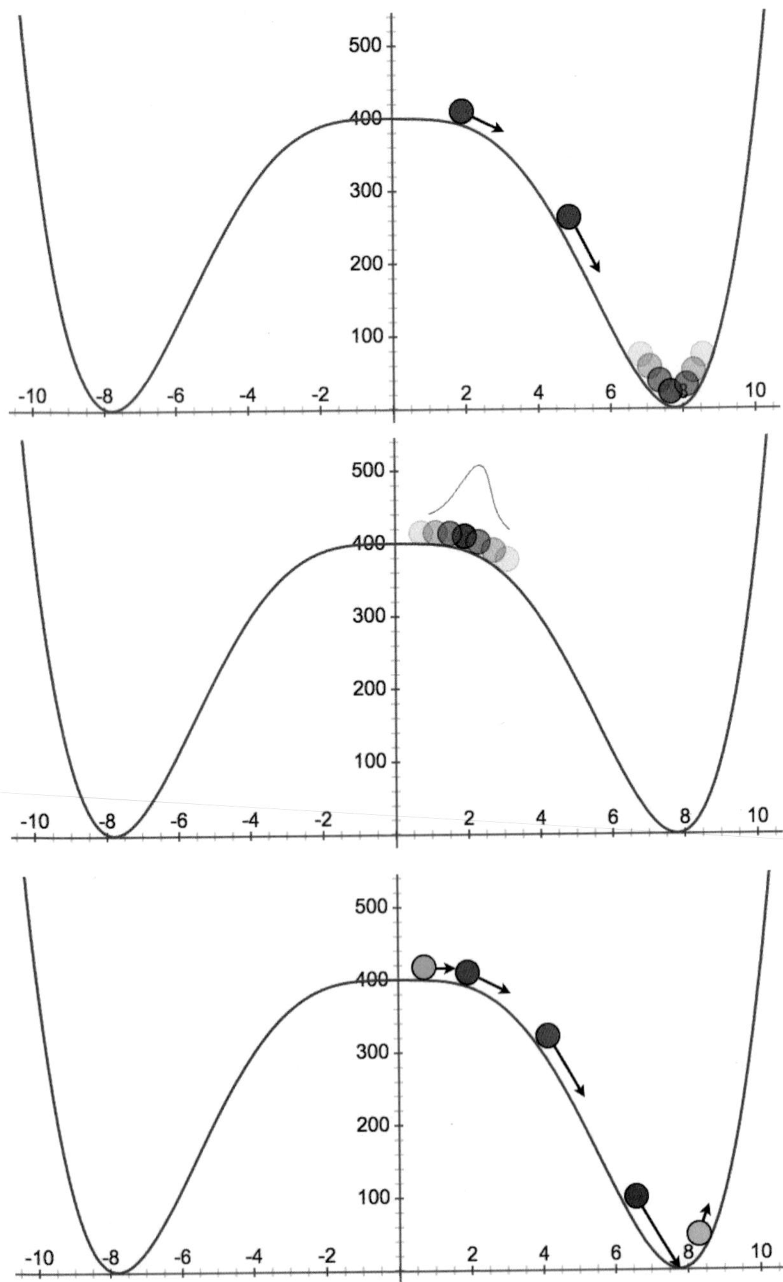

Figure 8.21. *(Continued)*

Figure 8.21 (*Continued*) In the top panel, inflation acts like a *classical* field, rolling slowly down the hill. No matter how long it takes to roll along the flat part, eventually you will reach the steeper part and roll into the valley, which takes you to the end of inflation, a reheating phase and the hot Big Bang. But if you are a quantum field, in addition to rolling slowly, your quantum wavefunction (and hence, your position atop the hill) also spreads out over time. Since you are rolling very slowly, the uncertainty in your position can be *greater* than the amount you would have rolled in a given amount of time, so the middle panel shows where your position will be likely to be as you move forward in time. In the bottom panel, five different regions are color-coded to show what has happened in them. While inflation has ended or is ending in some of them, some regions are not only still inflating, but are *farther away* from having inflation come to an end at all. Image credit: E. Siegel.

But hang on a moment. There were fluctuations stretching across the Universe — both density (scalar) fluctuations and gravitational wave (tensor) fluctuations — when you were at the top of this hill, because you *were not* acting just like a ball. In reality, you were a quantum field. As far as we can tell, at a fundamental level, *all* forces, particles and properties of the Universe must be quantum in nature, including whatever field is responsible for inflation. And one of the things that quantum fields do over time is develop an inherent uncertainty in their position. All other things being equal, they probabilistically *spread out* over time! (Fig. 8.21)

Why is this interesting? Because if we are in an inflationary state — at the top of this hill — where we are rolling slowly and creating new space at an exponential rate, we are not necessarily approaching one of the valleys. At least, not everywhere. In some of the regions of space, perhaps in as many as 50% of them, the value of the quantum field spreads out in such a way that it moves away from the nearest valley, rather than towards it. This continued state of inflation is most pronounced on the flattest part of the plateau, where the spreading effect of the field can dominate the rolling effect that takes you towards the valley. And if space is expanding exponentially, which means that new space is rapidly created wherever inflation is ongoing, the inflationary state can be eternal to the future. As we move forward in time, we are constantly creating new regions of space where inflation is no closer to ending (and in many cases, *farther away* from ending) than at earlier times.

In other words, it is tempting to have a classical picture of inflation in your head: that at some point in the past, you have a region of space that is exponentially expanding, and it inflates, inflates and inflates some more, uniformly, and then stops in all those regions all at once, giving rise to the Big Bang in all locations. But that picture is *inconsistent* with what the known laws of physics, when combined with what we know of inflation, tell us! Instead, we have a region of space that is expanding exponentially, inflating, and spawning many more new regions. In some of them, inflation ends (giving rise to a Big Bang) while others continue to inflate some more, spawning many more new regions. In some of those new regions, inflation ends (giving rise to a Big Bang), but others have inflation continue further, spawning many more new regions. So long as inflation is rapid enough, this process can occur for an eternity.

If we treat inflation as a quantum field, and we calculate the rate at which quantum fields spread out and compare it to the rate at which the mean value of the field rolls down into one of the valleys, what we find is that *in all physically viable scenarios*, there are always regions of space where inflation lasts forever. In other words, if inflation begins in even *one* tiny region of the Universe, it continues, somewhere, eternally into the future. Sure, there are infinitely many regions like ours, where inflation comes to an end at some point, giving rise to a matter, antimatter and radiation-filled Big Bang, but separating each of those individual regions are places where the Universe continues to inflate. As time goes on, it is true that more and more regions will see inflation end. But the exponential expansion creates enough new space, continuously, to ensure that we will never lack for hot Big Bangs in independent, disconnected locations in our Universe. When people speak of a **multiverse**, or the idea that our Universe and what we can observe is not all there is out there, it is the very good science of inflation that leads to this inevitable conclusion! (Fig. 8.22)

* * *

So after all this, what can we conclude? That for all the successes of the Big Bang — and there are a great many — we *cannot* extrapolate the Universe back to an arbitrarily hot, dense state. Yes, the Universe is huge, cold and still expanding even now. Yes, we can go back in time to when

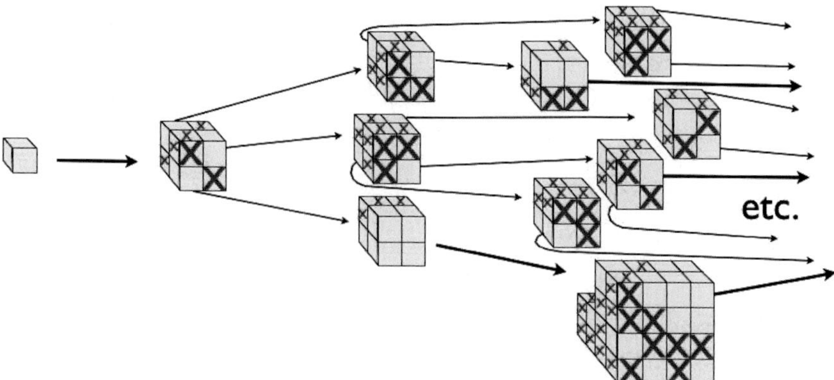

Figure 8.22 If even a single region of the Universe begins inflating, it will begin to create new regions of space. While many of these regions will have inflation end, giving rise to a hot Big Bang (marked with red X's), there are always going to be regions where inflation does not come to an end (blue cubes). In those places, inflation continues, giving rise to more exponentially expanding space. So long as — at any given time — there are *more* regions inflating than there were at a previous time, we are guaranteed that there will be locations in the Universe outside of our own, observable Universe where inflation continues for an eternity. Image credit: E. Siegel.

things were smaller, denser and hotter. Yes, there was a time when things were more uniform, when galaxies were smaller, more numerous and less evolved. Yes, there was a time before which any stars had formed at all. Yes, there was a time when it was so hot that neutral atoms could not stably form, and the entire Universe was an ionized plasma of electrons and atomic nuclei. Yes, there was a time even before that where it was too hot and energetic for even stable atomic nuclei to form, and all we had was a sea of free protons and neutrons. And yes, there was a time even before that, where matter-and-antimatter were spontaneously created in great abundance, where all the elementary particles (and their antiparticles) in the Universe could spontaneously create and annihilate solely from the energy available, and where energies far exceeded anything we have ever created here on Earth. But we cannot go back to a time where things were infinitely dense, to where all the matter and energy in the Universe was concentrated into a single point. As far as we can tell, our observable Universe never possessed such a state.

Instead, the Big Bang was preceded by a period of cosmic inflation, where instead of being filled with matter, antimatter and energy, the Universe was dominated by energy inherent to empty space itself. This was a *huge* amount of energy, enough that when inflation ended, that empty space provided the energy that gave rise to the hot Big Bang, as it converted into particles, antiparticles and photons. This period of inflation explains why the Universe is as flat as it is today: because no matter what the shape of space was beforehand, inflation stretches it to be indistinguishable from flat. Inflation also explains why the Universe is the same temperature in all locations: because despite the fact that these regions are causally disconnected from one another today, they *were* causally connected in the past, and were only carried apart across the Universe by the exponential expansion of space that occurred during inflation. And finally, inflation explains why there are no extraordinarily high-energy relics in the Universe: because the Universe never reached the incredibly high temperatures necessary to create them once inflation ends!

In addition, inflation explains why we see the sizes and scales of density fluctuations that the Universe was born with. This prediction was new to inflation, and has since been confirmed to an amazing degree of precision. Furthermore, it also predicts a spectrum of gravitational wave fluctuations that should imprint themselves on the polarization of the CMB. Although such a signal has not been detected yet, we know what it ought to look like, and as our satellites, experiments and observatories improve, we know exactly what we ought to see to verify the signal's existence (Fig. 8.23).

Finally, one bizarre but fascinating consequence of cosmic inflation is that our Universe represents a region of space where inflation came to an end. Although there are a great many such regions, *in between* all of them are regions of space where inflation continues eternally, something that has been ongoing for the last 13.8 billion years at least! This conclusion is arrived at simply by applying the known laws of physics to the physical condition of inflation, and the (seemingly necessary) assumption that inflation be quantum in nature. This leads to the existence of a multiverse, where our observable Universe is just one of a great many regions where inflation ended and gave rise to a Big Bang.

In the end, it is only the last 10^{-30} seconds or so of inflation that have any impact on our observable Universe, giving rise to the fluctuations that led to the formation of stars, galaxies, clusters and the great cosmic

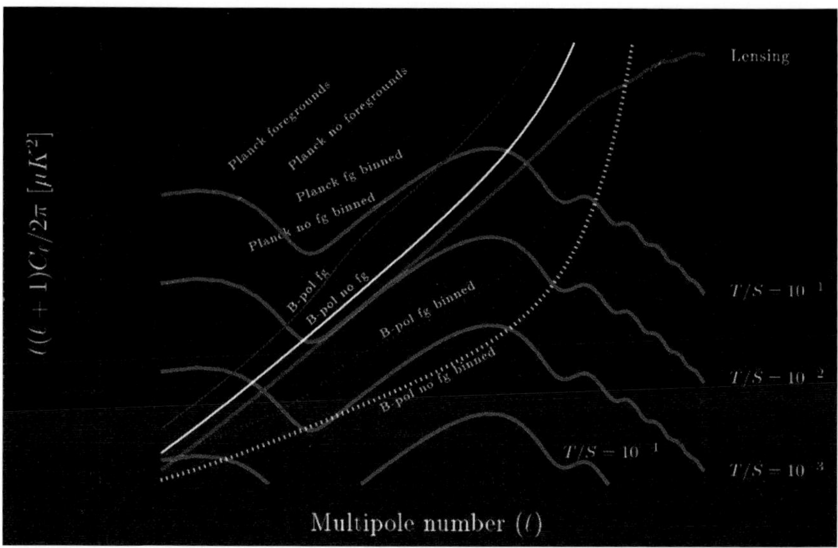

Figure 8.23 The spectrum of gravitational waves arising from inflation is independent of the type/model of inflation that occurs, but the *amplitude* of that spectrum is very sensitive to it. This signal, if present, will show itself in the B-mode polarization of the CMB. Depending on the particulars of inflation, this B-mode signature should, for many presently viable models, be potentially detectable with experiments that are presently either operational or in development. The different solid blue curves show the different amplitudes for various models of chaotic and new inflation. Image credit: Planck science team.

web. When it comes to the story of our Universe, we *do not know* what came prior to those final instants of inflation, how long the inflationary state lasted, whether it was eternal to the past, or whether an even earlier state gave rise to inflation. It also raises the disconcerting possibility that *we may never know* those answers, because the data necessary to determine the answers may not be present in our observable Universe. As vast and wondrous as the Universe we see is, it is finite both in terms of what is in it and how long it has been around, and so the information contained within it is finite as well. The Big Bang was not the beginning, as it was preceded by cosmic inflation. But as to what happened before inflation ended — before those last 10^{-30} seconds or so — we presently have no good ideas for how to find that out.

Chapter 9

Dancing In The Dark: Dark Matter And The Great Cosmic Web

With everything we know about the Universe from its early stages up through the present day, you might imagine that starting off with the right initial conditions and applying the laws of physics would be sufficient to reproduce the Universe as we know it. Any small discrepancies would merely be a matter of filling in the details. If we can start off from expanding spacetime itself, create the hot-and-dense state of the Big Bang, construct a matter–antimatter asymmetry, have the Universe cool to annihilate the leftover antimatter away, produce the first atomic nuclei and then form neutral atoms, it seems like gravitation acting on matter to clump it together to form stars, galaxies and clusters would get us the rest of the way there. In fact, we would be absolutely crazy if we did not try that first! After all, that is what our best theories predict ought to happen in the Universe after those first neutral atoms form.

But there is something important to look for: that what we predict and what we observe, for the large-scale structure of matter in the Universe, match up. If they do not, we are going to have some problems. Even though we know that those problems can often be omens of scientific advances, they can also remain without a definitive solution for many generations, which is one of the most frustrating situations to be in. If we want to understand how the Universe came to be the way it is today, complete with the great cosmic structures we observe, the place to start is with what we can see up close in our own cosmic backyard: the nearest and brightest galaxies and galaxy clusters to us. This represents simultaneously a test for both our theory of gravity and for what we think our Universe ought to be made of (Fig. 9.1).

Figure 9.1 The Virgo cluster of galaxies is the largest nearby cluster to us in the Universe, consisting of over 1,000 galaxies located some 50 to 60 million light years away. Although the largest elliptical and spiral galaxies are clearly visible, every "smudge" of light that isn't a single point source is actually a galaxy all unto itself. Nearby clusters like this one are essential for giving us insight into the formation of large-scale structure in the Universe. Image credit: Wikimedia Commons user Hyperion130, under a c.c.-by-s.a.-3.0.

* * *

We have two completely independent ways to measure the sum total of what is inside a huge structure like a galaxy or a cluster of galaxies: through starlight and through gravitation. We know how stars work, from studying not only our own Solar System but from measurements of the huge variety of stars out there, from dwarf stars to Sun-like stars to hot, young blue stars to giant, evolved stars. When we look at the light coming from either a single star or a population of stars, we know how to infer what the mass and age of the stars we are looking at are. This extends even to entire galaxies: by measuring the total amount of light coming in many different wavelengths are, we can determine how much of that galaxy exists in the form of stars. When we do that for a typical galaxy or for a cluster of galaxies — where galaxies consist of billions or trillions of stars and clusters contain up to many thousands of galaxies — we arrive at a number for the amount of mass that is locked up in the form of stars.

We can also use the motion of those same stars, and our knowledge of the laws of gravitation, to figure out how much *total* mass is present in those huge, bound systems. Just as knowledge of Newtonian gravity and the measurement of one single planet orbiting our Sun would be enough to tell us our Sun's mass, knowledge of General Relativity and the measurement of stars orbiting a galaxy are enough to allow us to infer the entire mass of the galaxy. Similarly, we can look at the different galaxies bound together in clusters, measure their velocities, and determine what the total mass of that cluster must be in order to keep it bound together as well. For both galaxies and clusters of galaxies, we can also arrive at a number for the amount of mass that is inferred due to gravitation.

You might think, based on our Solar System, where 99.8% of the mass is in our star, that measuring starlight would be an outstanding way to measure the mass of a huge structure like a galaxy or a cluster of galaxies. But our observations do not appear to support this at all. Instead, those two numbers we arrive at — the amount of mass in stars and the amount of mass inferred due to gravitation — differ by *a factor of fifty*. In other words, only 2% of the mass that we conclude must be there is actually seen in the form of stars. So what is going on with the other 98%?

* * *

This problem was first noticed by Fritz Zwicky, an eccentric and iconoclastic Swiss astronomer who did most of his famous work in the 1930s. Zwicky was a fan of many non-standard ideas, including the tired-light alternative to relativity, developed the idea that supernovae resulted in collapsed cores to stars that could condense into a ball of neutrons — a neutron star — and also (once he accepted relativity) that dense collections of matter could act as gravitational lenses, magnifying, distorting and creating multiple images of background galaxies, something finally verified in 1979! Known as a lone wolf in his research, Fritz Zwicky did all his own observing and calculations himself, and absolutely detested the mainstream astronomical community's resistance to new ideas and to entertaining wild speculation. There is even a story that one night, in an attempt to reduce atmospheric turbulence, he

Figure 9.2 The Coma cluster of galaxies, an even more massive, denser cluster than Virgo, located 330 million light years from Earth. The image shown is of the central, core region. It was the first cluster of galaxies where the amount of matter inferred from starlight and the matter inferred due to gravitation were shown to yield wildly different numbers. Although astronomers did not take the results seriously for a long time, this was our first clue that there was a new type of matter — dark matter — populating the Universe. Image credit: Adam Block/Mount Lemmon SkyCenter/University of Arizona, under c.c.-by-s.a.-3.0.

ordered a rifle be shot through the telescope slit towards the sky, in an attempt to reconfigure the air. Although the seeing did not improve at all, it goes to show how willing Zwicky was to try new things! (Fig. 9.2)

But perhaps most famously, in 1933, Zwicky was observing the galaxies in the Coma cluster, an extremely rich collection of galaxies containing thousands of members some 330 million light years away. By measuring the redshifts and blueshifts in their spectral lines, Zwicky could not only infer how quickly the Coma cluster was receding from us on average, but also how quickly each individual galaxy was moving relative to the cluster's mean speed. With the additional measurement of distance — including each galaxy's distance from the cluster's center — Zwicky could even determine (to an extent) how the individual galaxies were orbiting inside the Coma cluster. Because he knew how clusters worked, and he knew these galaxies all needed to be gravitationally bound together, he was able to calculate how much total mass must be present in the cluster to keep the galaxies from flying apart. When he compared the value with that inferred from starlight, he was alarmed; the numbers were *vastly* different! He coined a new term, *dunkle materie*, or **dark matter** in English, to explain this huge discrepancy (Fig. 9.3).

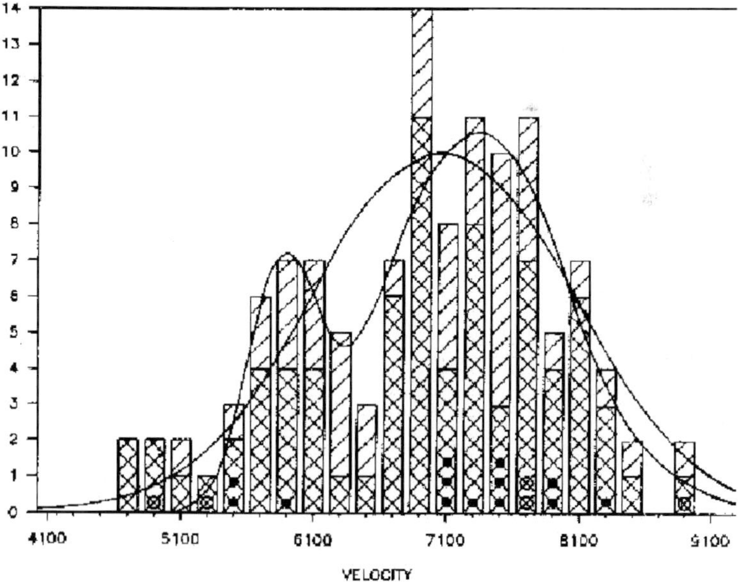

Figure 9.3 Despite being so distant, the galaxies in this cluster display a huge range of velocities, on the order of thousands of kilometers per second, indicating a tremendous mass for the cluster. By contrast, the amount of starlight emitted from the cluster indicates only a tiny fraction of that mass should be present. Image credit: G. Gavazzi, (1987). *Astrophysical Journal*, 320, 96.

This was a serious, legitimate problem that Zwicky noticed, one that should have caught the interest of a great many astronomers and astrophysicists of the day. But perhaps due to Zwicky's bristly reputation, and also perhaps due to political and personal reasons among the astronomical elite at the time, these findings were largely ignored for many decades. Although Zwicky's original estimate that the discrepancy between the amount of matter found in stars and the amount needed to keep this cluster bound were originally high by about a factor of three — Zwicky claimed a factor of 160 difference rather than the now-accepted factor of 50 — this was a serious enough problem that it should not have been so easily brushed aside.

Yet people did not believe this to be a serious issue, and came up with a number of ways to explain away Zwicky's results. Perhaps there were other forms of mass besides stars that accounted for these effects. Perhaps the combination of gas and dust could be responsible for the additional gravitation that was not present in stars. Perhaps there were smaller, fainter masses out there — more planets than we ever imagined, or perhaps large numbers of extremely dim or even failed stars — that accounted for this gravitation that did not emit detectable light. Unfortunately, little was done to test these ideas, and Zwicky's work largely languished in obscurity until the 1970s.

* * *

Nearly 40 years after Zwicky's original work, the investigation of another, largely independent phenomenon all of a sudden rekindled interest in the idea of dark matter. Telescope technology had progressed to the point where not only could redshifts (and blueshifts) be measured for individual objects such as stars within our galaxy or distant galaxies themselves, but for different locations within a single galaxy. Imagine a spiral galaxy, rotating about its center, with the inner portions spinning around rapidly and the outer portions taking longer to complete a revolution. If this galaxy is viewed face-on, like the Whirlpool Galaxy (M51), its rotational motion is *perpendicular* to our line-of-sight, and therefore does not create any observable redshift or blueshift. But if this galaxy is oriented at an angle to us, or in the extreme case, completely edge-on, like the Spindle Galaxy (NGC 5866, or M102), then as the galaxy rotates, one side of it will appear to have a relative motion *towards* us (an extra blueshift), while

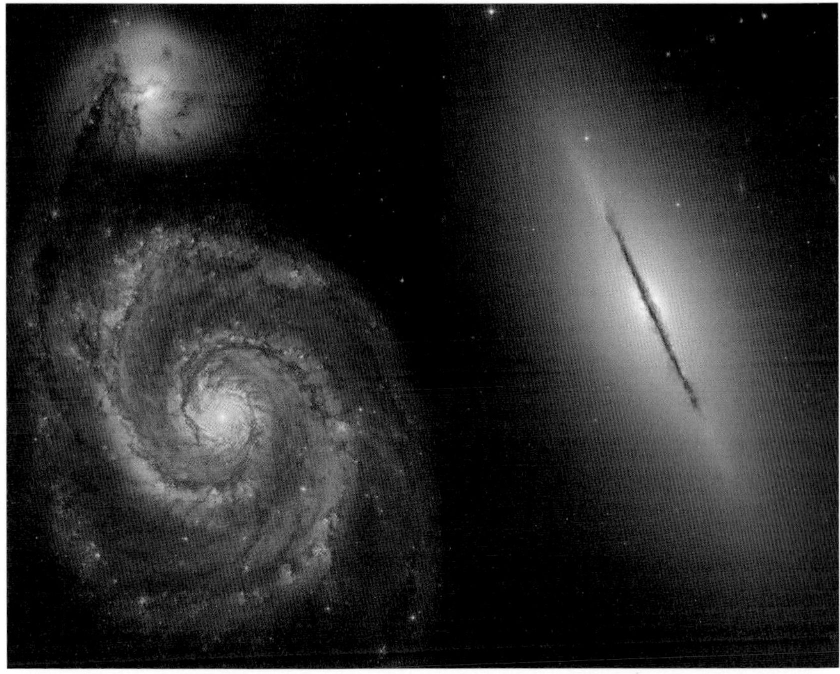

Figure 9.4 Face-on spiral galaxies like M51 (left) rotate in a plane perpendicular to our line of sight, and hence we cannot observe a shift in the light sources' spectra as the stars orbit the galactic center. But galaxies that are inclined at an angle — or in the most extreme case, completely edge-on like M102 (right) — have one side where the stars are moving towards us from our point of view, while the other side has the stars move away along our line-of-sight. This leads to a relative blueshift and redshift, respectively, for the stars on opposite sides. Images credit: NASA, ESA, S. Beckwith (STScI), and The Hubble Heritage Team STScI/AURA) (L); NASA, ESA and The Hubble Heritage Team STScI/AURA) (R).

the other side will appear to have a relative motion *away from* us (an extra redshift) (Fig. 9.4).

Based on the light we see in spiral galaxies, and how the underlying matter appears to be distributed, we would expect there to be a greater concentration of mass at the center, falling off to lower and lower densities as we move out towards the outskirts. A more extreme example of what we would expect is found in our Solar System; with 99.8% of the mass located at the center, it is no surprise that Mercury is not just the closest planet to the Sun, it is also the *fastest* planet in orbit around it. As we move out, away from the Sun, we find that the speeds of each planet drop the farther out we go. Mercury moves at an average speed of 47 km/s, while Earth moves at just 30 km/s, Jupiter at a mere 13 km/s and Neptune at a paltry 5.4 km/s. Since galaxies also have a far greater concentration of matter at their centers, we would expect that each galaxy would display a similar velocity profile when we measured its rotation.

Beginning in 1970, that is exactly what Vera Rubin — an astrophysicist who studied under George Gamow — sought to measure. She started with the Andromeda galaxy, focusing in on its gas clouds. Since Andromeda is close to edge-on, tilted at just 30° relative to our line-of-sight, it should have been easy to measure its velocity profile. Andromeda is rare in that it's a blueshifted galaxy, but what she found was quite surprising: not only *did* the rotational speed of the gas clouds *not* drop as her measurements moved farther away from the galaxy's core, but in some cases, they appeared to speed up!

Throughout the 1970s and 1980s, Rubin and other scientists went on to verify that this phenomenon robustly exists not only in Andromeda, but across practically every galaxy that could be measured. What they saw was that velocities did not drop as we moved away from the center, but rather stayed constant on average, a shocking result given the concentration of stellar matter towards the center! (Fig. 9.5)

When the cluster results of Zwicky were revisited and combined with the individual galaxy results of Rubin and those who built upon her work, it signaled the onset of a crisis in modern astrophysics.

* * *

On the one hand, we can take all the matter that we can observe in the Universe — made up of all the elementary particles known to exist — and add it up, obtaining a certain amount. On the other hand, we've got gravitation and the laws of General Relativity, which seem to be successful describing the Universe in a myriad of testable, verifiable ways. The fact that there is not only more gravitational force exerted than we know how to account for, but that gravity exerts this force in a region that seems to be inconsistent with the observed matter distribution, is what brought on this crisis. As far as we could tell, there were only three ways out of this:

1) Perhaps the particles that we know of have some bizarre property that we have not yet discovered. Perhaps they cluster, clump or behave in

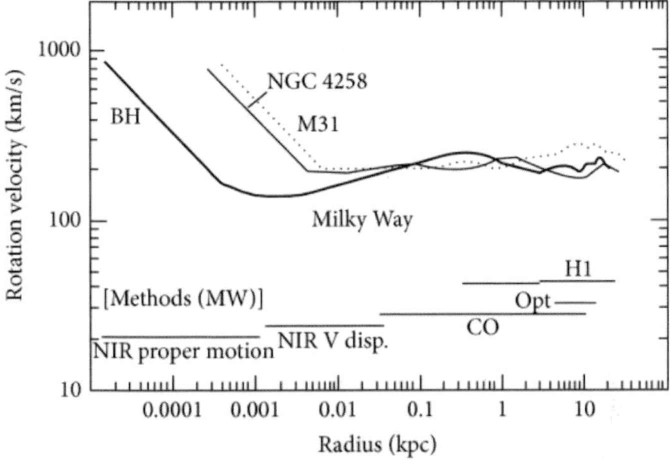

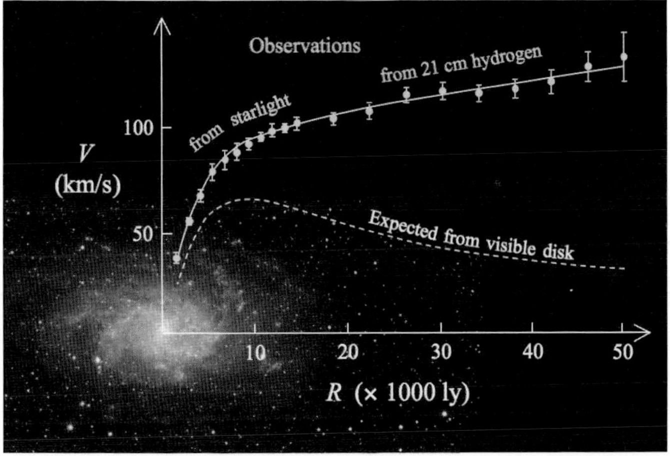

Figure 9.5 Based on the distribution of visible matter, we would fully expect the rotational speed of stars and gas around a galaxy's core to decrease as we reached the outskirts of the galaxy, as most of the matter is concentrated towards the center. But what we see instead — in the vast majority of galaxies — is that the rotational speed remains constant or even continues to rise as we probe stars and gas farther away from the central region. The rotation curves of the Milky Way, Andromeda and NGC 4258 are shown on top, along with the observational methods used to track the galaxy's motion, on a logarithmic scale. Below it, a pictorial representation of the Triangulum Galaxy, M33, along with its rotation curve is shown at right. Along with the cluster data, these galactic rotation curves are a compelling indication that something more is at play in the Universe than normal matter under the influence of General Relativity. Images credit: Paul Gorenstein and Wallace Tucker (2014). *Advances in High Energy Physics*, 2014, Article ID 878203 (top); Wikimedia Commons user Stefania.deluca (bottom).

certain ways that allow them to exist in states we have not been sensitive to or understood very well, and that this accounts for this apparent inconsistency. In other words, perhaps matter is not located where we think it ought to be.

2) Perhaps the laws of gravity are the culprit; perhaps it has been too big an assumption that General Relativity holds on the scales of a galaxy and larger. If we can find the right modification to the laws of gravity — something we did once before when Newtonian gravity did not work — perhaps we can explain all of General Relativity's prior successes, the new, observed phenomena, and also make novel predictions that we can subsequently either verify or refute.
3) Or, finally, perhaps we actually have everything right. Maybe the particles do behave as we understand them, and there simply are not enough of them. Maybe General Relativity is correct, and is the right theory of gravitation on all scales. And maybe, in addition to all of that, there is a new form of matter that exists that we have not discovered yet, a form responsible for all the apparent excess gravitation.

The first idea is a thought along the same lines as those of Zwicky's original detractors: there might just be normal matter we have not been very good at detecting responsible for this whole thing. The second idea is more revolutionary, but would have a lot of explaining to do, given how successful General Relativity has been and how difficult it would be to make such major modifications without disrupting the predictive successes that have already been verified. The third idea is perhaps the most outlandish, as the idea that there is not only a new form of matter out there, but that this form actually *dominates* the gravitation of the Universe, would require some extraordinary evidence (Fig. 9.6).

Thankfully, the past 40 years have brought forth *almost* all the evidence we could ask for in our quest to distinguish between these three very different possibilities.

* * *

The simplest explanation is — in all cases — the one we should consider first. If we can explain all the observations we can make *without* either

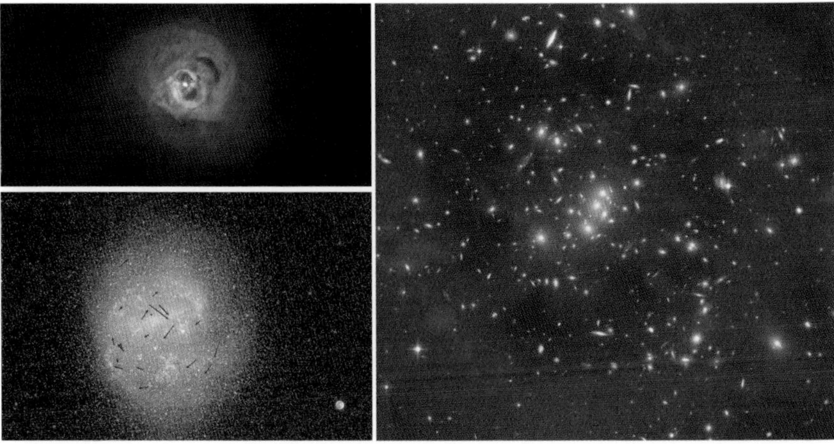

Figure 9.6 It is conceivable that there is plenty of matter that is simply not in the form of light-emitting stars present, such as copious amounts of gas, which might be visible in the X-ray, as shown in the Perseus cluster at top left. It is possible that the laws of gravity are in need of modification on the largest scales, and cause galaxies and clusters to behave with different dynamics than General Relativity predicts, in accord with observations like those shown at bottom left. Or, perhaps, a new form of matter exists that is different from the normal, known particles, and is distributed in a large, diffuse halo around galaxies and clusters such as CL0024 + 17, as shown on the right. Images credit: NASA/CXC/SAO/ E. Bulbul, *et al.* (top left); NASA, ESA, A. Feild and Z. Levay (STScI), Y. Beletsky (Las Campanas Observatory) and R. van der Marel (STScI) (bottom left); NASA, ESA, M.J. Jee and H. Ford (Johns Hopkins University) (right).

modifying the laws of gravity or introducing a new type of matter or particle, that would be the preferred explanation. But we have to consider that possibility honestly and rigorously, and that means considering everything conventional that not only *is* out there in the Universe, but everything that could be present! While stars are certainly the easiest form of matter to see, due to their light-emitting properties, there are plenty of other forms that ought to be out there. Thanks to the use of a number of clever techniques, we have come up with schemes to not only detect these components, but to measure their abundances as well.

What are the types of ways normal matter, made up of the constituents of atoms, could manifest itself in our Universe? There could be ionized particles like free electrons, protons and nuclei. There could be diffuse, neutral atoms that live in gas clouds, both within galaxies and clusters as well as between them, in the intergalactic/intercluster medium. There could be a large number of "failed stars," or massive clumps of matter simply too dim to give rise to nuclear fusion in their core. There could be huge numbers of planet-sized (or smaller) collections of matter that do not emit any visible light of their own, or even tiny grains of dust that could be obscuring significant amounts of light. And finally, there could be collapsed stellar remnants — things like white dwarfs, neutron stars and black holes — which emit little-to-no light, but could have tremendous gravitational effects.

One way to test the viability of this possibility is by measuring the abundances of all these types of components directly: by taking a cosmic census. We know how to detect each and every one of these types of matter, many via multiple techniques. For ionized particles, no matter how diffuse they are, we can look in the ultraviolet and X-ray portion of the electromagnetic spectrum, as even the most tenuous plasma should emit light in those frequency bands. Satellites such as XMM-Newton and Chandra have observed X-rays consistent with this, and there are *plenty* of particles that are present in this form. If we calculate how much of the Universe is present in the form of stars, there is up to *six to seven times* as much in this form of ionized plasma, which is known as the WHIM: the Warm–Hot Intergalactic Medium. It is not nearly enough to explain the factor-of-50 discrepancy between the observed matter and gravitation, however. In addition to that, these ionized particles are found almost exclusively *outside* of galaxies and clusters (Fig. 9.7).

When we look at clouds of gas, however, there are plenty of those as well, and they are found *inside* galaxies and clusters. We can measure this gas directly in a variety of infrared wavelengths. On top of that, when clusters or galaxies go through intense periods of star formation, we can measure the X-rays emitted from them to determine the fraction of gas that is present. Quite consistently, we find that somewhere between 11% and 15% of the total mass of galaxies and clusters are present in the form of gas: a *huge* improvement from the 2% present in stars. But again, this is not nearly enough to get us up to the 100% we would need to explain the gravitation that we see.

Failed stars — objects like brown dwarfs or dim, bigger-than-Jupiter gas giants — only give off very faint visible light radiation in the red part of the

Figure 9.7 We find that there is plenty of normal matter present in both intergalactic space, some of which can fall into galaxies and quasars (top), as well as in clusters and galaxies (bottom), which show up thanks to the signatures of X-ray light and neutral hydrogen. All told, however, these sources of gas and plasma only outweigh stars by a factor of seven or so, nowhere near the factor of 50 we would need to account for all of the missing mass. Images credit: Christopher Martin, Robert Hurt/Caltech (top); A. Chung/NRAO/ROSAT (bottom).

spectrum, or in some cases, no visible radiation at all. This makes them very difficult if not outright impossible to detect with conventional telescopes, but *infrared* telescopes can spot them via dedicated searches. As it turns out, there might be just as many objects which can only fuse deuterium — a heavy isotope of hydrogen — as there are stars. But since brown dwarfs are much lower in mass than conventional stars, it is thought that they can contribute at most a fraction-of-a-percent to the total mass of a galaxy or cluster. These objects exist, but not in the abundance we would need in order for them to make up a significant portion of this missing mass (Fig. 9.8).

Lower mass objects, such as moon-or-planet-sized ones, can be detected via two different methods: transits and microlensing. Most of the time, planet-sized objects are completely imperceptible to us: the light they emit or absorb at all wavelengths is too small in magnitude to detect by almost any means. But every once in a while, dependent on how many of them there are, they will randomly pass across the line-of-sight of a distant star. Distance-wise, if they are very close to the star, they will transit across its disk, blocking a portion of its light for a certain amount of time, while if they are very far from it, their gravity will act like a lens, magnifying the starlight for a brief amount of time. We have actually detected planets via both of these methods, and what we have been able to do is constrain that the amount of mass present in our galaxy due to these objects is again less than 1% of what is needed to explain the galaxy's gravitation. These objects exist — and are called MACHOs, for MAssive Compact Halo Objects — but there are not enough of them to account for what we need.

Dust grains are also interesting, since we can detect them by their absorption parameters. In particular, because dust comes in grains of a particular size, it will interact differently with light of various wavelengths; when we measure absorption features, we can therefore tell what the size and concentration of dust is. Despite how prominent it appears to be when we look at any galaxy that contains it, the dust fraction in even the most obscured galaxy is only about 1% of the gas present; dust exists in negligible abundances when it comes to gravitation.

And finally, there are collapsed objects: white dwarfs, neutron stars and black holes. Black holes are some of the most massive objects in

Figure 9.8 Brown dwarfs, shown here in circles alongside the star field centered on the Sun, may be nearly as numerous as stars, but are lower in mass. They are a form of very low-luminosity mass that is made of normal matter, and they do exist in modest abundance, but they make up a negligible fraction of the missing mass. Image credit: NASA/JPL-Caltech/WISE.

the entire Universe, with some objects reaching into the billions or even *tens* of billions of solar masses. Unfortunately, these are not only preferentially clustered towards the centers of galaxies, which means they could not explain the rotation features we see, but there simply are not enough of them overall; there is far less mass in these stellar remnants than there are in stars today. These high-mass, low-luminosity objects, too, are all detectable via microlensing as well as other methods, and all make significantly less of a contribution to the overall mass than stars do (Fig. 9.9).

Adding together all of these sources as well as the others that astronomers have been able to uncover accounts for somewhere between 13–18% of what is required to explain the gravitational effects that we see. As far as we understand the cosmic energy inventory, normal matter — stuff made up of protons, neutrons and electrons — simply is not going to resolve this puzzle.

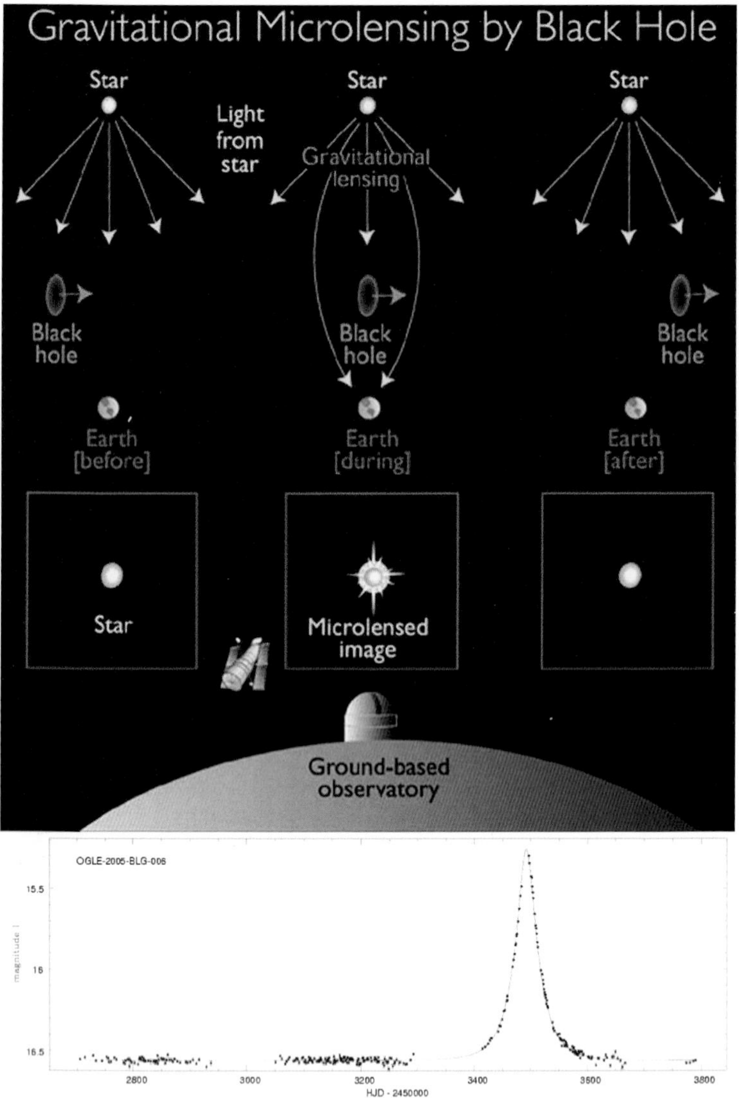

Figure 9.9 Microlensing occurs when a small, compact but significant mass passes between the line-of-sight of ourselves and an object. The bending of light around the object causes a temporary spike in the apparent brightness that then fades down to previously normal levels (top). Microlensing has been observed for objects ranging from the size of small planets all the way up to multiple solar masses, always leading to the same type of characteristic light curve (bottom). While an interesting phenomenon, none of its sources can account for the missing mass of the Universe. Images credit: NASA/ESA (top); OGLE/Jan Skowron (bottom).

* * *

Rather than count up all the different types of normal matter, there is another, more elegant way to approach this problem. You will remember that there was a time very early on in the Universe where it was too hot to form individual atomic nuclei; collisions with photons were energetic enough that any atomic nuclei that did form would be immediately split apart into individual protons and neutrons. When the Universe expanded — and the photons cooled — past a certain point, protons and neutrons then rapidly fused together into the light elements and their isotopes: deuterium, helium-3, helium-4 and lithium-7.

As it turns out, based on the laws of physics as we understand them, there is only *one* free parameter that determines what the ratios of these early isotopes will be: the ratio of the number of baryons (protons and neutrons combined) to the number of photons. Think about this for a minute. We can measure how much hydrogen, deuterium, helium-3, helium-4 and lithium-7 were present in the early Universe. We can accomplish this by measuring stellar populations that have undergone varying amounts of star formation and extrapolating back to a time when they had undergone *no* star formation at all. We can, in some rare cases, even measure pristine clouds of gas with bright sources (like quasars) behind them, determining things like the deuterium or helium-4 abundance directly. And just by observing the cosmic microwave background (CMB), we can measure the number of photons-per-unit-volume that were present back at any arbitrary time in the early Universe.

That leaves a whole set of data for a variety of independently measurable quantities and only *one* unknown: the abundance of baryons, or protons and neutrons, in the Universe. Since an electron is nearly 2,000 times lighter than a baryon, if we can measure the baryon density, we can determine what the total gravitational influence of normal matter is on the Universe! We can determine the density of normal matter, and see how close it comes to reproducing what we need gravitation to be in order to account for the galaxies and clusters we see (Fig. 9.10).

And to no one's surprise, consistent with our observations of all the individual components of the Universe, it falls well short, making up only about *one sixth* of the total amount of gravitational matter that needs to be there to account for the internal motions of galaxies and clusters. Although there is much more normal matter in the Universe than is

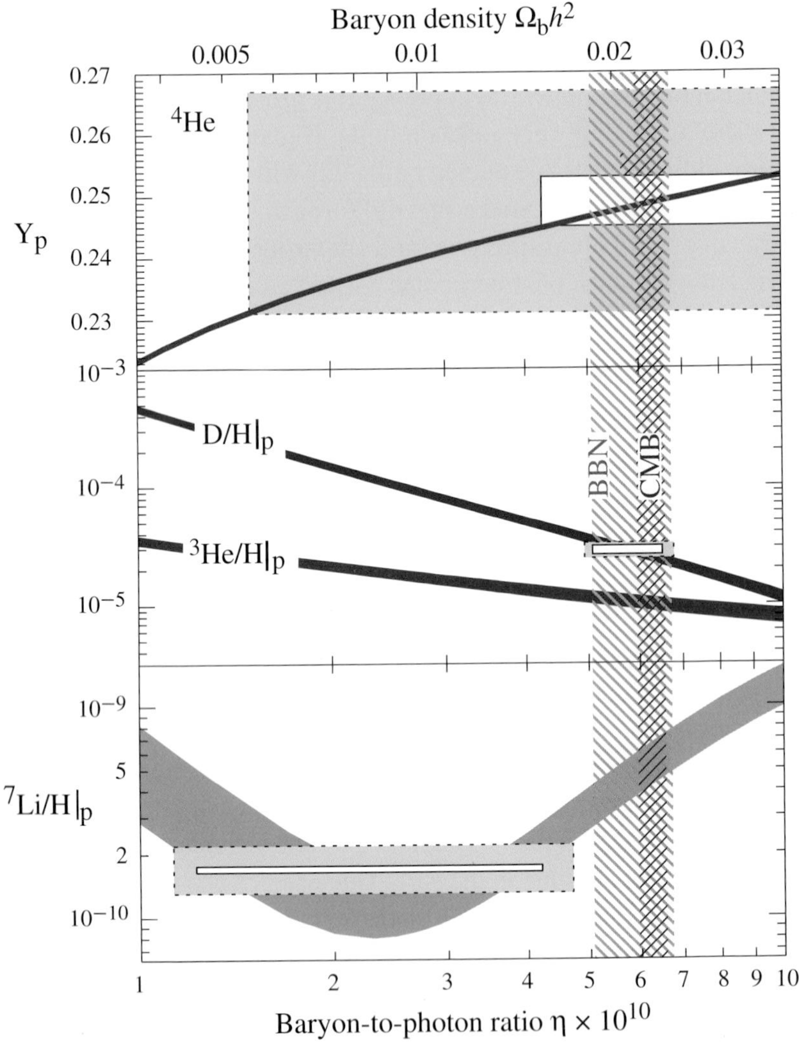

Figure 9.10 The astrophysically measured abundances for helium-4 (Y) and deuterium agree spectacularly with both the predictions of Big Bang Nucleosynthesis and observations from the WMAP and Planck satellites. The total amount of normal matter — that is, matter made of protons, neutrons and electrons — is too small to account for the mass observed in the Universe by approximately a factor of six. There is a slight discrepancy between predictions and observations of lithium-7, but this is thought to be due to its easily-destroyed nature. Image credit: Beringer *et al.* (Particle Data Group), *Physics Review* D86, 010001 (2012).

present in stars, there most definitely is not enough of it to account for the gravitational effects that we see. Something else must be at play in order to solve this puzzle.

* * *

One of the defining characteristics of normal matter is that it interacts with — by absorbing, emitting or generally colliding with — photons. This is incredibly useful for a number of scientific aspects, including:

- Observing an element's spectrum simply by exciting its atoms and watching what wavelengths it emits.
- Determining what elements are present in a cloud of gas simply by shining a light on them and seeing what wavelengths get absorbed.
- Detecting when two high-speed gas clouds collide by measuring the X-rays emitted from their excited atoms.
- And for transferring heat and kinetic energy from one hot system to an adjacent, cooler one.

If we go all the way back to the young Universe, where everything was much denser and hotter, matter's interaction properties with photons have an amazing application that you might not have expected.

You see, the Universe was *born* with overdense regions: places in space where there was just slightly more matter and energy than average. These overdense regions existed on all scales, from nanometers to meters to kilometers to light years to scales of millions or even billions of light years. If all you had was cold, motionless (or slow-moving) matter, you would expect that the overdense regions on the smallest scales would grow first, followed progressively by the larger ones, as gravitation — limited (like all things) by the speed of light — would take more time to grow the overdense regions on the largest scales. It is a reasonable thought, and it means that we would expect to have individual stars well before we had galaxies, and individual galaxies long before we had clusters.

But what happens if you throw radiation into the mix? What happens if you throw a sea of photons, and all its associated interactions, in with our matter? Whenever you create an overdense region, the photon pressure will increase, too, causing two things to simultaneously happen:

1) It causes the photons to push out against the overdensity, reducing the matter density back towards the "average" value.
2) The photons themselves stream out of the overdense region, reducing the energy density back towards the "average" value.

In short, a young Universe with more of its energy in the form of radiation *washes out* the small-scale overdensities, and brings everything back towards equilibrium (Fig. 9.11).

This phenomenon should show its effects in two different ways. First, when we look at the pattern of fluctuations in the CMB from the young Universe — a snapshot at 380,000 years of age — we should see a series of specific effects. We should see that the largest scales, the ones that are larger than the amount of distance the speed of light would allow gravity to travel in that time, should have fluctuations that appear truly *scale-invariant*. Without time for photons to push against the matter or stream out of the overdense regions, this effect will simply not occur on the largest scales. We should see that on scales smaller than that, there is a "peak" where gravity is pulling matter in towards the overdense regions, on scales that have not had time for that peak to hit a maximum and for photons to push it out, yet. And as we go to even smaller and smaller scales, we should see those fluctuations fall, with peaks and valleys that continue to decrease extremely rapidly each time we go to a smaller scale. Of course, this assumes that the Universe consists *only* of normal matter and of radiation. If there is a new type of matter there in addition to the normal matter — something that does not interact with radiation — it is going to elevate practically all of the later peaks, some from a near-zero level, and change the details of the spectrum of fluctuations in significant, measurable ways. If dark matter exists, and is something that does not interact with radiation, it should show itself to us in both the small-scale structure we can observe today and also in the fluctuations of the CMB at the earliest times.

With our current technology, we have been able to measure the spectrum of fluctuations in the CMB down to scales of 0.07°. (See Fig. 8.16.) Although we do see these peaking-and-falling features in the spectrum, the "falls" are much less dramatic than we would expect if there were only normal matter and photons present. In fact, there are additional peaks that

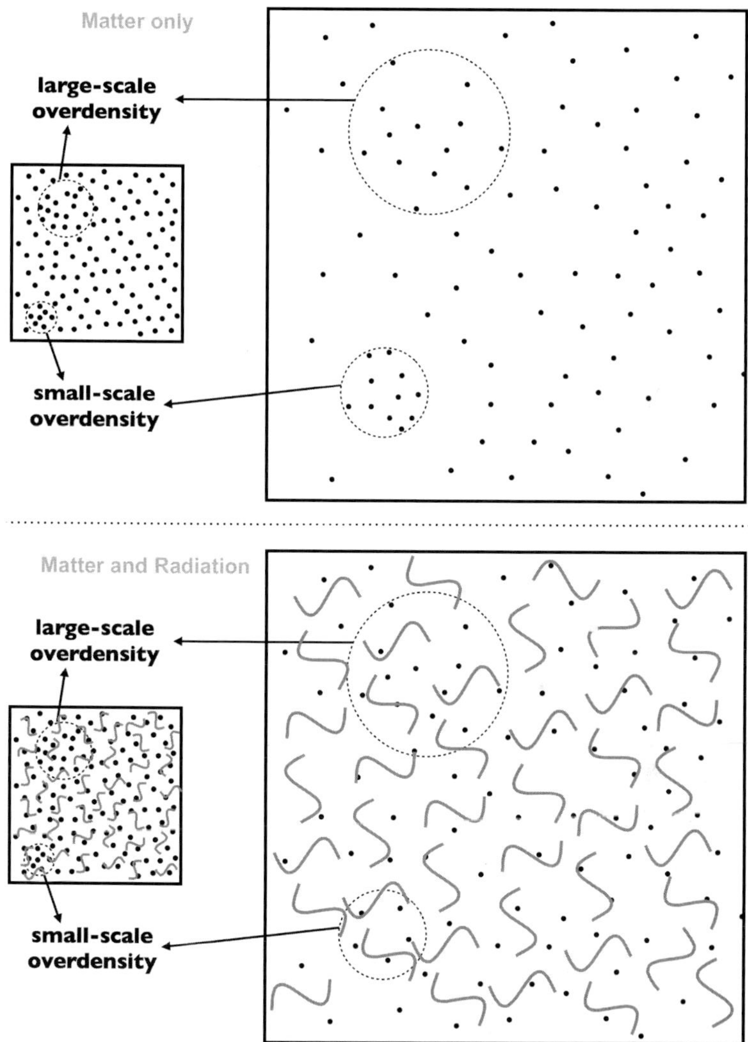

Figure 9.11 In an expanding Universe with only matter, the smallest-scale overdensities grow into clumps and clusters the fastest, since it takes less time for gravity to pull things that are already closer together in towards one another. Larger-scale overdensities take longer amounts of time, since gravitation requires more time to bring more distant masses close together. But in a Universe with matter and radiation, the hot radiation pushes back against the matter, washing out the smallest-scale structure and preventing it from growing. If this were all that were at play in our Universe, the smallest-sized galaxies in the Universe would have been unable to form. Image credit: E. Siegel.

would not have even appeared if all we had were normal matter and photons. What we see, instead, is evidence that there is an additional component to the energy in the Universe: there is some type of matter that is not being pushed out by the photons the same way that protons, neutrons and electrons are. And just like our other lines of evidence, we see that same suspicious figure: about 80–85% of that matter is some type of dark matter, completely distinct from anything in the Standard Model.

But there is another effect that should be even more dramatic. Not only will the initial, *small* fluctuations be affected by the interaction between matter and radiation, but the way the large-scale structure of the Universe forms and grows over time will be impacted tremendously! In particular, the smallest-scale structures would not exist if normal matter and radiation were the only major ingredients in our Universe. The radiation stays hot and its pressure remains significant for not only a few hundred thousand years but many millions of them, which will prevent star clusters and galaxies from forming early on. By using incredibly powerful telescopes like Hubble, combined with the magnifying effect of gravitational lensing, we can probe these great distances. When we do so, we find that these small, faint galaxies did, in fact, exist at early times (Fig. 9.12).

In addition to the smallest scales, a Universe consisting of normal matter and radiation alone will show huge suppressions of structure on even relatively larger scales, preventing galaxies from clustering together at specific distances. Our Universe — without a new kind of dark matter — is predicted to have predominantly large galaxy clusters of specific sizes and scales, with very few isolated groups of galaxies (like our own) and practically no small galaxies by themselves. These features — known as either Silk damping or Baryon Acoustic Oscillations — all stem from the interaction of normal matter with photons. The only way around it, again, is if there is some new type of dark matter in addition to the normal matter, which changes the story dramatically.

Our largest-scale surveys of structure in the Universe can discriminate between these possibilities, and can do so with tremendous precision. We can go back to the very beginning of the Universe and run simulations that reproduce the structure of the Universe on galaxy, cluster, and even larger scales. We can run simulations with various amounts of dark matter, with various types of dark matter and with no dark matter at all, and compare

Figure 9.12 The luminous red galaxy at the left of the image acts as a gravitational lens, magnifying and stretching the light from ultra-distant galaxies behind it. These small, dim but very distant galaxies show us that small galaxies *did*, in fact, form very early on in the Universe, on scales below what would be predicted if the Universe consisted of normal matter and radiation alone. Image credit: ESA/Hubble & NASA.

them to the Universe we actually observe. And we can do it in great detail and on a huge variety of distance scales.

When we do all this, we find that a Universe without dark matter would be *irreconcilably* different from the Universe we observe. The magnitude of the oscillations that we would get from a Universe whose matter was made up solely of baryons would be far too great when compared to what's observed. The amount of power (the technical term for numbers of galaxies we see) on small-scales would be suppressed well below what we actually observe, and the details of large-scale structure would be all wrong. And yet, if we are willing to throw dark matter into the mix, we find that our observations and our predictions line up perfectly if we add five times the amount of dark matter along with the normal matter in our simulations (Fig. 9.13).

On all these fronts, adding the same amount of dark matter seems to solve all our problems.

* * *

There is one more measurement we can make to get a handle on the total amount of mass in a galaxy or cluster of galaxies, and it is a measurement we can make *directly*. Rather than having to rely on simulations or iterative calculations, there is a very simple prediction that General Relativity makes: that when a distant source of light passes through a region of space with a significant amount of mass, that light gets magnified, distorted, and possibly either stretched into multiple images or a ring, depending on the mass configuration. This phenomenon of gravitational lensing, which goes all the way back to the aforementioned maverick of astrophysics, Fritz Zwicky, comes in two varieties: **strong lensing** and **weak lensing**. Despite being predicted as far back as the 1930s, it was not seen observationally for the first time until 1979! Since the late 1970s, however, gravitational lensing has been observed in a very large number of situations, enabling us to measure not only the total mass of an object simply through the physics of General Relativity, but also the mass distribution of certain galaxies, quasars and galaxy clusters.

Strong lensing is the easier type to understand and to detect, as its signatures are unmistakable to even an untrained observer. When we look out into the Universe, compact sources of mass are seemingly everywhere. In many cases — such as for individual stars — these sources emit significant amounts of their own light, but in other cases, the amount of light they emit is small, particularly compared to the large amount of mass in there. Obvious examples include neutron stars and black holes, but there are much larger, more diffuse cases as well. For example, consider the case of a dusty, edge-on galaxy, from which we might see a luminosity that is 10 billion times greater than our own Sun, but which might have a *trillion* solar masses of total matter residing inside. The overall luminosity might be huge, but the mass-to-light ratio is what matters here. With so

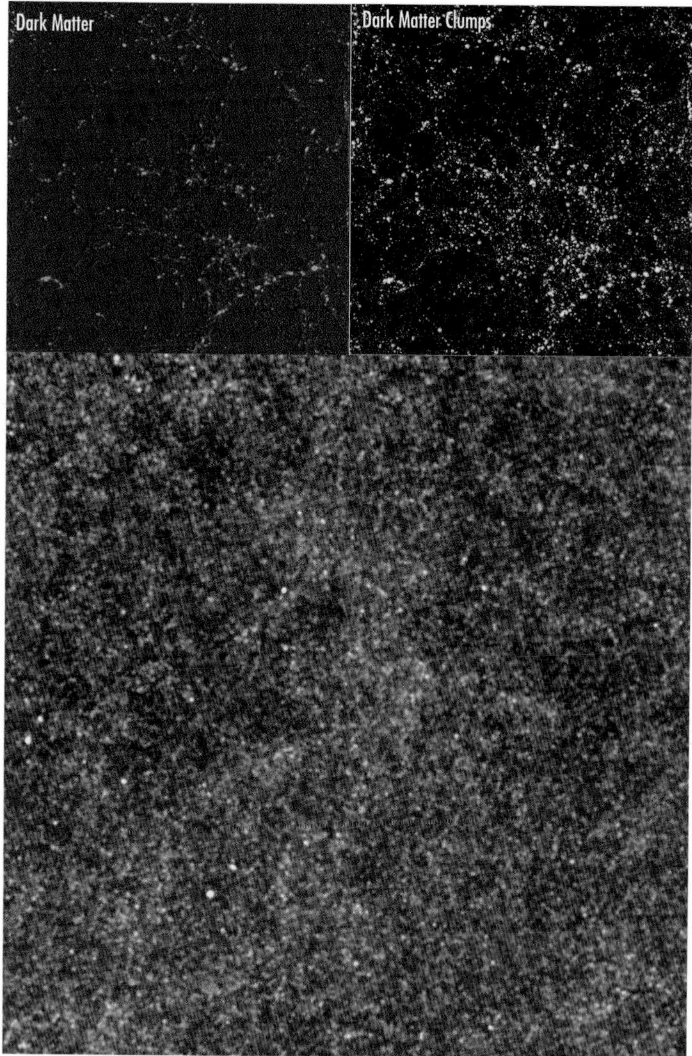

Figure 9.13 The top panels represent a simulation containing large amounts of dark matter of the Universe on the largest scales. All matter is shown in the top left panel. In the top right panel, the objects that ought to represent significantly large galaxies — those about 15% the size of the Milky Way and larger — are teased out of the simulated data in yellow. At the bottom, an actual infrared image of galaxies populating a region of space known as the Lockman Hole are shown, taken with ESA's Herschel space observatory. The statistical agreement between simulations and observations provides strong, conclusive evidence that a large amount of dark matter is necessary to accurately describe our Universe. Images credit: Virgo consortium/A. Amblard/ESA (top panels); ESA/Herschel/SPIRE/HerMES (bottom).

much more mass than its observed luminosity would indicate, the ability of such a galaxy to act as a strong gravitational lens is tremendous. For that matter, sometimes an entire galaxy cluster — particularly if it's distant enough — can function as a gravitational lens! (Fig. 9.14)

All you need for strong lensing is for that lens to be *in between* you, the observer, and whatever source of light it is that you are attempting to observe. If there is a distant star, galaxy or quasar behind your lens, you will see that observed object undergo a magnification, a stretching (or distortion) of the image along a circle centered on the lens, and the possibility of multiple images, where two (or more) light paths can reach your eyes via entirely different routes, thanks to the bending of space. Strong lensing is a remarkable tool, because it allows us to so accurately reconstruct what the mass of the object is that behaves as a lens, giving us a direct mechanism for measuring the mass of a distant object.

Less visually striking but perhaps even more important for astrophysics is the other major method: weak gravitational lensing. Rather than a compact, point-like source acting as a lens, weak lensing is when an extended source — a source whose mass is distributed over a large angular region of the sky — has luminous objects behind it that need to pass through that lens in order to reach our eyes (and telescopes). The weak lens is not capable of significant magnification, nor of stretching the background objects into a circular path, nor of even creating multiple images. What it can do is distort the background galaxies into elliptical shapes in a way that is dependent on the mass distribution of the weak gravitational lens. This distortion, for a single galaxy, would do us no good, since galaxies come in all sorts of shapes naturally. But so long as there are large numbers of objects in the background, we can take advantage of this method, since we know that statistically, distant galaxies are likely to be randomly oriented. This allows us the ability to reconstruct not only what the total mass is of the cluster acting like a gravitational lens, but to determine how that mass is distributed (Fig. 9.15).

For both weak and strong gravitational lenses, we find that the total amount of mass present must be about five to six times more than normal matter — particles in the Standard Model — can account for.

* * *

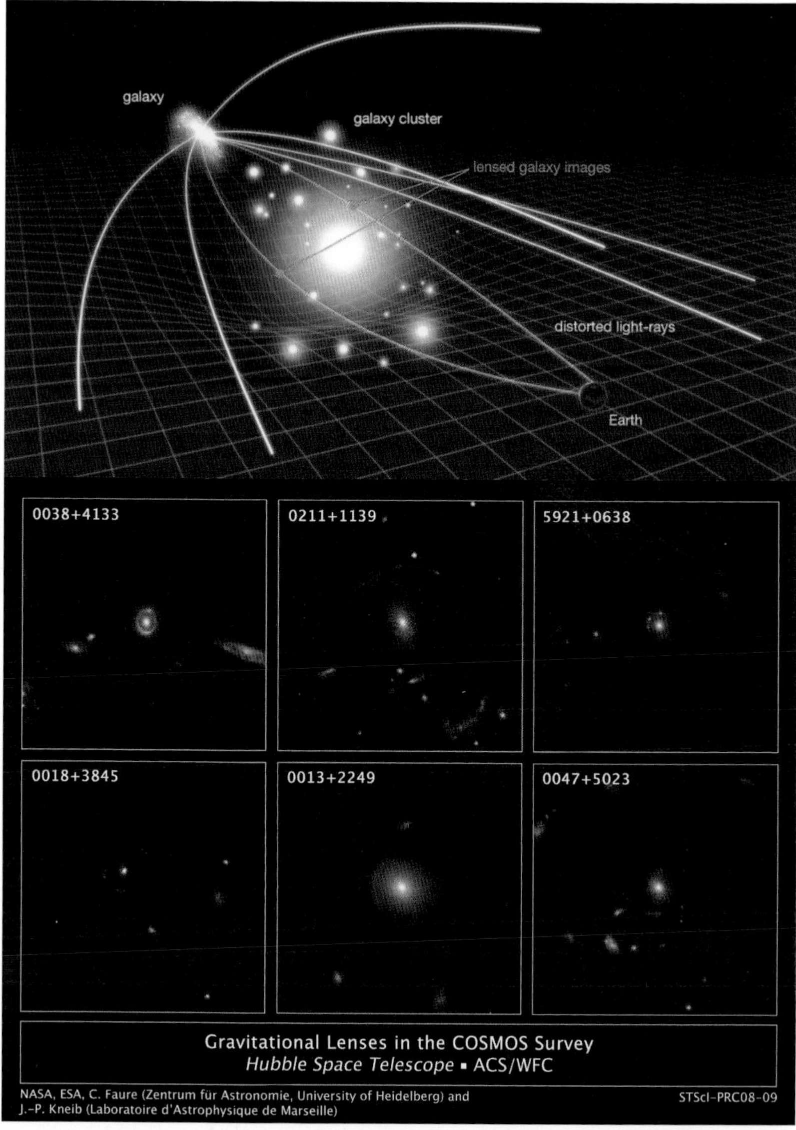

Figure 9.14 A very strong gravitational source can bend spacetime so significantly that a background galaxy or quasar can often arrive at your eyes via multiple light paths, sometimes as multiple distinct images or sometimes bent into a ring or an arc (top). At bottom are six (of many) strong gravitational lenses discovered by the Hubble space telescope. Images credit: NASA/ESA (top); NASA, ESA, C. Faure (Zentrum für Astronomie, University of Heidelberg) and J.-P. Kneib (Laboratoire d'Astrophysique de Marseille).

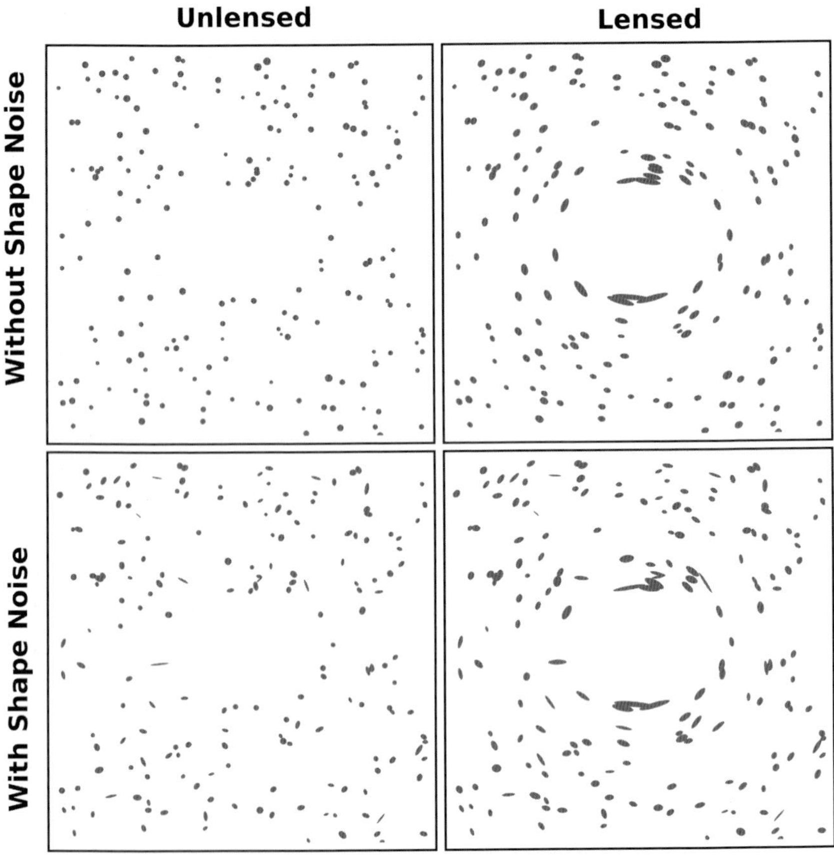

Figure 9.15 Galaxies should be distributed in some fashion behind a large, extended galaxy cluster (upper left), whose gravity causes their shapes to become distorted thanks to weak gravitational lensing (upper right). Although real galaxies are more complicated than simple points (lower left), the lensing signature is still unmistakable, and can be teased out of the data so long as large enough numbers of background galaxies are available (lower right). Image credit: Wikimedia Commons user TallJimbo, under c.c.-by-s.a.-3.0.

These multiple lines of evidence, originating from several completely independent observations, all indicate that normal matter and the laws of gravity (General Relativity) cannot explain the full suite of observations that we have. It seems that simply adding dark matter, a new kind of matter that does not collide with matter or radiation, in about five times the abundance of normal matter, can explain everything we observe. But hypothesizing that most of the matter in the Universe is *invisible* is a great leap indeed. There is another possibility worth considering: what if it is not the addition of a new type of matter that is the solution to all of these puzzles, but a further modification to the laws of gravity? After all, General Relativity has only been directly tested on scales of the Solar System and smaller; it could well be that on scales the size of a galaxy or larger, it is the laws of gravitation that need modification, not the matter content of the Universe!

By the end of the 1970s, the galactic rotation problem was the one puzzle that had the most data behind it, with the rotation profiles of dozens of galaxies having been measured. As the stars, gas and dust clearly indicate, the density of normal matter is much greater towards the *center* of spiral galaxies than in the outer regions. If this were the only matter present — if the normal matter determined gravitation — we would expect the observed rotational speeds to *fall* as we moved away from the centers of galaxies towards the outskirts. Not only does this not appear to be the case, but the speeds do not fall even as we measure out to the observable limits of these galaxies! If we expect dark matter to account for this, each galaxy would require a halo of dark matter around it that is far larger than the extent of the luminous matter itself. But there came another idea: what if all the matter that was present was the normal, atomic matter, but with a small tweak to the law of gravity? As it turned out, there was a very suggestive — and brilliant — correlation that was noticed: rather than allowing the acceleration of any system to solely be determined by the matter-and-energy present, what if a very small *but non-zero* value was added to the acceleration of each part of the system?

This would alter *Newtonian* gravity, not to mention Einstein's, unless the extra value of acceleration were so small that it had no effects on Solar System-scale measurements. This single addition of a minimum, non-zero acceleration would not only explain the observed rotational motion of every galaxy, but the value of the acceleration needed to do so turns out to be *the same* for each and every galaxy we had measured. This idea — of MOdified Newtonian Dynamics (or MOND) — became the start of the major alternative idea to dark matter: modified gravity. First

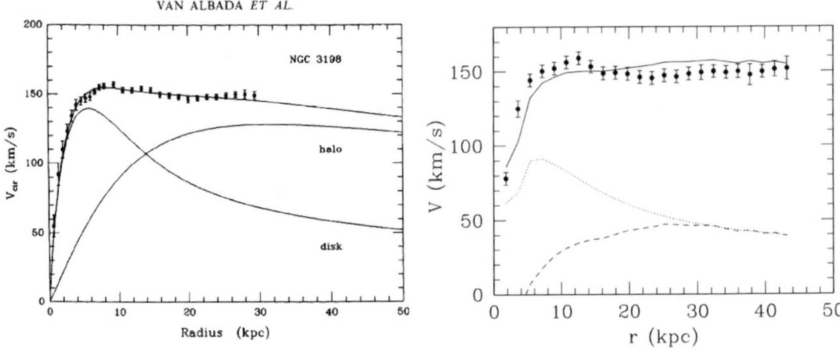

Figure 9.16 While the phenomenon of galactic rotation curves can be fitted by adding a diffuse halo of dark matter (left), they can also — and often *better* — be fitted by making a slight modification to Newtonian gravity at very small accelerations (right). In the figure at left, the normal matter (in the disc) and the dark matter (in the halo) are clearly marked, while at right, the gas in the thin disk and beyond the plane of the galaxy is shown by the dashed line, the dotted line shows the stellar disk, and the solid line shows the predicted curve due to MOND. The data points extend farther out in the graph at right due to additional measurements published in 2008. Images credit: T. S. van Albada, J. N. Bahcall, K. Begeman and R. Sancisi, (1985). *Astrophysical Journal*, 295, 305 (L); G. Gentile *et al.* (2013). *Astronomy & Astrophysics*, 554, A125 (R).

put forth in 1980 by Moti Milgrom, MOND was never intended to be a complete "alternative" to General Relativity with dark matter, but rather a compelling starting point that might lead to a more complete theory (Fig. 9.16).

The idea is sound: surely hypothesizing that 80–85% of the matter in the Universe is of some hitherto undiscovered type, different from all the particles known to exist, represents a *greater* leap than making a tweak to our theory of gravity. After all, tweaking our theory of gravity to explain Mercury's orbital motion was what led to General Relativity in the first place! But it has proven to be very difficult to modify General Relativity in a way that is still consistent with the full suite of our observations, and so translating MOND into a viable theory of gravity has so far proven elusive.

People have tried adding new fields to the way gravity works; some of the best attempts have been the Modified Gravity (MoG) theory put forth by John Moffat and the Tensor-Vector-Scalar (TeVeS) theory put forth by

Jakob Bekenstein, but both theories fail to reproduce the large-scale successes of dark matter with General Relativity. In particular, gravitational lensing, the cosmic web of structure and CMB observations all go unexplained in all the modified gravity theories put forth so far, with the mismatch between predictions and observations of large-scale structure being particularly egregious. The only way, thus far, to save such gravity-modifying theories *is to introduce large amounts of dark matter*, which defeats the idea's purpose altogether! MOND (and the exploration of an alternative theory of gravity) remains an attractive avenue of investigation, as it is still more successful at predicting the rotation curves of individual galaxies, overall, than the theory of dark matter is. But its failure to meet the criteria of reproducing the successes of the already-established leading theory means it has not yet risen to the status of scientifically viable.

Nonetheless, the idea that our theory of gravity may be incomplete and in need of modification remains an attractive one, and will persist as the leading alternative until dark matter is detected directly. But back in 2006, an *indirect* test provided the strongest evidence to date that dark matter must be real.

* * *

Imagine what tests you'd perform if the Universe were your own experimental laboratory, rather than being something we could only make passive observations of. What would you force it to do to detect the presence or absence of dark matter? If you were clever, you might come up with the bright idea of taking two *huge* collections of matter and slamming them together at very high speeds. The largest collections of matter we see are giant galaxy clusters — vast regions of space with hundreds or even thousands of large galaxies inside — all bound together by their mutual gravity.

Even in a large cluster, the regions containing galaxies are tiny; the vast majority of the volume in a cluster is devoid of stars. There are big, massive elliptical galaxies near each cluster's center, with a smattering of spirals that become more and more prevalent as we move towards the outskirts. Within each cluster, there is also a large amount of neutral gas: matter that will form future generations of stars. This gas is diffuse,

Figure 9.17 The visible matter in a galaxy, a galaxy cluster or set of clusters comes from the individual stars within galaxies, but the dark matter can be revealed through the weak gravitational lensing of all the objects behind it. Here, four different galaxy clusters located very close together have their dark matter mapped in pink (along with the dark matter in the surrounding regions) via imaging taken from the Hubble space telescope. Image credit: NASA, ESA, C. Heymans (University of British Columbia, Vancouver), M. Gray (University of Nottingham, U.K.), M. Barden (Innsbruck) and the STAGES collaboration.

existing in sparse clouds surrounding each of these galaxies, and in many cases, in the intergalactic medium between the galaxies. Now, if dark matter is present, there should be about five times as much of it as all the normal matter combined, distributed not only in halos around the individual galaxies, but in one large, diffuse halo around the cluster itself (Fig. 9.17).

Imagine, now, taking *two* of these clusters and smashing them together at very high speeds. What would you expect to see? Remember, irrespective of dark matter, there is about six times as much matter in neutral gas as compared to stars, and — if dark matter is correct — there is about five times as much dark matter as there is normal matter overall. There are three major things to consider:

1) **The stars, and the individual galaxies that house them**: a star is a rather large entity, with each one of the roughly hundreds-of-billions

in each galaxy averaging a little over ten million kilometers in diameter. The galaxies themselves are even larger, averaging around 100,000 light years in diameter. But the galaxies are spread out over tremendously huge volumes of space in a cluster, which typically spans around *ten million* light years across. If you collide two such clusters at high speeds, even assuming there are thousands of galaxies in each, you're likely to have only a handful (much less than 1%) of the galaxies actually collide; the vast majority will pass right through one another. For objects such as these, it is like taking two guns filled with bird shot and shooting them simultaneously from across a field at one another. Yes, occasionally you will get two pellets that hit each other, but the overwhelming majority of them will simply pass by one another, and that is what happens to individual galaxies and the stars within them inside a cluster.

2) **The gas, both surrounding each galaxy and in each cluster overall**: this is a vastly different story from the stars. The stars and galaxies are compact entities, taking up a very small amount of volume in relation to the entire cluster. When two clusters collide, most galaxies and practically all the stars simply pass by one another, unnoticed. But the neutral gas is *much* more diffuse, and fills pretty much the entire volume of the cluster, albeit at a very low density. When two clusters collide, the gas from the two clusters mutually interacts. Because these clusters are moving at high speeds relative to each other, the gas within them has large amounts of kinetic energy, and so the molecules collide energetically, heating up, *slowing down* and causing the emission of X-rays. If two clusters collide, we should see the gas lagging behind the galaxies, and we should be able to detect this gas through an X-ray signature.

3) **Dark matter, both surrounding each galaxy and for the respective clusters overall**: this starts off as a similar story to the gas, except with the halos being even larger in size and more diffuse. The dark matter fills the entire volume of the cluster, and extends out beyond even what the gas fills. But unlike the gas, because dark matter does not have the same type of interactions (specifically, electromagnetic interactions) that normal matter does, dark matter does not collide, either with itself or normal matter. That means it does not heat up, it

does not emit light, and most importantly, *it does not slow down*. The dark matter — much like the individual stars and galaxies — ought to pass right through itself, the gas and the galaxies, winding up clear on the other side (Fig. 9.18).

The reason this is such an important scenario to consider is not because we can orchestrate such a collision; of course we can do no such thing. But considering that there are literally expected to be many tens of millions of galaxy clusters in our Universe, there are bound to be examples

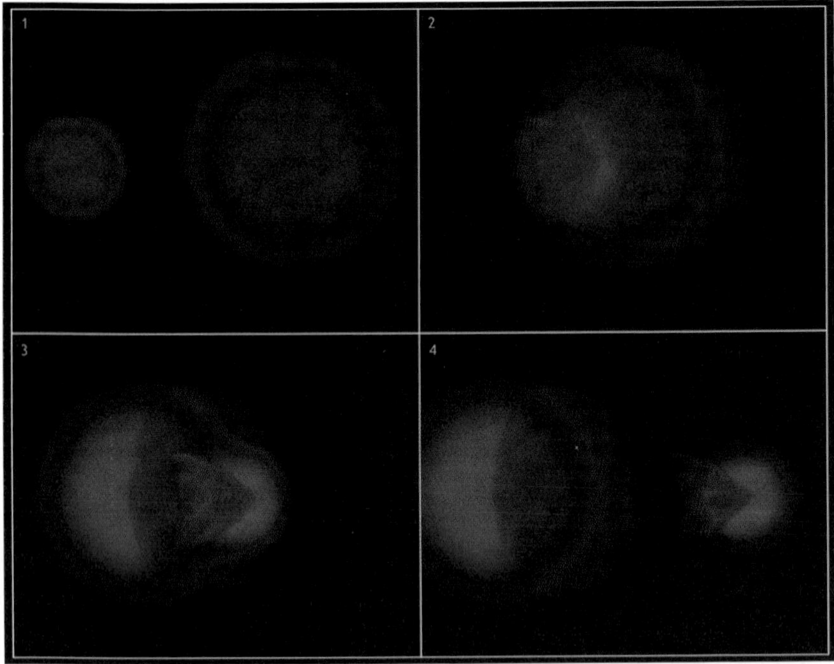

Figure 9.18 If you had two large galaxy clusters, you would expect them to consist of a mix of dark matter (in blue) and normal matter, mostly in the form of gas-and-plasma (in red). When they collide with one another, the dark matter would be expected to pass right through the other cluster's dark matter and normal matter, while the normal matter should interact, heat up, slow down and emit X-rays. If we can observe this separation between the light emitted from normal matter and the gravitational effects of the dark matter, that would be an ideal astrophysical test of dark matter's existence. Image credit: NASA/CXC/M. Weiss.

of clusters colliding that are visible to us today. The overwhelming majority of the normal matter in clusters is found in the form of gas, but dark matter — if it exists — will vastly outweigh the normal matter, giving us an unambiguous sign to seek out!

If we can find clusters of galaxies that have collided recently, sometime in the past, we would expect to see an X-ray signature that is displaced from the galaxies in the colliding clusters. The gas is going to follow a different trajectory than the individual galaxies, and so the light (from the galaxies) would not match the X-ray light from the gas. We would also expect the gravitational signature — which we can reconstruct from weak gravitational lensing — to be displaced from the X-rays if there *is* dark matter. On the contrary, if there *were no* dark matter, we would expect the gravitational signature to follow the X-rays and be displaced from the galaxies!

In 2006, the first gravitational lensing map of the colliding galaxy cluster 1E0657-558, colloquially known as the "Bullet cluster" after the high collisional speeds between the two clusters involved, was completed. As anticipated, a comparison of the X-ray map (in pink) from the Chandra X-ray telescope showed a significant offset from the stars and galaxies composing the visible portion of the cluster. But what about the lensing map? It revealed (in blue) that the gravitational signal from this cluster was concentrated in the two clumps *that passed right through one another*. The vast majority of the mass appeared to be located where the individual galaxies were, well-separated from the X-rays and the gaseous normal matter. For the first time, we had direct, empirical evidence that the presence of normal matter alone — on scales of galaxy clusters — did *not* determine the majority of a galaxy cluster's gravitational effects (Fig. 9.19).

Subsequently, other galaxy clusters in various stages of major mergers have been discovered, and have shown a similar mismatch between their X-ray emissions (indicating the abundance of gas) and their gravitational lensing signals (indicating the overall mass). Famous examples include clusters Abell 520, DLSCL J0916.2+2951 (the Musket Ball Cluster), MACS J0717 and MACS J0025.4-1222, all located billions of light years away from Earth. These sets of observations pose a colossal obstacle for modified gravity theories, as they would not only need a mechanism to displace gravitation from the location of the matter, but they would have to discriminate between the way a gravitational signal should look

Figure 9.19 The Bullet cluster, shown here, is actually two separate clusters that have recently collided at a high velocity. Overlaid atop the optical image is the reconstructed gravitational lensing signature in blue, which shows the mass distributed in two well-separated clumps, and the X-ray signature in pink, which shows the highly shocked gas lagging behind the gravitational source. Clusters like this, where there is a clear difference between the physical location of the normal matter and the location of the (inferred) gravitational matter, provide the most direct evidence for dark matter to date. Image credit: X-ray: NASA/CXC/CfA/M. Markevitch *et al.*; Lensing Map: NASA/STScI; ESO WFI; Magellan/U. Arizona/D. Clowe *et al.* Optical: NASA/STScI; Magellan/U. Arizona/D. Clowe *et al.*

immediately preceding, during, and subsequent to a cluster collision. Such a satisfactory explanation has yet to arise (Fig. 9.20).

* * *

But there is still evidence that we require before the idea of dark matter can be taken as a certainty. For all of the successes of dark matter in predicting large-scale structure, in explaining the fluctuations in the CMB, and in accounting for astronomical phenomena in general, we have yet to directly detect and isolate the particle responsible for it! We know it

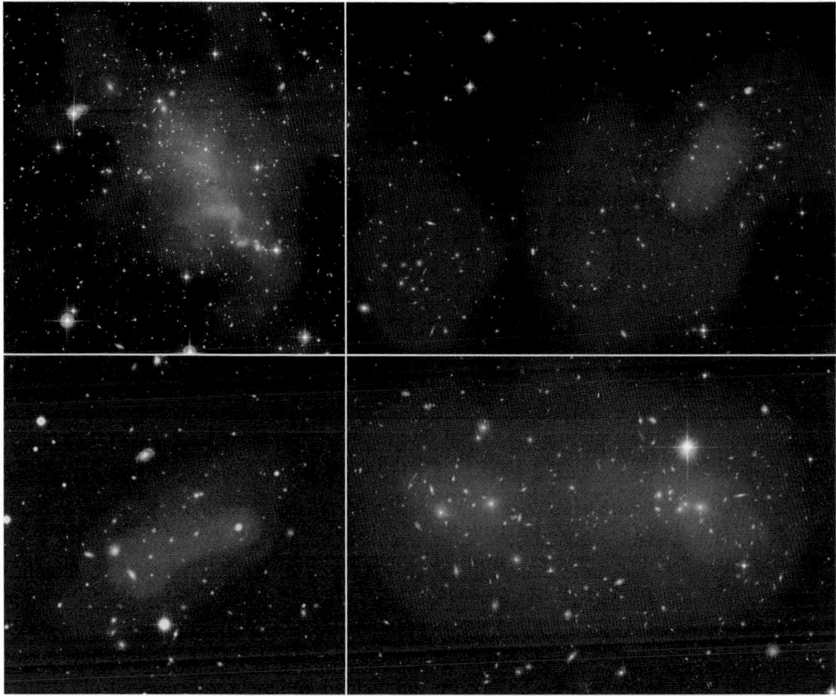

Figure 9.20 In each case, these four colliding galaxy clusters show a separation between the optical signal and the dark matter (in blue) from the X-ray gas (in pink). If the gravitational signal were due to the normal matter, you would expect that the lensing and X-ray signatures would align; the fact that they do not is very strong evidence for the existence of dark matter, and that the dark matter is something other than the known particles in the Universe. Clockwise from top left, the colliding clusters are Abell 520, the Musket Ball Cluster, MACS J0025.4-1222 and SL2S J08544–0121, the last of which is the smallest cluster ever discovered to display this effect. The geometries of these clusters are all so different from one another due to the fact that they are in different stages of a major merger, where the youngest cluster collisions began only a few hundred million years ago, while the oldest have endured nearly two billion years since their major collision. Images credit: X-ray: NASA/CXC/UVic./A.Mahdavi *et al.* Optical/Lensing: CFHT/UVic./A. Mahdavi *et al.* (top left); X-ray: NASA/CXC/UCDavis/W.Dawson *et al.*; Optical: NASA/STScI/UCDavis/ W.Dawson *et al.* (top right); ESA/XMM-Newton/F. Gastaldello (INAF/IASF, Milano, Italy)/CFHTLS (bottom left); X-ray: NASA, ESA, CXC, M. Bradac (University of California, Santa Barbara), and S. Allen (Stanford University) (bottom right).

cannot be any of the particles in the Standard Model, as they all either interact through the electromagnetic force (quarks and the charged leptons), are unstable (the heavy bosons), or are far too light (neutrinos) to account for the missing mass. If we want to convince ourselves beyond any reasonable doubt that dark matter is real, we have to find this new particle (or, possibly, sets of particles) directly.

Because of how good our observations are, however, the particle responsible for dark matter cannot just have *any* sorts of properties. For example, dark matter particles cannot have a charge under either the strong or electromagnetic force, otherwise their interactions would be powerful enough that they would have been directly detected. Dark matter needs to be cold enough, meaning that it has to either have a rest mass that is at least 2% of the electron's mass or a mechanism for it to have been created with practically zero kinetic energy, otherwise the formation of the small-scale structure we see would have been too greatly suppressed. Dark matter cannot "stick" to itself the way that nuclei or atoms do, meaning that there are huge constraints on its self-interacting cross-section. And yet it *must* have a rest mass, it must interact according to the gravitational force, and if it at all couples to the weak force — the force responsible for neutrino interactions and for radioactive decay — it has to be *billions* of times weaker than any other particles experiencing the weak force.

That said, there are two main classes of particles that, theoretically, ought to be seriously considered (and searched for) as cold dark matter candidates. The first is a type of stable particle that was created along with all the normal ones during the hot Big Bang, when there was enough available energy that all interactions happened readily and spontaneously. In short order, as the Universe expanded and cooled, interactions for this new particle ceased, and if there were other, unstable particles created as well, many of them could have decayed into the new, stable dark matter candidate. The particle would have to be massive enough that it would support (and not suppress) the formation of small-scale structure, and could interact through the gravitational force and possibly, *very* slightly, through another force like the weak force. We call this class of particles, generically, Weakly Interacting Massive Particles, or WIMPs.

There are a number of theoretical candidate particles that could be WIMPs, and they arise naturally in models with supersymmetry, in models with extra dimensions, or in models with new types of heavy neutrinos. There are presently a number of experiments searching for dark matter of exactly these types, including CDMS, XENON, Edelweiss, DAMA/LIBRA, LSND, Mini-BooNE and LUX. They are variously searching for telltale nuclear recoils, signs of "missing energy" of a specific amount in colliders, and novel interaction pathways. Despite a number of suggestive signals that have arisen from a variety of experiments, the other experiments operating in that range have failed to confirm the existence of any dark matter candidates. Two improved-sensitivity experiments — SuperCDMS and LZ (successors to CDMS and LUX, respectively) — will push the limits of the possible dark matter cross-sections down even further, or possibly even detect and measure them! As of today, however, no dark matter particle (or WIMP in general) has yet been discovered (Fig. 9.21).

The second class of dark matter candidate particle is similar, in the sense that it would be a form of dark matter, it would have a non-zero

Figure 9.21 The LUX detector (at left) and the ADMX detector (at right) are two of the many experiments attempting to detect dark matter particles directly. While many independent teams are searching for new particles, no direct detection experiment has given a confirmed, positive signature of dark matter. Whatever its nature is, we still have not found it just yet. Images credit: Wikimedia Commons user Gigaparsec (L); Wikimedia Commons user Lamestlamer (R), both under c.c.-by-s.a.-3.0.

rest mass, it would be cold (with little kinetic energy compared to its mass), it would interact through the gravitational force, and any other interactions would be very weak via any other force. But this type of dark matter candidate was not born with the other particles in the Big Bang, but rather came about as the Universe cooled through a certain transition. There are many types of symmetries that are restored at higher energies — the electroweak symmetry, for example, a threshold above which the electromagnetic and weak forces behave the same as one another — that are *broken* at lower energies. We can hypothesize a new type of symmetry that leads to the spontaneous production of a new class of particles at rest once that symmetry is broken: the Peccei–Quinn symmetry, which produces very light, massive particles known as axions. If axions exist, they should couple extraordinarily weakly (but in a very specific, detectable fashion) to photons, meaning that an electromagnetic cavity with the right properties should be able to detect it. There is an experiment actively searching for exactly this signal — ADMX (the Axion Dark Matter eXperiment) — and is presently gearing up for an enhancement that will make it even more sensitive. Thus far, these types of experiments have only constrained axions; none have ever been detected.

Many other types of even more exotic candidates abound, including WIMPzillas, gravitinos and Q-balls, among others. Direct searches from dedicated experiments are ongoing, and teams are looking for signals both from colliders and from astrophysical observations that would indicate the existence of any new, novel particle. So far, however, no new fundamental particles outside the standard model have yet been discovered. The exact nature of dark matter remains a mystery.

* * *

What we have learned from all of this is that when we take stock of all the luminous matter in the Universe — matter like stars that emit their own light — it cannot come close to accounting for the gravitational effects we see on scales of galaxies, clusters, and even larger cosmic structures. Even if we include all other possible sources of normal matter, including gas, dust, plasma, black holes, planets, asteroids and more, raising the total amount of normal matter that must be present significantly, all of that still cannot account for the observed effects. Furthermore, the synthesis of the light elements in the early stages of the Universe places a strong constraint on the amount of normal matter that can be present overall. The Universe seems to require an additional ingredient: either a modification to the theory of gravitation or a new type of matter whose mass abundance greatly outstrips all the "normal" matter combined.

If we apply the same three criteria we always have to new ideas, we would find that they would have to do the following:

1) Explain equally well all the phenomena that the previous leading theory (General Relativity without dark matter) could.
2) Provide explanations for the full suite of new phenomena that the prior theory could not: galactic rotation, individual galactic motions within clusters, the CMB fluctuations and large-scale structure formation.
3) Make new predictions that had never been tested but could be verified, such as an observed physical separation of gravitational effects from the majority of normal matter.

The dark matter explanation can satisfy all three of these criteria, while — thus far — modifications of gravity cannot even fully satisfy a single one. Although we are not certain what the nature of dark matter actually is, we can be incredibly confident of its existence, particularly when we look at how galaxies cluster *on average* on the largest cosmic scales (Fig. 9.22).

Still, the quest to detect the particle responsible for the Universe's missing mass is the next logical scientific step to take, so searches continue both at colliders and through dedicated direct detection experiments. If nature is kind to us, dark matter may reveal itself to us during the next generation of searches. However, it could just as easily be the case that dark matter's cross-sections for doing things like annihilating with itself, getting produced in colliders and scattering off of normal matter are well below what even the loftiest practical plans are for detection. For the time being, we have to accept that we simply do not know everything we would like about the source of the majority of mass in the Universe.

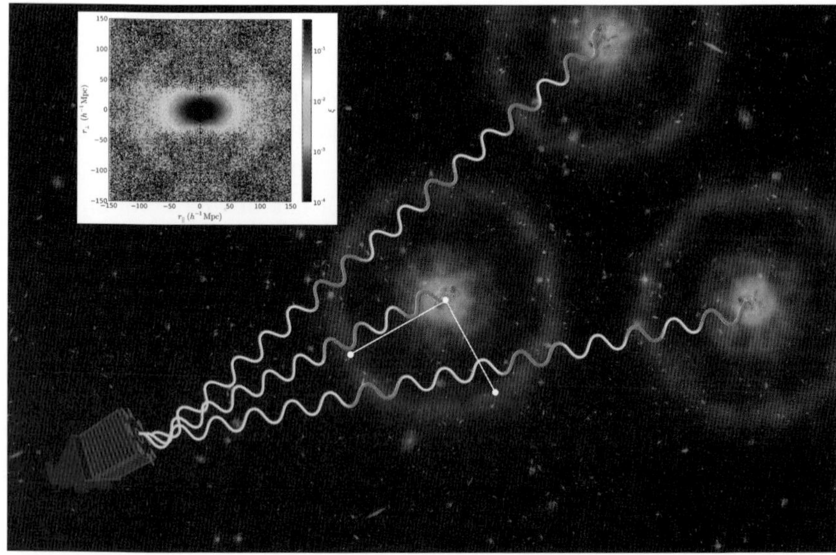

Figure 9.22 One additional prediction that comes with a Universe that is composed mostly of dark matter is that wherever you have a galaxy, you are slightly more likely to have another galaxy at a distance of 500 million light years from that galaxy than at either 400 million or 600 million light years. This bizarre prediction is due to the combination of Baryon Acoustic Oscillations in the presence of dark matter. Observations (inset) show an increased likelihood where the faint circle appears, and once again points to a Universe whose matter content is roughly 85% dark matter and 15% normal matter. Images credit: Zosia Rostomian (Lawrence Berkeley National Laboratory) and Andreu Font-Ribera (BOSS Lyman-alpha team, Berkeley Lab), with an inset image by L. Samushia *et al.* (2014). Monthly Notices of the Royal Astronomical Society, 439(4), 3504.

Chapter 10

The Ultimate End: Dark Energy And The Fate Of The Universe

The story of where our Universe came from is a remarkable one, and the fact that we have reached the point where we understand as much of it as we do is perhaps equally remarkable. But up until very recently, the *fate* of our Universe — how it would continue to evolve in the future — was very much unknown. We have already explored (back in Chapters 4 and 8; see Fig. 4.2) the idea that there are three possible fates for the Universe, dependent on the relationship between the amount of matter-and-energy present and the rate of the initial expansion. These fates are:

1) **The Big Crunch:** an initially large expansion rate causes the young Universe to grow incredibly quickly, but the density of matter and energy is also tremendously large, slowing the expansion rate down over time. Eventually, there is *just* enough gravitation to overcome the expansion completely, causing the Universe to reach a maximum size and cease expanding altogether. Once it reaches this turning point, the gravitation from all the matter and energy present causes the Universe to begin contracting. After the same amount of time passes again that it took the Universe to expand from the Big Bang until reaching maximum size, it will recollapse into an arbitrarily hot, dense state again: the Big Crunch. A Universe that ends in a Big Crunch will have a positive (closed) spatial curvature.

2) **The Big Freeze:** the initial conditions of the Universe — the initial expansion rate and the matter-and-energy density — are practically

indistinguishable from the Big Crunch conditions early on. The Universe expands incredibly rapidly and slows down to a minuscule fraction of its initial value. But rather than there being just enough gravitation to stop and reverse the initial expansion, it is the other way around: the initial expansion is just slightly too great for the amount of matter and energy in our Universe. As a result, the expansion rate remains positive at all times, and the galaxies that are not gravitationally bound to ours recede farther and farther away as time goes on. Inevitably, the Universe keeps expanding and cooling forever and ever. A Universe that ends in a Big Freeze will have a negative (open) spatial curvature.

3) **A Critical Universe:** at most, the departure from a perfect balance between the early matter-and-energy density of the Universe and the initial expansion rate must be less than one part in 10^{25}, so why couldn't the balance be exactly perfect? Perhaps the Universe really is right on the border that separates an eternal expansion from a recollapse, where only one more proton in the Universe would change our fate. A critical Universe will have the expansion rate drop, asymptotically, towards zero, but the expansion rate will never reverse itself. A Universe that possesses this exact critical density of matter-and-energy will have zero (flat) spatial curvature.

As we have learned from measurements and observations of the cosmic microwave background (CMB), the Universe indicates that — to the best of our measurement capabilities — it is indistinguishable from having zero spatial curvature. For a long time, the critical case appeared to be the scenario that described our Universe precisely (Fig. 10.1).

In hindsight, this should not have come as a surprise, since the theory of inflation explains why the Universe appears to be flat: the very process of cosmological inflation takes whatever shape the Universe had and stretches it so that it appears to be indistinguishable from perfect flatness! The reason the expansion rate and the energy density balance so extraordinarily well is because the Universe went through a phase where the expansion rate was determined by a huge amount of energy intrinsic to space itself. When that spatial energy decayed into matter and radiation,

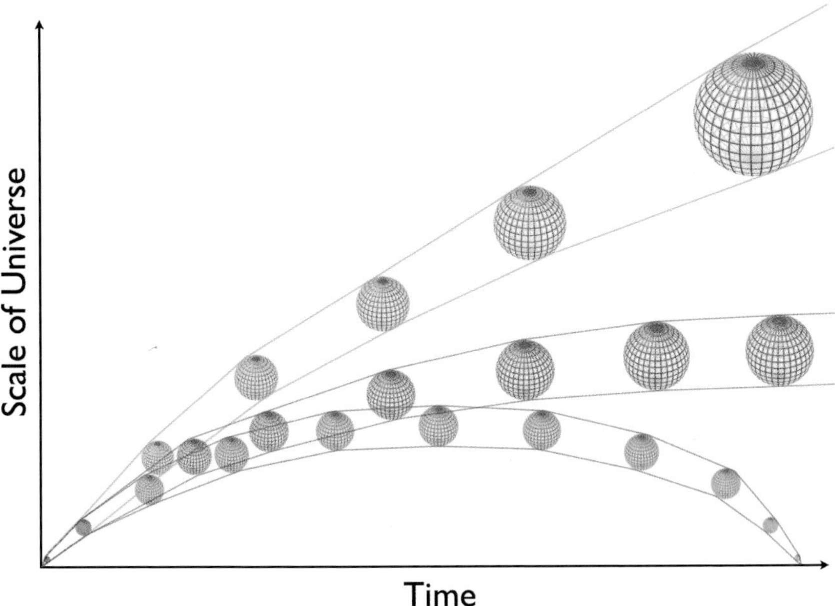

Figure 10.1 The three basic cases for the Universe are schematically shown here as they evolve in size over time: a closed, recollapsing Universe (blue), an open, forever expanding Universe (green) and a critical Universe (red). Note how, at all times, the size of every Universe grows less quickly than it was at any given prior time. In all three scenarios, the slope of the line describing the Universe's scale decreases over time. Image credit: E. Siegel.

the expansion rate began dropping in accordance with how matter and radiation dilute in an expanding Universe.

Therefore, a critical Universe really ought to be what we find ourselves living in. But the Universe has another surprise in store: even knowing that the Universe is spatially flat does not tell us everything about its future fate.

* * *

When it comes to the Universe's expansion rate, we often talk about matter and radiation in the same breath, as they both exist in abundance and both have the same conceptual effect: the Universe expands, the matter and radiation densities go down, and hence the expansion rate drops as well. But the way these forms of energy behave, individually, causes significant differences in the Universe depending on which form of energy is more important.

Matter, as you will remember, behaves very simply: as the Universe expands, the number of particles stays the same, but since the volume of space increases, the energy density drops. In particular, it drops in proportion to the scale of the Universe *cubed*. This means when the Universe was half its present size, the matter density was 8 times larger; when the Universe was one tenth its present size, the matter density was 1000 times larger. This analysis applies to both normal matter and dark matter; the types of interactions a massive particle experiences have no practical effect on how its energy density evolves as the Universe expands.

But things are different for radiation; it sees its density drop a little bit faster than matter does. Not only does the number density of particles dilute as space expands, but unlike matter, radiation has its wavelength stretch as the Universe expands as well. Combined, these effects mean that its energy density drops proportional to the size of the Universe *to the fourth power*. When the Universe was half its present size, the radiation density was 16 times larger; when the Universe was one tenth its present size, the radiation density was 10,000 times larger. Changes in the radiation density are more dramatic than changes in the matter density. Interestingly enough, at high enough energies, when matter starts moving very close to the speed of light, its kinetic energy is so large that it starts behaving as radiation, too! (Fig. 10.2)

Another way to examine this is from the perspective of *pressure*, or how much force a given form of energy exerts on whatever surface area it pushes up against. For absolutely stationary matter, it has no pressure at all; this is why if you dunk an inflated balloon into liquid nitrogen, the balloon deflates, as the molecules inside transition from moving very quickly (which they do at room temperature) to hardly moving at all, causing the pressure to plummet. But take that balloon out again, and the molecules heat up and the balloon reinflates. The reason is that the faster any form of matter moves, the greater the pressure it exerts. This increase continues all the way up to a maximum value that would be identical to radiation's, exerting a strong outward (positive) pressure. A stellar example are the very low mass neutrinos, which behave as radiation when the Universe is very young and hot, but which act as matter by the time the Universe is large and cool. Dark matter, since it is either born with

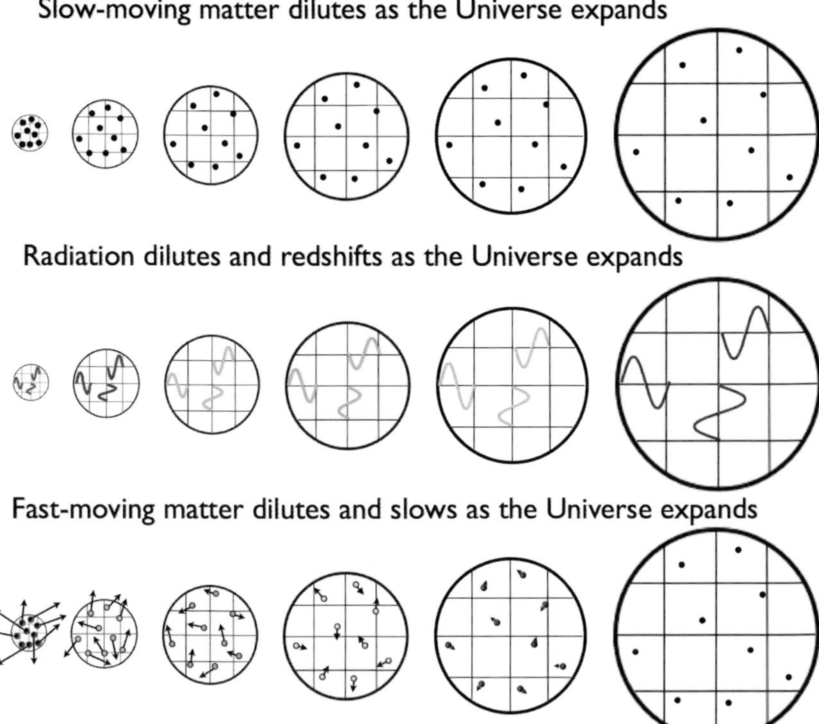

Figure 10.2 When matter moves slowly (or not at all) compared to the speed of light, its energy density drops as the Universe's volume expands. For radiation, which always moves at the speed of light, its energy density drops by an additional factor of the scale of the Universe, due to its wavelength stretching. But for matter that starts off moving very quickly, it loses energy as radiation does early on, slowing precipitously, and then loses energy as slow-moving matter does later on, once it is no longer moving close to the speed of light. Radiation and fast (relativistic) matter both have positive pressures, while slow (non-relativistic) matter has a negligible pressure. Image credit: E. Siegel.

very low energies (like axions) or is extremely high in mass (like WIMPs) pretty much always is expected to behave as slow-moving matter.

But normal matter, dark matter, radiation and neutrinos are only some of the possibilities that various forms of energy can take in our Universe. There are plenty of other more exotic forms of energy that can exist, and many of them do not exhibit the same types of effects we are accustomed to. In particular, there are forms of energy like cosmic strings, domain walls, cosmological textures and energy inherent to space itself, all of which dilute *less quickly* than any form of matter or radiation. In addition, they have a *negative* pressure inherent to them, rather than a positive or zero pressure. If the Universe consisted of even a tiny amount of any of these forms of energy, they would eventually come to dominate the energy content of the Universe, the same way that matter eventually came to dominate radiation, since the latter dropped in energy density more quickly as the Universe expanded.

It is important to keep these other possibilities in our minds as we observe the Universe. It is all too easy to let our preconceptions about what we expect to find color what we look for. Imagine how much more quickly science could have advanced if people had taken Fritz Zwicky's observations (and his suggestion of dark matter) more seriously back in the 1930s; we would have had a 40 year head-start on the problem! So when we try and measure the expansion history of the Universe — and then reconstruct what the different forms of energy present are — we have to keep *all* the theoretical possibilities in mind, even if they seem far-fetched or might raise additional unanswered questions (Fig. 10.3).

This means, when we are calculating what ought to be present in our Universe, we should consider that our Universe might contain:

- Radiation, whose energy density dilutes proportional to the size of the Universe to the fourth power, possessing a strong positive pressure,
- Normal and dark matter, whose energy density dilutes proportional to the size of the Universe to the third power, possessing a negligible pressure,
- Neutrinos, which behaves like radiation at early times and then transitions to behave like matter at late times,
- Cosmic strings, whose energy density dilutes proportional to the size of the Universe to the second power, possessing a strong negative pressure (just as strongly negative as radiation's is positive),
- Domain walls, whose energy density dilutes proportional to the size of the Universe to the first power, possessing a negative pressure twice as strong as that of cosmic strings,
- Cosmological textures or energy intrinsic to space itself (a cosmological constant), whose energy density *does not dilute at all*, possessing a negative pressure three times as strong as that of cosmic strings.

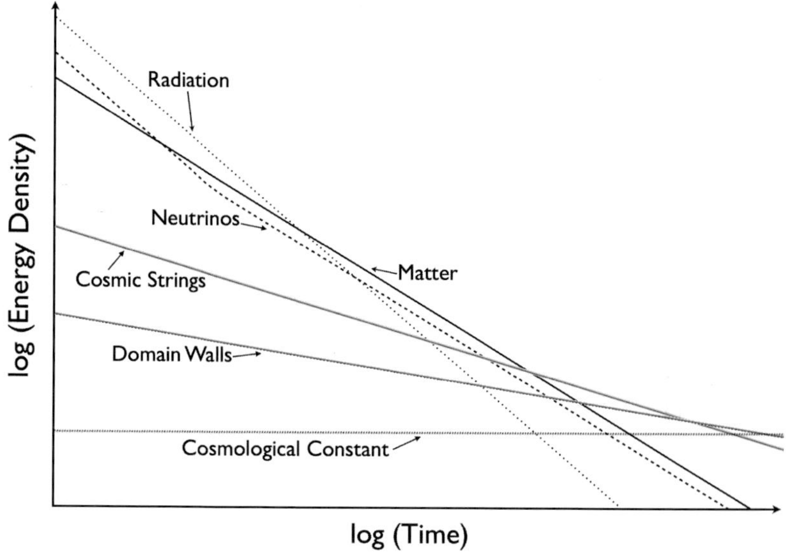

Figure 10.3 This logarithmic graph represents how energy density would evolve over time, assuming the Universe consisted of a mix of radiation, neutrinos, matter, cosmic strings, domain walls and energy intrinsic to space itself (a cosmological constant). Notice how, if your Universe contains even a tiny amount of any of the latter three cases (the ones with negative pressure), those components will eventually come to dominate the energy content of the Universe. Furthermore, the component with the most negative pressure will always come to dominate your Universe, given enough time. Image credit: E. Siegel.

Keep in mind that — in addition to any of these — the Universe might consist of some other component that we have not yet considered. It is often the most surprising and hard-to-come-by thoughts and ideas that turn out to reflect the Universe that nature actually gives us.

* * *

With all of these possibilities in mind, it is time to turn to the actual data. When Hubble first discovered the expanding Universe, he only had one tool in his arsenal to measure cosmic distances: a certain class of individual star. In particular, he was able to observe Cepheids, which are luminous, hot blue stars that vary in brightness, not only in our Milky Way, where their behavior had been understood, but in *other* galaxies as well! By measuring an intrinsic property of those stars (such as the period of pulsation), he was able to determine their intrinsic brightness, and then by comparing that with their observed brightness, he could determine their distances from us. Combine that information with the host galaxy's redshift — or apparent recession speed — and you know that galaxy's distance and relative velocity to you. Take these measurements for enough objects, and you can infer the expansion rate of the Universe!

Since Hubble's time, however, our toolkit has improved dramatically. No longer are we limited to measuring galactic distances by the individual stars that we can see within, but we've developed a whole host of methods that function as cosmic standard candles. Recall that the "standard candle" moniker refers to the fact — with a candle as an example — that if you know how intrinsically bright an object is, and you can measure how bright that object appears to be, you can figure out its distance from you, as the astronomical relationship between distance and brightness is well-understood. In addition to the Cepheids that Hubble used and many other classes of individual stars, we can use properties like the width of emission lines in spiral galaxies (the Tully–Fisher relation), the radius, brightness and velocity dispersion of elliptical galaxies (the fundamental plane relation), how light systematically shifts to higher frequencies due to the presence of hot gas (the Sunyaev–Zel'dovich effect) and how the surface brightness of an old galaxy fluctuates from one region to another (surface brightness fluctuations) as standard candles as well. In all these cases, there are particular classes of galaxies that exhibit a relationship between these easily-measurable quantities through observations here on Earth and their intrinsic brightnesses. Even though they are too far from us to measure the individual stars inside, we can measure properties that allow us to deduce the intrinsic brightness of a galaxy through these other correlations. If we also measure the apparent brightness of that galaxy, we can know its distance as well. Combine that with the galaxy's redshift, and we can measure the expansion history of the Universe out to hundreds of millions (or even billions) of light years (Fig. 10.4).

* * *

The most limiting factor when it comes to the use of standard candles is that they become harder and harder to see at large distances. This means that measuring the expansion history far back into the past is very difficult, and

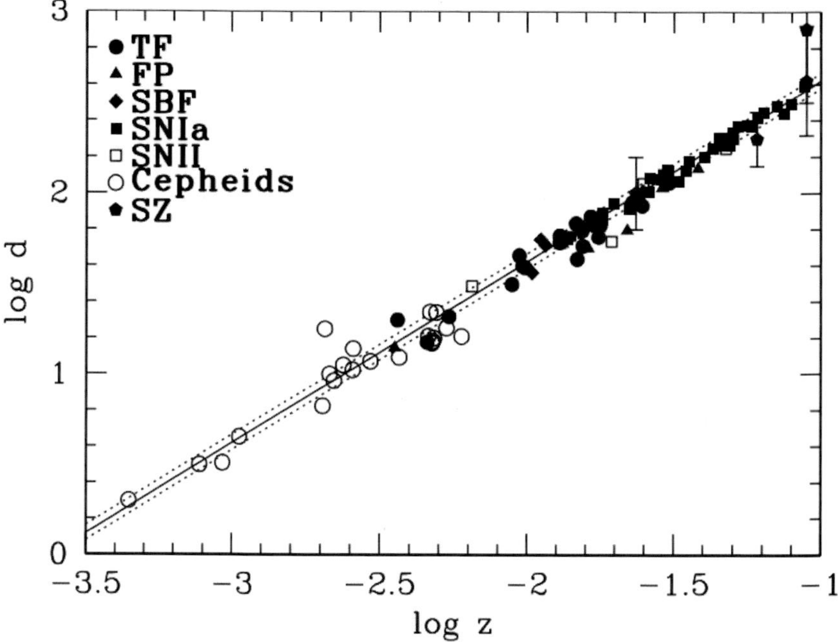

Figure 10.4 Many different methods are used in concert to arrive at the redshift (x-axis) vs. distance (y-axis) relationship. The final results of the Hubble space telescope key project led to an accurate measurement for the Hubble expansion rate (solid line) with very small errors (dotted lines), but were unable to go far enough back in the Universe's past to distinguish between expansion histories that might have been dominated by different energy components. Image credit: W. L. Freedman, (2001). *The Astrophysical Journal*, 553(1), 47.

relies on an intrinsically very bright source to fill that need. Luckily, we have one that is better than anything else we have mentioned so far. The most robust standard candle out to the largest distances is a specific type of stellar explosion: a type Ia supernova. All stars in the Universe are powered by nuclear fusion in their cores, a process that releases energy by combining light nuclei together to form heavier ones. Protons are fused into deuterium, which is then fused with additional protons and deuterons in a chain reaction that leads to helium. In most stars — including the Sun — this is the dominant path by which fusion occurs, and hence by which a star's energy is created. As stars burn through their fuel, their cores increase in temperature and density, increasing both the rate of fusion and the size of the region in which it occurs. But stars fall into three categories as respects what happens next, depending on their mass.

1) The lowest mass stars, the ones with less than 40% the mass of our Sun, will never reach high enough temperatures to begin fusing helium into heavier elements. When these stars run out of fuel in their cores — which will not happen for many times in the present age of the Universe — they will contract down and cool into a white dwarf composed of helium.
2) The intermediate mass stars, ones with between approximately 40% and 400% (but perhaps as high as 800%) the mass of our Sun, will go through a second stage of fusion. When the inner core runs out of hydrogen, it will contract and heat up, reaching sufficient temperatures to ignite the helium in its core, enabling fusion into heavier elements like carbon and oxygen. When this phase of fusion exhausts its fuel, the (mostly) carbon-and-oxygen core will contract down to form a white dwarf, with the outer (mostly hydrogen-and-helium) layers blown off into the interstellar medium in a planetary nebula.
3) The highest mass stars, the ones born as O-stars or massive B-stars, will not only fuse hydrogen and then helium, but will then go through a carbon-burning phase, producing even heavier elements all the way up to iron, nickel and cobalt in their cores. These stars, when they reach this end-stage, will die in a spectacular type II supernova, producing a magnificent supernova remnant with either a neutron star or black hole at the center.

The lowest mass stars that were formed in our Universe are all still hydrogen-burning, M-class stars, as the Universe is too young for hydrogen to have burned to completion in these stars. They offer no hope for a supernova explosion. The highest mass stars are incredibly important for the formation of heavy elements in the Universe and for the origin of rocky planets and organic molecules, and they do die relatively quickly in a supernova explosion. But the types of supernovae they generate when they die — type II supernovae — are extremely varied, and most importantly, *non-standard* in their brightnesses.

But the white dwarfs that intermediate-mass stars leave behind are of particular interest. For one, they are all primarily made out of the same elements: carbon and oxygen. For another, they all have the same physics preventing them from collapsing further: the quantum rule that no two identical particles can occupy the same quantum state. This rule, the Pauli exclusion principle, allows the electrons to hold up individual atoms against the force of gravity that tries to pull everything together. And this would not be sufficient if it were not for one other property: all of the white dwarfs that these stars form are below a very specific mass threshold — about 1.4 times the mass of our Sun — known as the **Chandrasekhar mass limit**. As it turns out, this limit is critical for understanding what makes type Ia supernovae such an important standard candle in the Universe (Fig. 10.5).

White dwarfs are very unusual entities. Try and imagine an object that is the physical size of Earth, and made out of atoms, again, quite similar to the ones on Earth. But instead of having our planet's density, imagine that it had a density *hundreds of thousands* to *millions* of times greater than our own world. You would have an object where a human being would be literally crushed to death in seconds if they stood on its surface, with their bones breaking under the weight of only their bodies. If that's what happens at the surface, imagine how much force must be on the atoms at the *center* of a white dwarf! The more and more mass you pile onto your white dwarf, the smaller in size it gets, as the additional gravitational force is so strong that it makes your object take up less space overall. It is only the quantum degeneracy pressure — caused by the Pauli exclusion principle, which forbids identical particles (like electrons) from existing in the same place with the same properties — that allows white dwarfs to exist at all. Without it, the high temperatures and densities at the core would simply enable a runaway nuclear fusion reaction, destroying the white dwarf altogether.

Even with this important quantum mechanical property, a runaway fusion reaction can still occur if a white dwarf gets too massive. The pressure of the electrons can hold the atoms up against the force of gravity, but only to a limit: that is what the Chandrasekhar mass limit is (Fig. 10.6). As it turns out, there are two major mechanisms by which a white dwarf that was previously *below* the mass limit can come to exceed it. Both of the

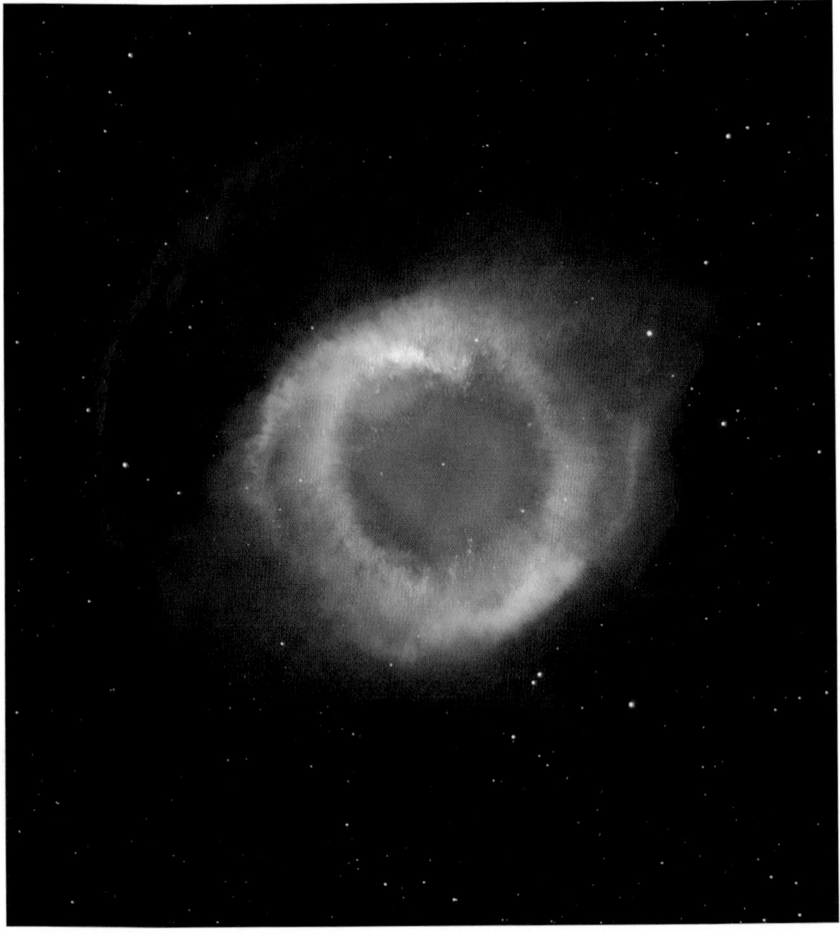

Figure 10.5 The Helix Nebula (shown above) is the remnant of a Sun-like star that has run out of nuclear fuel and expelled its outer layers in a planetary nebula. At the very center of this nebula is a white dwarf: a stellar remnant that consists of a significant fraction (10–50%) of the original mass of the star. Although these objects may be as massive as up to 1.4 times the mass of the Sun, they are only about the physical size of Earth. Image credit: NASA, ESA, C.R. O'Dell (Vanderbilt University), M. Meixner and P. McCullough (STScI).

scenarios lead to the start of a fusion reaction that culminates in a tremendous explosion that destroys the white dwarf itself: a type Ia supernova.

1) The white dwarf can accrete matter from an orbiting companion star. While our Sun is the only star in our Solar System, many star systems

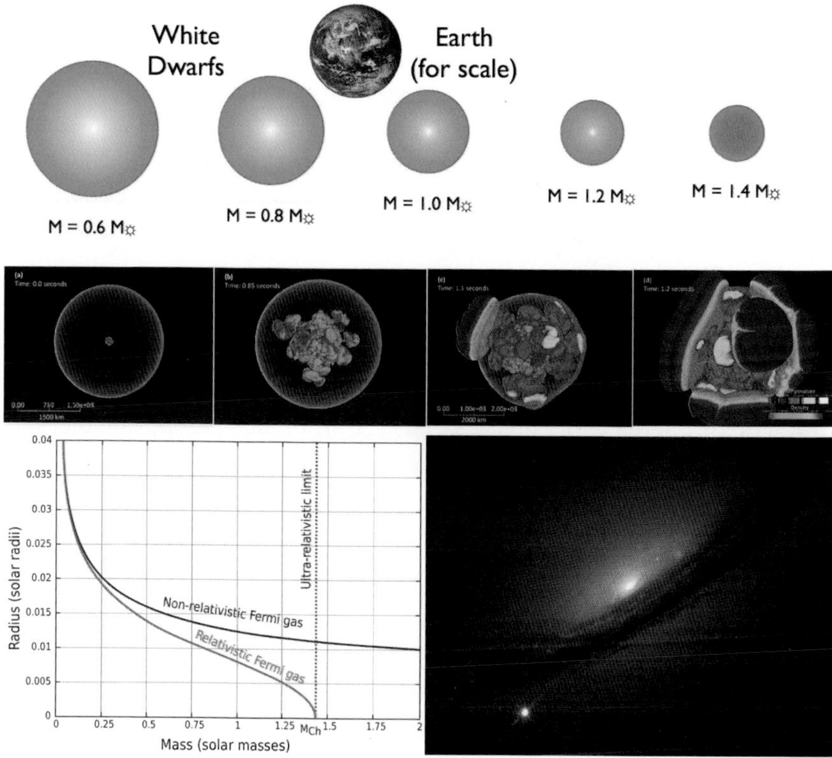

Figure 10.6 Most white dwarf stars are significantly lower in mass than the Chandrasekhar mass limit. As they get more and more massive, they shrink in size and their cores become denser. Above some mass threshold — about 1.4 times the mass of our Sun — nuclear fusion ignites in the core, destroying the white dwarf in a runaway fusion reaction and leading to a type Ia supernova. Images credit: E. Siegel (top); Argonne National Laboratory/U.S. Department of Energy (middle); Wikimedia Commons user AllenMcC (lower left). NASA/ESA, The Hubble Key Project Team and The High-Z SNe Search Team (lower right).

have two or more large, nearby masses. Due to a white dwarf's incredible density, it can often siphon off mass from a less dense companion star's outer layers. In most cases, this siphoning is gradual and results in periodic bursts of fusion at the white dwarf's surface: a nova. But over enough time, it can accrete enough mass to exceed the Chandrasekhar mass limit, which is catastrophic for the atoms at the white dwarf's center.

2) The white dwarf can encounter another white dwarf, and the two can potentially merge together. This can happen either violently, by a chance collision, or slightly less violently, by having the two masses spiral in a decaying orbit, eventually merging together. In either one of these situations, the additional gravitational force provided by these two massive objects merging together causes the total system to pass the Chandrasekhar mass limit, with catastrophic results (Fig. 10.7).

When we say "catastrophic results," the atomic nuclei at the centers of these objects are compressed into such a small region of space (and at such high temperature) that their quantum wavefunctions overlap, and carbon fusion begins in the white dwarf's center. Carbon nuclei fuse together into even heavier elements, releasing energy in the process. Because of the white dwarf's incredible density, the released energy has nowhere to go, and simply slams into the surrounding particles, heating them up tremendously. This increase in temperature leads to an increased rate of carbon fusion in the surrounding regions, and so more of the white dwarf's central atoms fuse, further heating up its interior. In a rapid chain reaction, the rate of fusion increases and the temperature goes up everywhere, until the entire white dwarf is destroyed in a supernova explosion.

This class of supernova is known as a type Ia supernova, and they have very particular properties common to all events in this class. Because they originate from the same types of object (a white dwarf) at right around the same mass threshold (about 1.4 solar masses), these type Ia supernovae all exhibit very similar light-curves, which is how much light they put out over time. In all cases, they brighten rapidly, reach a maximum, peak brightness, and then fade gradually over time. Because of the consistent brightnesses and shapes of these light-curves, these events make *excellent* standard candles. All we have to do is measure the shape, apparent brightness and duration of the light curve, along with the redshift of the supernova's host galaxy, and we can infer the distance to the galaxy from our knowledge of these supernova events (Fig. 10.8).

What makes these type Ia supernovae uniquely useful as distance indicators (or standard candles) is not simply their consistency, but rather their consistency combined with how incredibly bright these events are! Whereas a star like the Sun might put out a total of 4×10^{26} watts of power

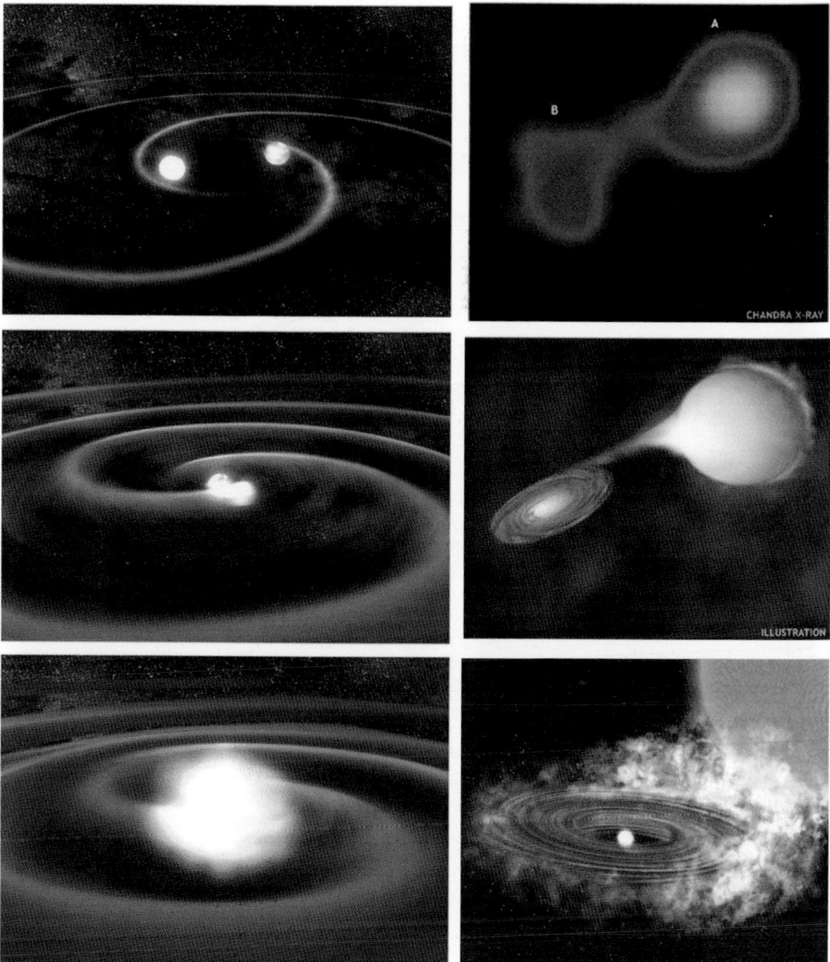

Figure 10.7 On the left, two massive white dwarfs that are both below the Chandrasekhar mass limit exist in a binary orbit around one another. Over time, this orbit will decay, leading to an eventual merger. If the sum of the white dwarf masses exceeds the Chandrasekhar mass limit, a runaway fusion reaction will occur, destroying both stellar remnants. On the right, actual X-ray observations (top) indicate a white dwarf accreting matter from a much larger, more diffuse binary companion (middle). Over time, a sufficient amount of matter will accumulate on the white dwarf to exceed the Chandrasekhar mass limit (bottom), causing a type Ia supernova to be triggered. Images credit: NASA/Dana Berry, Sky Works Digital (left sequence); NASA/CXC/SAO/M. Karovska *et al.* (top right); CXC/M. Weiss (middle right); P. Marenfeld/NOAO/AURA/NSF (bottom right).

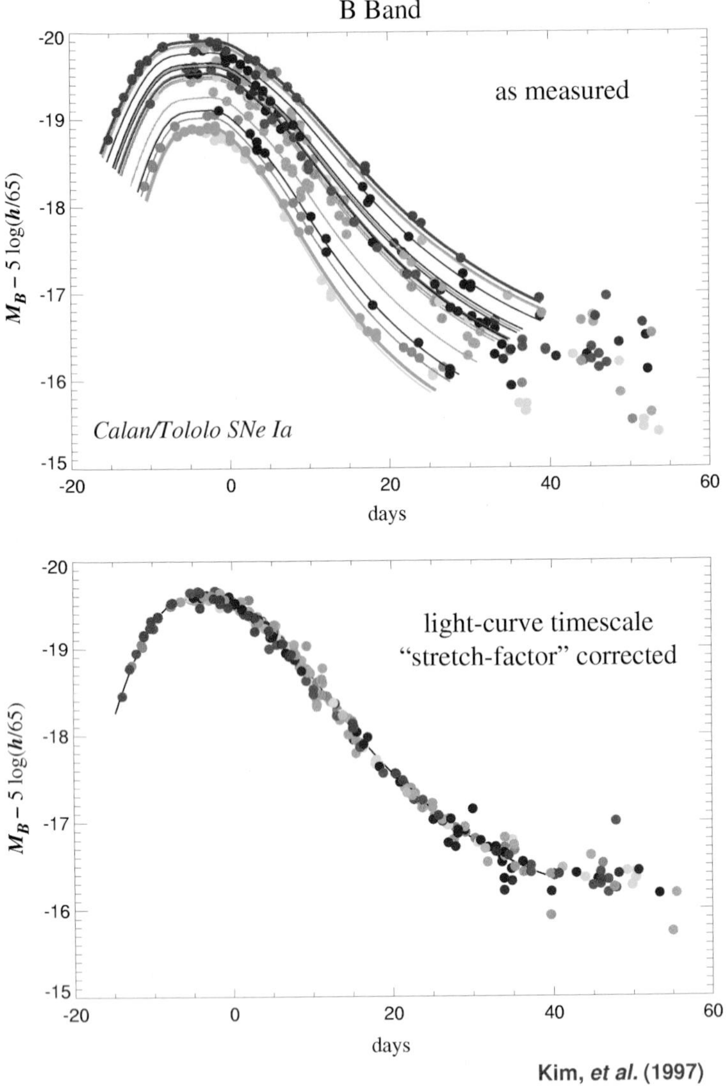

Figure 10.8 Type Ia supernovae do have minor intrinsic differences between them (top), but the shape of their light curve is well understood, and based on parameters such as peak brightness and the decay rate from peak brightness, they can be fitted to a universal light curve (bottom). By measuring the light curve of a distant supernova and correcting for its redshift (and hence, its time dilation), it is possible to determine the supernova's intrinsic brightness, and hence, its distance. These properties, as well as the universality of type Ia supernovae, make them ideal standard candles. Image credit: Saul Perlmutter, Supernova Cosmology Project, of the Stretch Corrected Hamuy Supernovae.

(or 4×10^{26} joules per second), for a total of around $\sim 10^{34}$ joules of energy per year, a type Ia supernova produces a total of about $\sim 10^{44}$ joules of energy in just one explosion, or as much energy as a Sun-like star emits over its entire lifetime! Instead of measuring distances out to hundreds of millions or even a few billion light years, we can use type Ia supernovae to probe distances out to many billions of light years, with the present record-holder (discovered in 2013) reaching a measured distance of more than *16 billion* light years from Earth (Fig. 10.9).

* * *

Being able to measure the expansion rate out to these incredible distances not only allows us to measure the Hubble expansion rate more accurately, it provides us with a way to measure how the expansion rate changes over time. This is incredibly important for telling us what makes up the Universe, because the rate of change in the expansion rate is determined by all the different components that contribute to the energy density. A Universe that is dominated by radiation but that also contains matter will eventually — thanks to the physics of expansion and cooling — have the radiation density drop below the matter density. Similarly, a Universe that is dominated by matter but that contains either cosmic strings, domain walls, a cosmological constant or any form of energy with a negative pressure, will eventually see that latter form of energy come to dominate matter.

All of this can only be probed, observationally, at large distances and large redshifts: where there is a *transition* from one form of energy dominating the Universe to another. To find out what makes up the Universe, all we have to do is look far enough back in time and space. The more distant an object we observe, the farther into the Universe's past we are looking in order to see it. Since the Big Bang occurred a finite amount of time ago, when we look back at a galaxy millions or billions of light years distant, we are looking back millions or even billions of years in time. The distant galaxy, as we see it, is situated when the Universe was much younger, hotter and had undergone only a fraction of its expansion. This last point is extremely important, because as we observe light from a very distant object, that light is affected by how the Universe has been expanding from the moment of its emission until the present day.

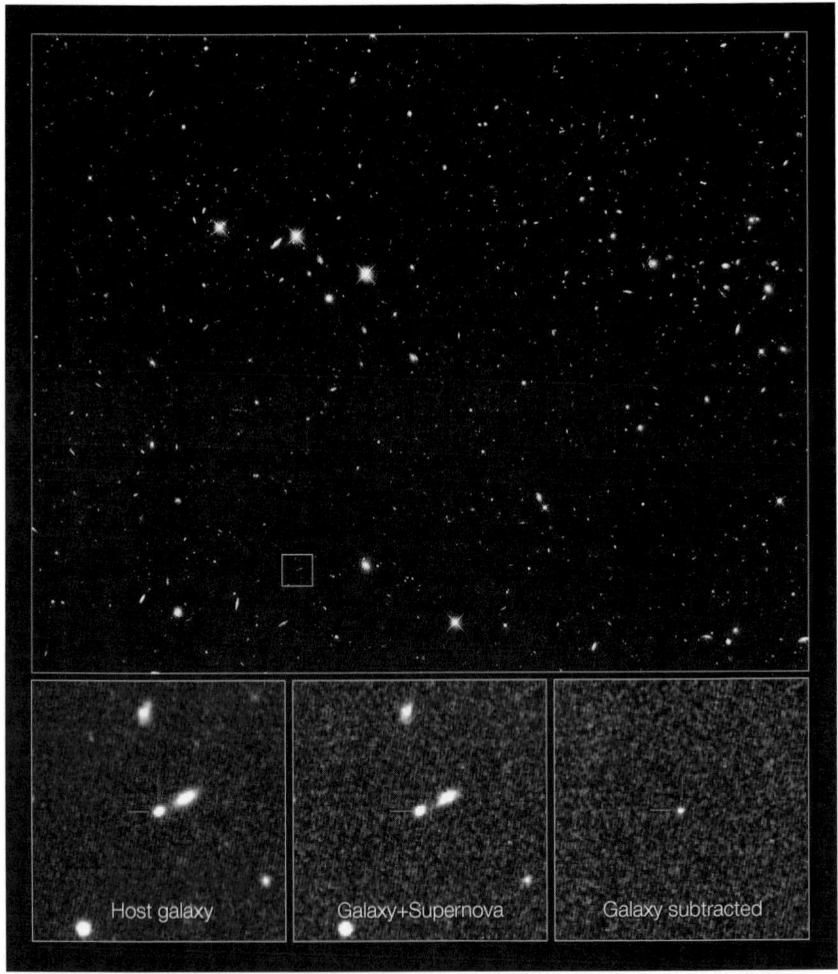

Figure 10.9 In 2013, the most distant type Ia supernova ever recorded was discovered: SN UDS10Wil, at a redshift of $z = 1.914$. This was spotted in data taken by the Cosmic Assembly Near-infrared Deep Extragalactic Legacy Survey (CANDELS) field in December, 2010. Follow-up data was taken by Hubble's Wide-Field Camera 3's spectrometer and the ESO's Very Large Telescope, verifying the supernova's distance and determining that it was a type Ia supernova. Image credit: NASA, ESA, A. Riess (STScI and JHU) and D. Jones and S. Rodney (JHU).

For emphasis, consider what it would be like to follow a photon from the moment of its emission directly to our eyes: when you observe light emitted from a distant object, that light experiences the effects of the Universe's expansion at every instant along the way until it reaches our eyes. What this means for our observations is that when we look at an object whose light arrives at our eyes after journeying for a billion years, that object is now *more* than a billion light years distant, because the Universe has been expanding all that time. That object was less than a billion light years distant from us when the light was emitted, the light traveled a total of a billion light years in order to reach us, and today, that object is more than a billion light years distant, thanks to the expansion. In addition to that object, if we also look at an object whose light was emitted two billion years ago, we can learn how the Universe expanded at even earlier times (Fig. 10.10).

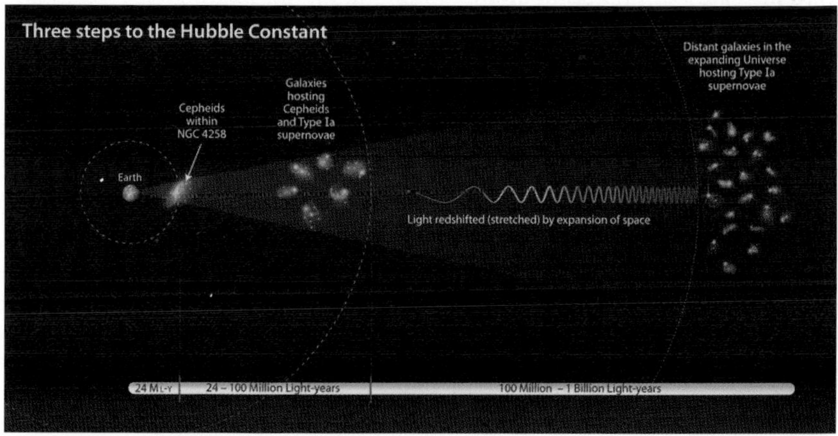

Figure 10.10 The more distant an object is — such as a type Ia supernova — the more that light gets redshifted by the expansion of the Universe. In order to reach our eyes, that light needs to travel huge distances in space, while the space it travels through is simultaneously expanding. Depending on what forms of energy are in our Universe, light will redshift by a particular amount. By measuring the amount that light redshifts from sources at varying distances, we can reconstruct how the expansion rate has changed over time, and come to understand what all the different components contributing to the energy density of the Universe are. Image credit: NASA, ESA and A. Felid (STScI).

Once the properties of type Ia supernovae became well understood, scientists attempted to take advantage of the new observational capabilities of telescopes such as the giant Keck Telescope in Hawaii and the Hubble Space Telescope. Telescopes such as these could, in principle, be used to not only search for these distant supernovae, but to measure the spectra of their host galaxies. (Other, smaller ground-based telescopes could be used as well, and were instrumental in the early successes of type Ia supernova-hunting.) Starting in the mid-1990s, there were two independent teams measuring large numbers of type Ia supernovae with the goal of determining the expansion history of the Universe: the Supernova Cosmology Project and the High-Z Supernova Search Team. Although it took many years of observations, both teams were able to find and measure type Ia supernovae out to great distances and early times: to when the Universe was less than *half* its present age. By measuring the light curves of each of these supernovae, they reconstructed their intrinsic brightnesses. Once the intrinsic brightness was known, the distance to those galaxies could be determined by comparing intrinsic with observed brightness. By then combining that information with the measured cosmic redshift of their host galaxies, they could map out what the expansion history of the Universe was.

The incredible thing about the Universe's expansion history is that every unique combination of various ratios of radiation, matter, and anything with negative pressure lead to an expansion history that is also unique. Every individual combination of matter, radiation, neutrinos, cosmic strings, cosmological constant and more comes complete with its own set of unique relationships between distances and redshifts that apply to all objects in the Universe. By January of 1998, both teams had announced that a Universe with matter and radiation alone could not account for the observations, and that there must be something else whose energy density contributed significantly to the Universe. Some other component — something with negative pressure — had become important at late times. In March of 1998, the first bombshell dropped: the High-Z Supernova Search Team submitted a paper where they plotted their supernovae and did the best possible fit to the data. The conclusion? That in addition to radiation and matter, the Universe not only consisted of but *was dominated* by some form of energy with significantly negative

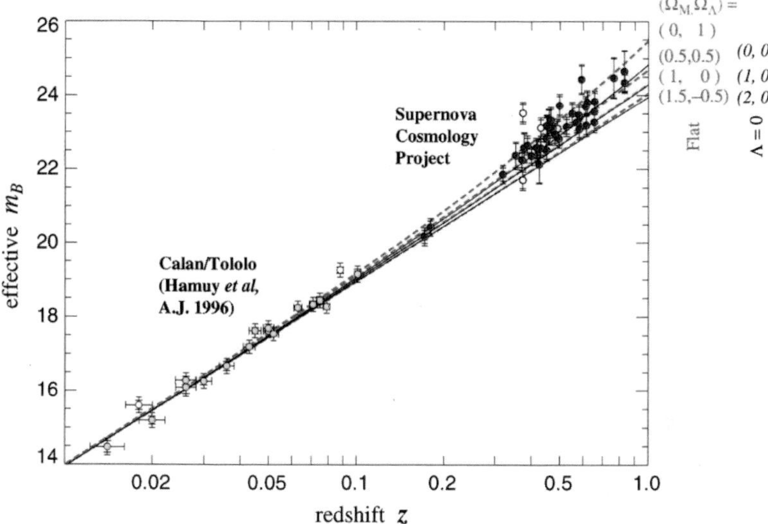

Figure 10.11 If our Universe was a critical Universe in the way we had expected — dominated by matter — it would have followed the second solid black line from the bottom. Instead, the data from the highest redshift supernovae came in significantly above that line, indicating that there was a component to the Universe with significantly negative pressure. The conclusion drawn from this was that the Universe must be accelerating at late times. Image credit: S. Perlmutter *et al.*, (1999). *Astrophysical Journal*, 517, 565.

pressure, and that the best fit to that data was as severe as energy intrinsic to space itself! Later that year, in September, the Supernova Cosmology Project released their results. The data from both independent teams led to the same conclusion: the Universe was dominated by some form of energy that was not only more important than matter was, but had a strongly negative pressure and was causing distant galaxies to recede from us at an *increasing* rate. Counter to the expectation of generations, the Universe's expansion was not slowing down, but was accelerating (Fig. 10.11).

* * *

What does it mean that the expansion of the Universe is accelerating? It *does not* mean that the Hubble rate of expansion is rising; that is probably the biggest misconception out there. To better visualize this, imagine that there are only two galaxies in the Universe: ours, and a very distant one, caught up in the expansion of space. If all that existed in the Universe was matter and radiation, whether our Universe's fate was to recollapse, expand forever or anywhere in between, the speed of that galaxy's recession would appear to slow down. It might be getting farther and farther away as time goes on, but the gravitation acting on everything in the Universe — even if the distant galaxy and our own are the only two present — will still fight against the expansion. Over time, its recessional speed will drop: to zero and then reversing in a closed Universe, asymptoting towards zero but never quite reaching it in a flat Universe, and towards some finite, non-zero value in an open Universe. But in all cases, the Hubble rate itself will continue to drop, since the Hubble rate is a speed-*per-unit-distance*, with distances to the galaxies continuing to increase as time goes on.

All of these cases, though, considered a decelerating Universe: one where a distant galaxy, as time goes on, will see its apparent recession velocity drop over time, with respect to us. In an **accelerating Universe**, however, the recessional speed actually increases over time, as a distant galaxy will appear to move away from us at a faster and faster speed as it moves farther away. It means that the Hubble rate itself need not necessarily drop over time, since even though it is a speed-per-unit-distance, the increasing distances could be matched or even exceeded by the increasing speeds. The Hubble rate can drop, remain constant or increase in an accelerating Universe, but it must not drop fast enough to allow a galaxy's recessional speed to even asymptote to a constant. In an accelerating Universe, all distant galaxies, clusters or other structures that are *not* gravitationally bound to one another at the time acceleration begins will never become gravitationally bound, and will instead speed away from one another. Eventually, if the acceleration continues, the recession speed of these objects relative to one another will exceed the speed of light, so that they effectively become unreachable beyond a certain point (Fig. 10.12).

A Universe dominated by radiation, neutrinos, normal matter or dark matter would not accelerate, as the gravitational properties of these types of energy can only result in a deceleration, or a slowing down of distant galaxies. Even if we consider cosmic strings or a Universe with

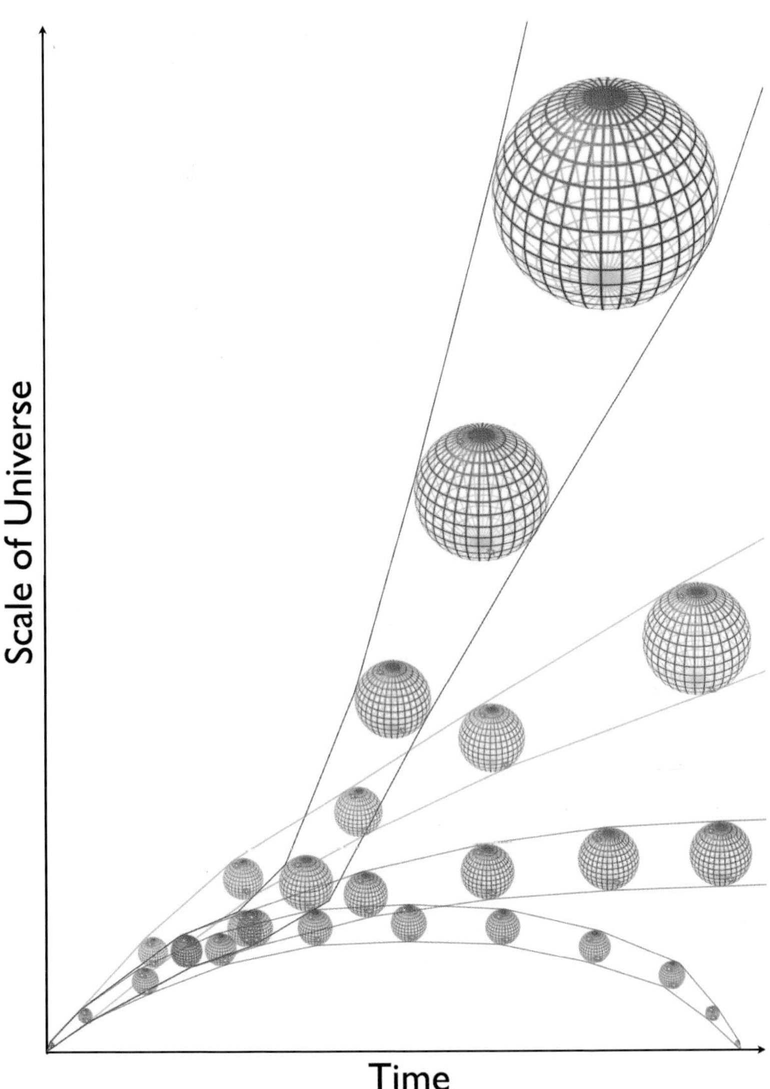

Figure 10.12 In the three decelerating Universe scenarios dominated by matter and radiation — a closed, open, or critical Universe — any galaxy will see the recession speed of another decrease over time, even though they continue to grow more and more distant. However, in an accelerating Universe (purple), the expansion matches the critical case for a time, but once the acceleration phase begins, the individual galaxies appear to speed up, moving away from one another more quickly than before. The acceleration only becomes important at late times, after the matter density has dropped by a significant enough amount to allow this new form of energy to grow to dominance. Image credit: E. Siegel.

sufficiently large spatial curvature, we still cannot get acceleration; the best we can do is get a Universe that expands forever with a constant recessional speed to distant objects, right on the border between deceleration and acceleration. (This is also what you would get in an expanding but completely empty Universe.) But if we consider a Universe with domain walls, cosmic textures, a cosmological constant (i.e., energy intrinsic to space itself), or any other energy component with a sufficiently negative pressure, it eventually *must* cause the Universe to accelerate.

The big questions, of course, are about what the exact effects of this acceleration are, as well as what its root cause is. While we do not know how to measure this form of energy directly, we can measure properties such as the expansion rate at a variety of times, and reconstruct a parameter (that we call w) that relates the energy density of this type of material to its pressure. In principle, w can be anything ranging from $-\infty$ to $+\infty$, but only a few values are physically motivated. In particular:

- Radiation, fast neutrinos and matter moving close to the speed of light has $w = +\frac{1}{3}$.
- Slow-moving or stationary matter (normal, dark and slow neutrinos) has $w = 0$.
- Cosmic strings or intrinsic negative spatial curvature has $w = -\frac{1}{3}$.
- Domain walls have $w = -\frac{2}{3}$.
- A cosmological constant (or energy intrinsic to space itself) has $w = -1$.

Even though the simplest models all come in increments of $\frac{1}{3}$, it is conceivable that not only could w take on any possible value, it does not even need to be constant; it could change over time.

To account for all of these possibilities, we gave this new form of energy — the one responsible for the observed acceleration to the Universe — a generic name: **dark energy**. If we could measure exactly how the Universe expanded over its history, we could better understand the phenomenon of this acceleration and the properties, and possibly the nature, of the dark energy that dominates it.

* * *

At present, there are literally thousands of type Ia supernovae that have had their distances accurately measured out to a billion light years or more. The farther away we can accurately measure these events, the more pronounced the differences between various forms of dark energy will appear to be. Continued work by scientists on those two major teams — the High-Z Supernova Search Team and the Supernova Cosmology Project — led to superior measurements and a suite of evidence that, when taken together, overwhelmingly supports the existence of dark energy, an accelerated expansion, and can rule out many confounding factors such as dust or disappearing photons. In fact, to about an uncertainty of 30%, they were able to establish that the preferred value of $w = -1$, a tremendous (if surprising) accomplishment! For their discovery of the accelerated expansion of the Universe, three of the leading scientists from those teams, Brian Schmidt, Saul Perlmutter and Adam Riess, were awarded the 2011 Nobel Prize in Physics (Fig. 10.13).

The supernova data is an incredibly strong piece of evidence in favor of dark energy's existence, but a single piece of evidence, no matter how strong, should not be enough on its own to change our conception of what makes up the Universe. After all, what if it turns out that type Ia supernovae are not as standard as we think; what if they have evolved over the past few billion years? What if the supernovae remain unchanged, but the environments in which they occur have evolved? There are numerous effects that could mimic what we perceive as the relationship between the expansion of space and an accelerating Universe. Rather than attempting to rule them all out individually, we can seek a more robust solution: to obtain multiple, independent lines of evidence that support the existence of dark energy.

Although it is the oldest method and possibly the most intuitive, measuring the expansion history by using a standard candle is not the only possible route. Instead, we can use a standard *ruler*, where we know the physical size of something at a certain distance, and can measure how large its angular size appears to be. Just as there is a well-known relationship between distance and brightness, the relationship between physical size and angular size in an expanding Universe is equally well-known, and also dependent on the Universe's expansion history. Only, instead of using individual galaxies, we can use what we know about the

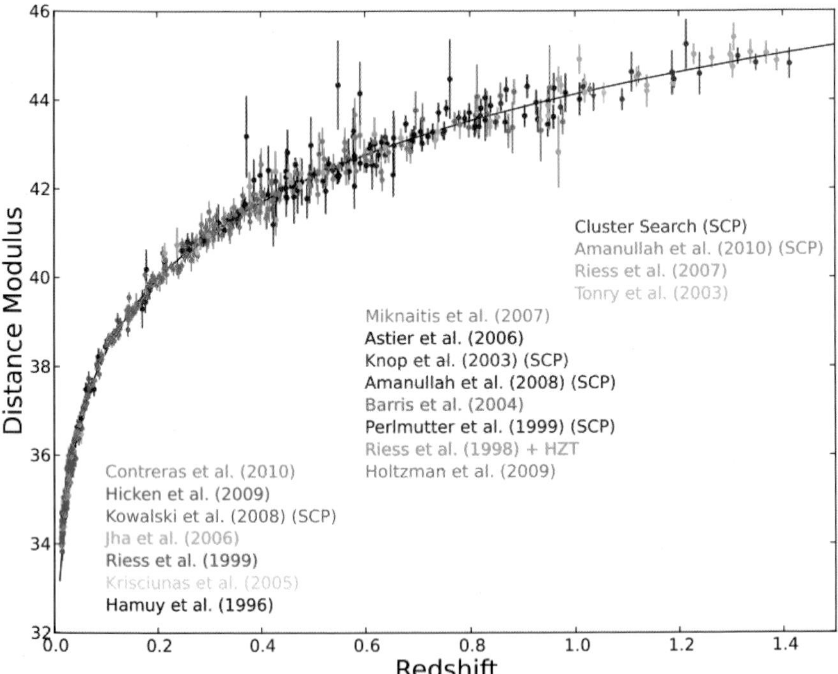

Figure 10.13 This up-to-date plot includes all the known type Ia supernovae that apply to this analysis, along with their brightness/distance measurements. The best-fit line in this plot represents a Universe that, today, consists of approximately 28% matter (both normal and dark, combined) and 72% dark energy. There is a small uncertainty on these numbers — about 4–5% in either direction — as well as a small uncertainty (of around 20–30%) on the best value of w, but it is consistent with $w = -1$ (a cosmological constant), and not with any of the other well-motivated options such as cosmic strings or domain walls. This data does not rule out an evolving or composite form of dark energy. Image credit: N. Suzuki (2012). *Astrophysical Journal*, 746(1), 85.

history of structure formation in the Universe: the scale at which galaxies cluster together. The ratios of normal matter to dark matter do not only impact the scales at which fluctuations form in the CMB, but imprint themselves in the likelihood of finding two galaxies separated by a particular distance. By taking large-scale surveys of the Universe out to great distances, we can look for evidence of this particular clustering pattern and how it evolves with redshift. (See, for example, Fig. 9.22.)

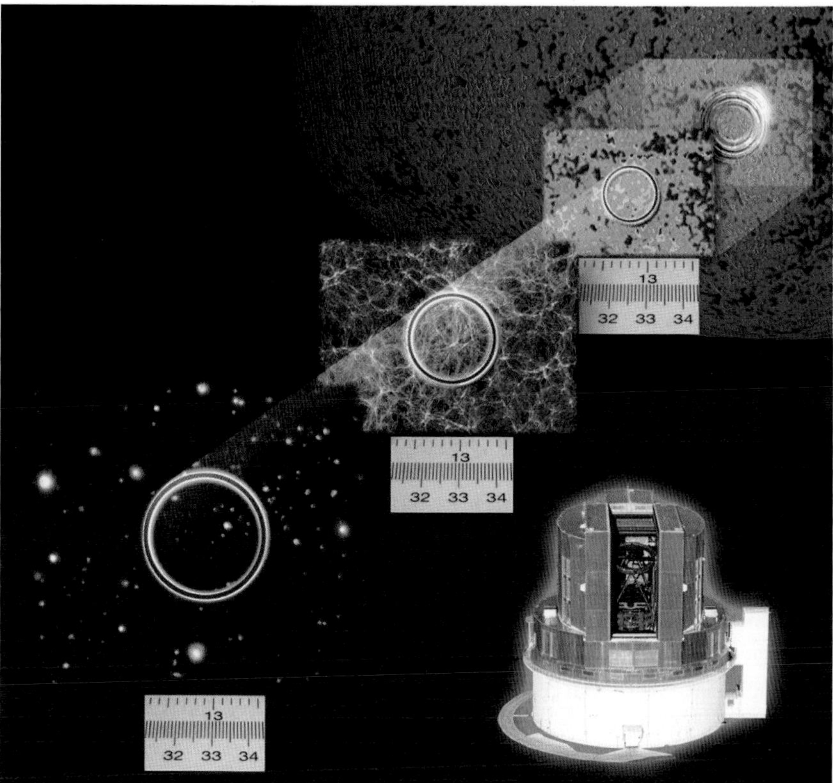

Figure 10.14 The scale at which normal matter begins to preferentially cluster together is determined back when the Universe is an ionized plasma, before neutral atoms formed, thanks to the interplay between matter and radiation. Once neutral atoms do form, those clustering patterns remain, and imprint themselves not only on the CMB, but also on large-scale structures and the clustering of galaxies at late times. By observing galaxy clustering both today and in the distant Universe, we can measure the expansion history of the Universe by using this clustering scale as a standard "ruler," an independent complement to a standard candle. Image credit: Gen Chiaki, Atsushi Taruya, for the SuMIRe Project.

These large-scale structure features are known as Baryon Acoustic Oscillations (BAOs), and give us an independent method of measuring the Universe's expansion history (Fig. 10.14).

In addition, the fluctuations in the CMB itself give us a constraint on the amount of dark energy that's present. The magnitudes and correlations of the fluctuations on different scales are very sensitive to not only the amount of radiation, normal matter and dark matter present, but also to any other forms of energy, including dark energy. The fluctuations might not be very good at constraining what the value of w is for this new form of energy (the constraints are quite weak), but they are very good at measuring the total amount of dark energy. The most up-to-date constraints come from the Planck satellite, and give us a Universe that is 68.6% dark energy, with an uncertainty of only ±2.0% on that value.

When taken all together, these three major, independent lines of evidence — supernova data, BAOs, and the CMB fluctuations — all point towards a single consistent picture of the Universe: one where about 5% of the energy density is due to normal matter, where 27% is due to dark matter and the remaining 68% is due to dark energy (Fig. 10.15).

Since 1998, the observations of not only supernova teams but also of huge galaxy surveys such as the two-micron all-sky survey (2MASS), the two-degree-field galaxy redshift survey (2dF GRS) and the Sloan Digital Sky Survey (SDSS) and multiple experiments measuring the fluctuations in the CMB (BOOMERANG, WMAP and, most recently, Planck) have tightened down the uncertainties on these numbers so stringently that they are known to a precision of around 2–3% apiece. There are other pieces of evidence, as well, that point towards the same conclusions:

- Measurements of structure formation, galaxy pairs, gravitational lensing and large galaxy clusters show that only about 25–35% of the Universe's critical density exists in the forms of normal and dark matter combined.
- Measurements of spatial curvature (through, for example, BAOs and the CMB) show that the Universe is spatially flat, meaning that the sum total of all the contributions to energy density in the Universe must add to 100%.
- Other distance indicators that are less "standardized" than type Ia supernovae but still give important measurements — indicators like gamma-ray bursts, quasars/active galactic nuclei and distant individual galaxies — also support a Universe with ~70% dark energy.

At the present time, even if there were *no* supernova data supporting it, there is enough evidence from other sources favoring of the presence of dark energy that the conclusion of an accelerating Universe would be all but inescapable.

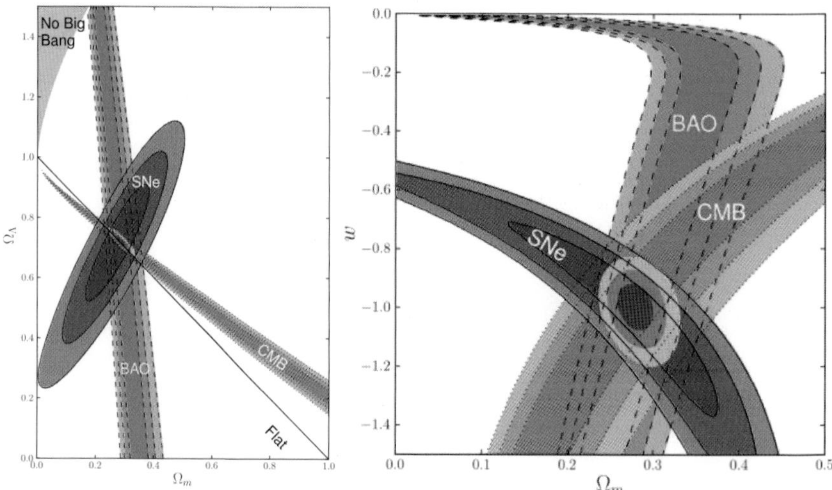

Figure 10.15 The combination of supernova data (SNe), CMB data (CMB) and data derived from galaxy clustering (BAO) lead to a single picture where all the observations agree: one where the Universe is made of approximately 30% matter and 70% cosmological constant. The graph at left shows this breakdown from all three sources combined (with error/uncertainty contours), where Ω_m and Ω_Λ are the matter fraction and dark energy fractions of the Universe, respectively. At right, all three sources are combined to place constraints on the equation of state of dark energy (w) and the matter fraction of the Universe (Ω_m) combined. The fact that these three disparate sets of observations all point towards a single, consistent picture is an incredible development that points to the existence and dominance of a new form of energy in our Universe: dark energy. Image credit: N. Suzuki (2012). *Astrophysical Journal*, 746(1), 85.

Interestingly enough, all the measurements of dark energy that are sensitive to it, with the greatest sensitivities coming from BAOs, also indicate that dark energy is most consistent with a cosmological constant. To the best of our measurements, we find that $w = -1.00$, that it does not appear to change over time, and that the value of -1.00 comes with an uncertainty of only around 10%. This rules out all the other conventional sources of dark energy, leaving behind contrived options as the only viable alternatives. Based on what the evidence indicates, it looks like there is a small but significant, and *positive*, energy inherent to space itself! (Fig. 10.16)

* * *

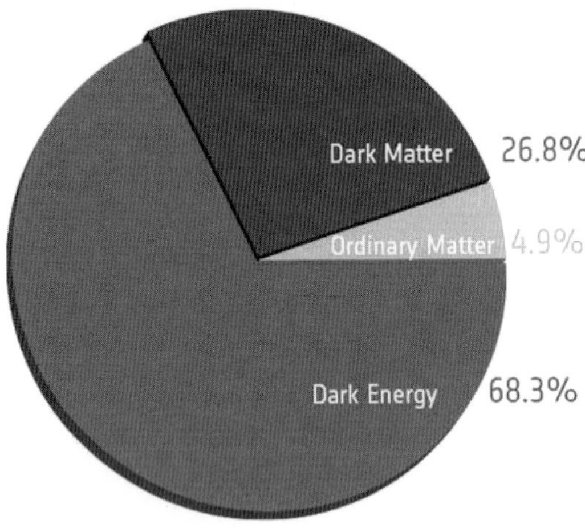

Figure 10.16 Thanks to data from a combination of sources, including the Planck satellite that measured the CMB, the SDSS-III collaboration that measured the BAOs, and various supernovae measurements, we have come up with a picture of what makes up the energy content of the Universe. With an uncertainty of only 2% or 3% on each number, we now know the Universe consists of approximately 4.9% normal matter, 26.8% dark matter and 68.3% dark energy. Surprisingly, the data indicates that 95% of the Universe is *nothing* like we expected! Image credit: ESA and the Planck Collaboration.

This sure does not look like the Universe we expected! Up until the 1970s, it was generally accepted that all of the energy in the Universe would be due to radiation and normal matter (with radiation being just a tiny fraction of a percent by the present day), and that the Universe's expansion would be decelerating. The only open question was that of the Universe's fate, something that we expected to determine by measuring the expansion rate's deceleration and calculating whether we would recollapse, expand forever or asymptote to a case where even the most distant galaxies eventually see their recession speeds drops to zero. Instead, we find that deceleration ruled for a time in the Universe, but around six billion years ago, all the objects that were not already gravitationally bound to us suddenly saw their recession speeds cease to slow down. Instead, those galaxies' recession speeds passed through a moment where they remained constant, and then began speeding up as they moved to greater and greater distances. By four billion years ago, the dark energy density had surpassed the total matter (normal plus dark matter) density, and at present, the dark energy density is more than double

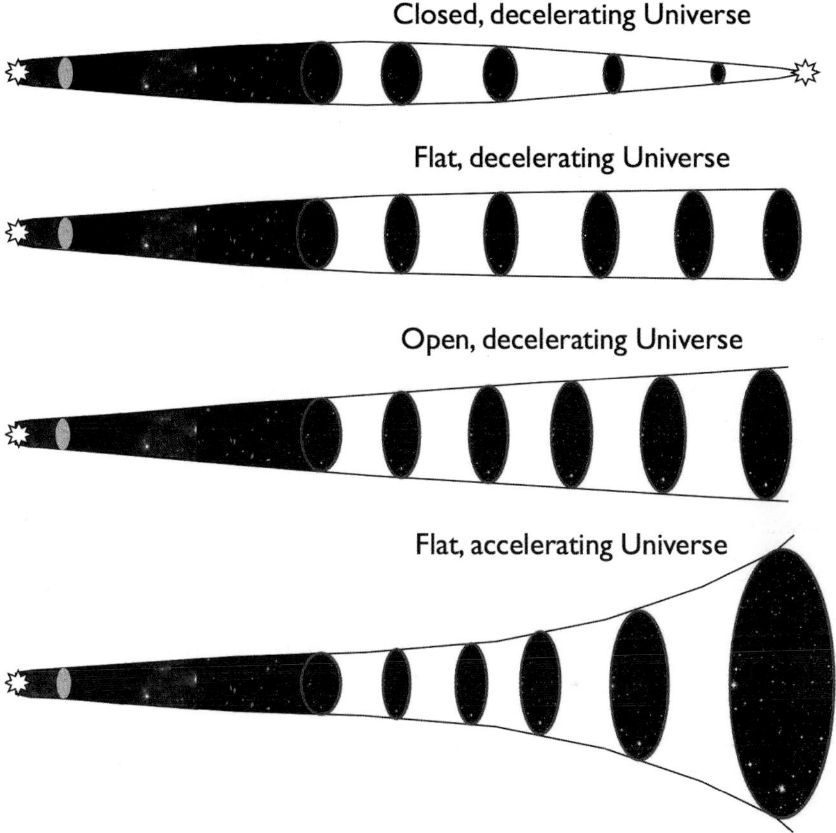

Figure 10.17 No matter what the Universe consisted of, we know it began at the end of inflation with the hot Big Bang, expanding and cooling to form light nuclei, neutral atoms, the first stars and then galaxies by the present day. If the Universe were dominated by matter and radiation alone, whether it was a closed, flat or open Universe, the rate of galaxy recession would have slowed down. But in an accelerating Universe, one that consists of some type of dark energy, once that component comes to dominate, all the galaxies in the Universe that are not bound together at the onset of acceleration will be driven apart. Image credit: E. Siegel.

the matter figure, which continues to dilute. As time goes on, the recession speeds continue to increase, and the Hubble rate — which appeared to have been dropping towards zero for the first few billion years — instead starts to asymptote to a *non*-zero value: approximately 46 km/s/Mpc, using the best values available today (Fig. 10.17).

Knowing the energy content of the Universe has enabled us to figure out a number of important facts about our Universe today, including its age: it has been **13.8 billion years** since inflation ended and the hot Big Bang began. For the first 7.8 billion years, the Universe was dominated by radiation and matter, with dark energy an undetectably small component of the cosmos. For all that time, galaxies that were expanding away from one another slowed down relative to one another, and in many locations, the local expansion started to reverse, as galaxies became bound together in groups and clusters. In our neighborhood, there was a very large cluster of galaxies — the Virgo Cluster — that formed not so far away from us, while the Milky Way became gravitationally bound to what we call our Local Group, consisting of Andromeda, ourselves, the Triangulum galaxy and approximately 40 other small, dwarf-sized galaxies. If not for dark energy, eventually we would have merged with all the nearby groups and the Virgo Cluster as well, as thousands of giant galaxies would have become our neighbors.

But instead, as the matter density continued to drop, dark energy — the energy inherent to space itself — became dominant, and all those hitherto unbound galaxies began to see their recession velocities rise, as seen from our perspective. Galaxies that were receding from us at only a few hundred km/s saw those speeds rise, eventually crossing into the thousands of km/s, where the Virgo Cluster is now. Worst of all, the most distant galaxies in the Universe — the ones located tens of billions of light years away — experience the most severe consequences of an accelerating Universe. As it stands now, the most distant particles of matter and radiation contained in the observable Universe are located 46 billion light years from us. But owing to cosmic acceleration, any galaxy that is more than about 14.5 billion light years distant has already obtained an effective recessional speed that is faster than the speed of light! This does not violate Einstein's relativity, as it is not a true motion of two things at the same location relative to one another; there is simply new space getting created in between us and these distant galaxies. Moreover, the space between us is expanding at such a rapid pace that, even if we left in a rocket ship today that could move arbitrarily close to the speed of light, we would never be able to reach galaxies beyond that mark. Frighteningly enough, that corresponds to *97% of the galaxies in the*

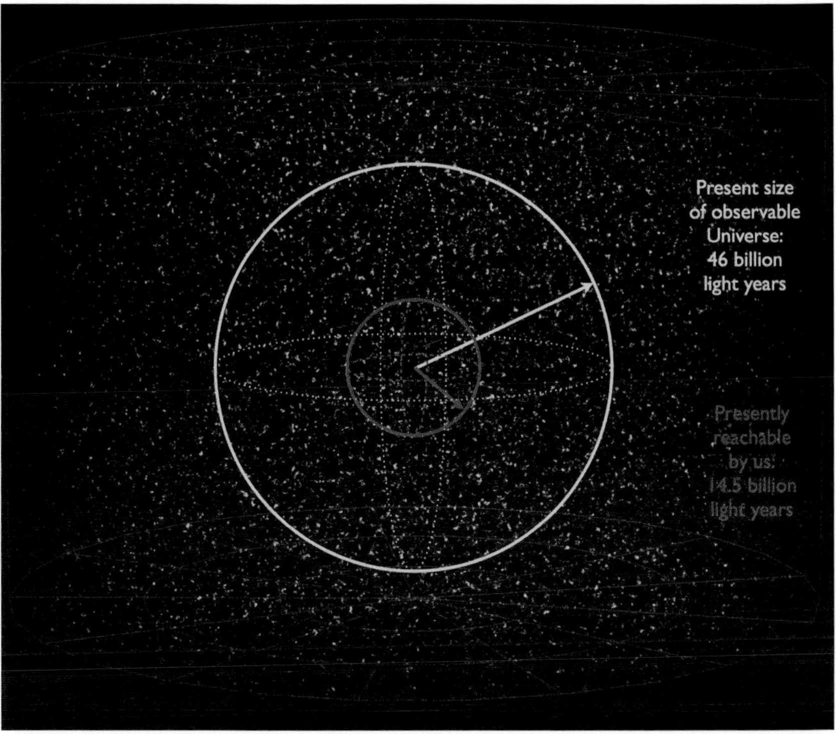

Figure 10.18 After 13.8 billion years in a Universe where space itself is expanding, the galaxies that are observable to us now fill a volume that is 46 billion light years in radius. But if we left today at the speed of light, we would only be able to reach a galaxy that is a maximum distance of 14.5 billion light years from us today, encapsulating only about 3% of the observable Universe. The rest are already forever beyond our reach. Image credit: E. Siegel, based on work by Wikimedia Commons users Azcolvin429 and Frédéric MICHEL.

observable Universe being unreachable. That is right: of all the galaxies we can, in principle, observe in the night sky, only 3% of them are still within our reach, with a new galaxy slipping out of our potential reach, on average, every three years (Fig. 10.18).

* * *

What does this mean, then, for the fate of our Universe? Assuming we have the story correct, that the Universe is composed of about two-thirds dark energy that takes the form of a cosmological constant, or energy inherent to the fabric of space itself, our Universe is headed towards a cold, empty fate, and we are headed there very quickly. When the Universe was half its present age, it seemed that all the hundreds of billions of galaxies in the observable Universe would still be reachable; today, that number is down to only a few billion. When the Universe reaches *double* its present age, the number of reachable galaxies will drop by another factor of ten, down to only a few hundred million. By time the Universe is perhaps ten times its current age, even the galaxies of the Virgo Cluster — just 50 to 60 million light years away at present — will have been pushed by the Universe's expansion to be hundreds to even thousands of times more distant, and will be receding too fast for even light to reach.

What will remain, then? Only the objects that were gravitationally bound to us before the Universe began accelerating! This tells us there is only a short window, just under eight billion years, for gravitation to do its work by attracting matter towards the overdense regions, before dark energy's acceleration takes over to drive the Universe apart. This is more than enough time to form structure on small scales, giving rise to star clusters, globular clusters and galaxies in great numbers. The larger scales need a bit more serendipity, though. While even small fluctuations on those tinier scales can grow rapidly enough to form bound, galaxy-scale structures, we need significant initial fluctuations to cause the larger scales to grow into galaxy groups or clusters. Our local group is one such success, as all the galaxies within two-to-three million light years of us are all bound together, including our larger sister galaxy, Andromeda. There are many nearby groups of similar size to ours, including the M81 group (which will be one of the last groups to disappear from our view) and the Leo Triplet, but when we start looking to larger scales, we find that the more massive structures are relative rarities. The Leo I Group (also known as the M96 Group) is many times as massive as our Local Group, but there are far fewer structures of this size than there are smaller groups or isolated galaxies. Very large clusters like Virgo, Coma or Pisces-Perseus are even rarer. Despite containing a thousand or more galaxies each, it appears there is fewer than one of these for every 10^{24} cubic light years of space! And when it comes to the largest structures in the Universe — great filamentary walls — it appears that there are only perhaps a few dozen of them in the entire visible Universe. There is a simple reason for this: larger, more massive structures take a longer time to grow and attract matter! (Fig. 10.19)

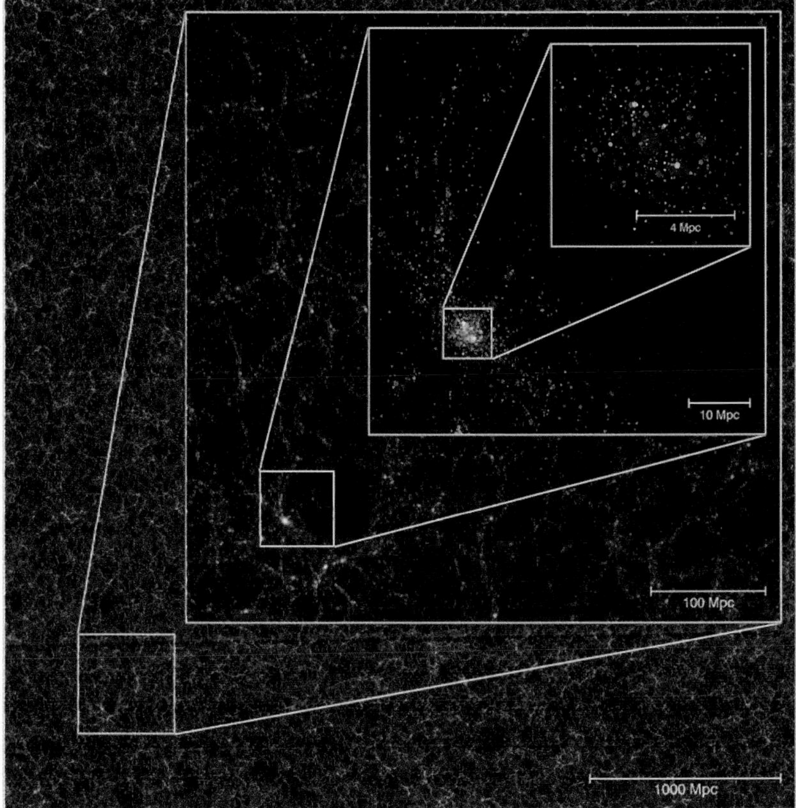

Figure 10.19 While the Universe today has structures that appear on all scales — including the very largest — it is only the smaller ones that have any hope of being gravitationally bound. Thanks to dark energy, the large filaments will stretch and break apart over time, as the expansion of the Universe will be too great for the force of gravitation to overcome. The structures that have formed on the smallest cosmic scales, like individual galaxies, groups and dense clusters, will remain gravitationally bound, but they will accelerate away from one another once dark energy becomes dominant. The Universe may appear to have larger structures like filaments and superclusters, but most of these are not gravitationally bound together, and will dissipate given enough time. Image credit: R. Angulo and S. White. Millennium XXL simulation, Max-Planck-Institut für Astrophysik.

Whatever the largest, gravitationally-bound structure is that you become bound to before dark energy takes over, that is the upper limit to the structure you'll become part of in the far future of the Universe. The unbound (to you) parts of the Universe that are closer to you when dark energy starts dominating the Universe's expansion will remain close and reachable (by spaceship and by light) for longer periods of time, but your home galaxy will never join them. As time goes on, the more distant galaxies will disappear from our cosmic horizon, a phenomenon known as **redding-out**, as their redshift and recession speed increases, eventually making all signals from them unobservable beyond a certain point. As still more time passes, even the once-nearby galaxies will disappear as well. In another 200 billion years or so, the Universe will be extremely cold and empty, with only the remaining stars, gas and dust from our now-merged-into-a-single-galaxy local group illuminating our skies.

How odd, then, that in the far future, our Universe will look much like we *thought* it looked just a century ago! Before we discovered that the spiral and elliptical nebulae in the sky were actually galaxies like the Milky Way — island Universes, far beyond our own local neighborhood of stars — we assumed that the extent of our own galaxy encapsulated everything that we saw in the night sky. It turned out that we were wrong: we live in a Universe full of matter, radiation and a whole lot more, with the observable part extending for some 46 billion light years in all directions and an unobservable part likely extending far beyond that, where a period of cosmic inflation came to an end some 13.8 billion years ago, giving rise to the hot Big Bang. But in our far future, all of those extra-galactic signals will be long gone! The last of the light from distant galaxies that could reach us will have already done so a long time ago, with no further signals forthcoming. The CMB will have dropped in both energy and photon density so fantastically that it will now be a cosmic *radio* background, and would take a radio telescope the size of a small planet to detect it. Unless an intelligent being somehow envisioned this possibility and managed to build such a large-scale detector to search for evidence of such an event, there would be no foreseeable way to know of our cosmic origins. Instead, we would conclude that our home, an elliptical galaxy (that we have already named *Milkdromeda*), consisting of all the matter making up our local group today, was the only source of matter in the Universe. The rest of it — expansion, acceleration, our Big Bang origins — would be obscured by a seemingly infinite abyss (Fig. 10.20).

This cold, empty Universe we envision is based on the assumption that dark energy truly is made up of energy inherent to space itself, with an

Figure 10.20 The Milky Way/Andromeda merger will happen roughly four billion years in the future, with the other 40 or so galaxies in our local group merging around that time or shortly thereafter to form a single, giant elliptical galaxy: Milkdromeda. Thanks to the presence of dark energy, however, all the other galaxies in the Universe, *even* the other galaxies of the Virgo supercluster, will recede from our view, eventually disappearing from our reach altogether. Image credit: NASA, ESA, Z. Levay and R. van der Marel (STScI), T. Hallas, and A. Mellinger.

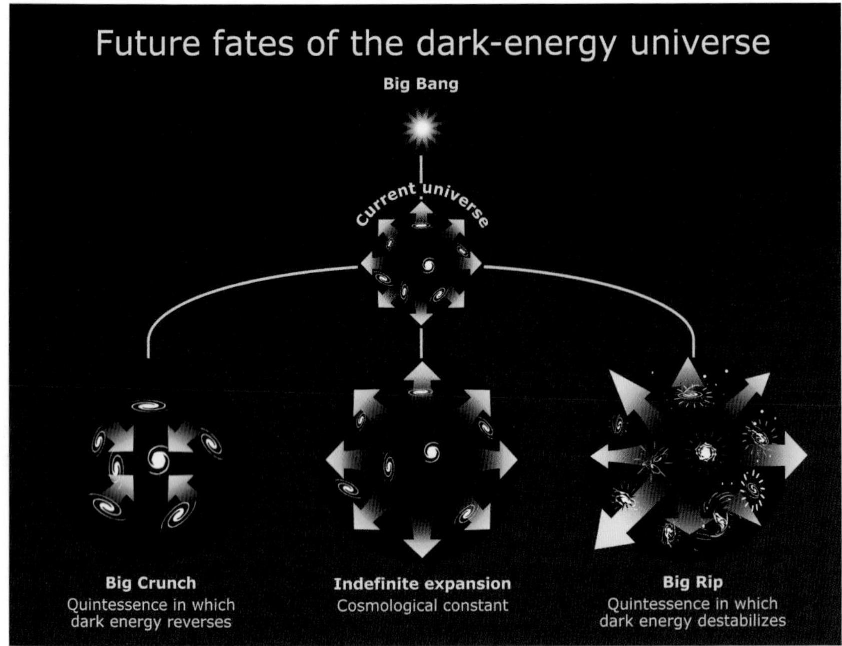

Figure 10.21 If dark energy is truly a cosmological constant, then the expansion will continue indefinitely, giving rise to a cold, empty Universe. But if dark energy changes its strength or properties over time — something theoretically possible but observationally without support — it could yet end in a Big Crunch or a Big Rip. The evidence we have today, however, overwhelmingly supports a "Big Freeze," the condition of expansion continuing at an accelerated rate indefinitely, and perhaps forever. Image credit: NASA/ESA and A. Riess (STScI).

equation of state of $w = -1$, exactly. It could, of course, *not* be our true fate if it turns out that dark energy is not a cosmological constant, and does in fact change over time. It could decay into another form of energy, perhaps leading to something similar to our Big Bang that occurred at the end of inflation, albeit at much lower temperatures. Dark energy could decrease in magnitude and eventually grow to have the opposite effect it has now, causing the Universe to recollapse after all. Some additional physics could even result in a cyclical Universe, where recollapses are followed by Big Bangs all over again, like a Universe stuck on an endless, repeating loop. Alternatively, dark energy could continue to increase in strength, leading to a "Big Rip" scenario in which structures like the galaxy are torn apart, with eventually even solar systems, stars, planets and the very atoms and subatomic particles themselves being ripped away from one another. All of these remain possibilities if dark energy is anything different than a cosmological constant, which it may well be (Fig. 10.21).

But right now, there is no evidence that points towards any of these exotic scenarios. If we follow where the evidence leads, the conclusion is that dark energy is a true cosmological constant where the Universe will continue to accelerate in its expansion uniformly and for all times. Inevitably, all the galaxies beyond our local group will continue to speed away from us, creating a far future where individual bound structures like groups and clusters will form one giant, elliptical galaxy at their center and will all disappear from one another's view. The Big Freeze will win after all, and will do so more quickly than anyone imagined just a generation ago.

Chapter 11

Past, Present And Future: All We Know About All There Is

In the span of the past 100 years, or a rather long human lifetime, our conception of the Universe has changed forever. A Universe assumed to be static, ruled by Newton gravity and spanning thousands of light years, within which all the stars — past, present and future — are contained, has been superseded in every way. The Universe is not static, but is rather expanding, cooling and evolving. Newtonian gravity is only an approximation of Einstein's General Relativity, which brings with it a whole host of observable consequences, nearly all of which have been validated. Our galaxy alone has been determined to be around 100,000 light years in diameter, and yet is only one out of hundreds of billions populating our observable Universe, which extends for some 46 billion light years in all directions (Fig. 11.1).

Moreover, our Universe has not been around *forever* in this form, but only came to be this way — filled with matter and radiation — some 13.8 billion years ago. Rather than being eternal to the past, our Universe as we know it had a birthday: a moment where it came into existence from a previous, inflationary state. It is also made of much more than we might have imagined; rather than being dominated by normal matter and radiation, those two components make up only around 5% of the total energy content of the Universe, with dark matter making up about 27% and dark energy, a form of energy intrinsic to space itself, making up the remaining 68%. And finally, our Universe will not remain this way

Figure 11.1 This is the Hubble eXtreme Deep Field, where the Hubble space telescope pointed its camera at a tiny, dark region of the sky for a total of 2 million seconds. What it found was a total of 5,500 galaxies in a region of the sky so small it would take 32 million of them to cover the entire Universe. Assuming that the rest of the Universe does not look so different from this — even though we know Hubble cannot reach the limits of the absolute most distant galaxies — we can conclude there are, at minimum, 176 billion galaxies in the observable Universe. Image credit: NASA; ESA; G. Illingworth, D. Magee and P. Oesch, University of California, Santa Cruz; R. Bouwens, Leiden University; and the HUDF09 Team.

forever, as our galaxy-filled skies will see all but a few dozen of our nearest neighbors eventually recede beyond our observable horizon, as they accelerate away from us. Our fate is a cold, lonely one, destined to be populated only by the matter bound together in our local group.

* * *

Our cosmic history is a remarkable one, and not one that could have easily been guessed a century ago:

- Our Universe began with an inflationary phase, expanding at an exponential rate, where a tremendous amount of energy was inherent to space itself. There was no matter or radiation present, only quantum fluctuations in the energy density and the gravitational field of space itself. Inflation may have gone on for as little as 10^{-32} seconds or as long as an infinite amount of time (or anywhere in between those two values) before it came to an end.
- In at least one region of space, inflation did come to an end, with the energy inherent to space itself getting converted into matter, antimatter and radiation. (Also mixed in there is a small amount of dark matter and a tiny amount of energy still inherent to space itself, although these do not become important until later.) This conversion process at the end of inflation is known as cosmic reheating, and results in — for the first time — our observable Universe being accurately described by the hot Big Bang. Our best understanding of inflation indicates that other regions exist where inflation did not come to an end, and in the majority of regions of space today, *outside* of our observable Universe, inflation continues eternally into the future.
- With our Universe consisting of matter, antimatter and radiation at incredibly high temperatures, yet rapidly expanding, it finds itself incredibly well-balanced between this initial expansion rate and the gravitational pull of all the different forms of energy within. It neither expands too quickly into a nearly empty state nor recollapses into a singularity; instead, the expansion is balanced by gravitation in a nearly critical Universe, resulting in the Universe gradually cooling down as it expands through a number of important transitions.
- At some point in the Universe's hot, early stages, a process took place that allowed the creation of *slightly* more matter than antimatter, resulting in an asymmetry of a little less than one extra particle of matter per *billion* particles of antimatter. When the Universe cooled through the critical phase so that the creation of matter–antimatter pairs was no longer as energetically favorable as the annihilation of matter–antimatter pairs into two photons, a process taking no more

than a second, the vast majority of matter and practically all of the antimatter annihilated away. This left behind an expanding, cooling Universe that was dominated by radiation, yet contained a small (but important) amount of protons, neutrons and electrons left over. The amount of dark matter left over, negligible until now in terms of energy density, winds up being some five times more important than the normal matter that remains.

- The protons and neutrons initially find themselves surrounded by particles too energetic to allow them to stably fuse together. Despite being in an environment more than hot and dense enough to initiate nuclear fusion, the first step in any nuclear chain reaction is to form deuterium, a nucleus composed of one proton and one neutron, which is easily dissociated by high-energy photons. Since photons outnumber protons and neutrons by more than a billion-to-one, the Universe needs to cool substantially before the first stable nuclear reactions can proceed, a process that takes a little more than three minutes. When it finally cools so that fusion can proceed, it leaves us with a Universe consisting (by mass) of about 75% protons, 25% helium-4 nuclei, with about 0.01% of each deuterium and helium-3, and a trace amount of lithium.
- The Universe attempts to form neutral atoms, but is still too hot for that, as photons are energetic enough to immediately reionize them. The Universe needs to cool for 380,000 years before electrons can stably bind to atomic nuclei, during which time the wavelength of radiation stretches so much that the Universe's energy density becomes dominated by matter: about 84% dark matter and 16% normal matter. When neutral atoms finally form, the photons left over from the Big Bang are free to travel in a straight line, unimpeded by free electrons and other ionized particles. This is the origin of the light, left over from the Big Bang, that is seen as the cosmic microwave background today.
- Thanks mostly to dark matter, the regions of space that are slightly denser than average preferentially attract more and more matter, both normal and dark, from their surroundings. As matter increasingly clumps together, the rate of gravitational growth increases, giving rise to dense clusters of matter, where the normal matter collides with

itself and sinks to the center. As the densities continue to increase, the most matter-rich regions undergo gravitational collapse, leading to the central regions heating up and igniting nuclear fusion. After tens-to-hundreds of millions of years, the first stars are born.

- While most of the stars that are born are relatively low in mass and will live for billions or even trillions of years, the most massive ones burn through their fuel extremely quickly and are very short-lived. After only a few million years, the most massive stars have burned through all the fuel in their core, dying in catastrophic supernovae and returning these now-processed, heavy elements back to the interstellar medium.
- As the individual clumps of matter form stars, gravitation causes structure to form on larger and larger scales. Smaller clumps of matter merge together or collapse to form the first proto-galaxies, which in turn merge together over time to form larger galaxies. On even larger scales (and at even later times), galaxies become bound together in groups, clusters and along vast, cosmic filamentary networks. All the while, the Universe continues to expand and cool.
- Within each galaxy, gravitational processes spur the continued formation of new stars, incorporating all the material available from the interstellar medium. This includes not only the pristine hydrogen and helium gas left over from the Big Bang, but the heavy elements created in and recycled from earlier generations of stars. In addition to the supernovae created by the heaviest stars, Sun-like stars blow off their outer layers in planetary nebulae, white dwarfs can either accrete matter or merge together, resulting in supernovae themselves, and neutron stars can merge, causing gamma-ray bursts and resulting in the creation of the heaviest elements known in the periodic table. With each subsequent generation of stars, more and more of these massive stars enrich the Universe with these heavy elements, and hence each new generation of stars is born with greater amounts of polluted atoms than the ones before.
- After enough generations of stars are born, the heavy element content of new star-forming regions results in not only the formation of new stars and gas giant worlds, but also of rocky planets, complete with the ingredients for complex chemistry, organic processes and — in places that get extremely lucky — life.

- In the meantime, the matter density, dropping all this time as the Universe continues its expansion, finally sees its effects on the expansion rate superseded by dark energy. At this moment in time, 7.8 billion years after the Big Bang, distant galaxies and clusters begin accelerating away from all the structures they are not presently bound to through gravity. A few billion years later, the dark energy density comes to exceed the matter density, and the formation of cosmic structure on the largest scales becomes so greatly suppressed that no new structures will form.
- Meanwhile, here in the Milky Way, some 9.2 billion years after the Big Bang, a run-of-the-mill star cluster is formed from a cloud of interstellar gas that has seen many generations of stars live and die. Only about 2% of the matter in it is heavier than hydrogen or helium, but this is more than enough to give rise to rocky planets around most of the newly formed stars. In the inner solar system around one of these stars, a planet forms at the right distance and with the right atmospheric content for water to exist in the liquid phase on its surface. Although it is not clear exactly how or when it happened, by time this world was only a few hundred million years old, life had taken hold on its surface. By time the Universe had reached 13.8 billion years of age, this life had evolved to the point where it could piece this entire cosmic story together, figuring it out from the pieces of evidence the Universe had left behind.
- Moving forward, dark energy will continue its dominance of the expansion rate, causing everything beyond our local group to accelerate away from us. While Andromeda, the Triangulum Galaxy and all the other 40-or-so dwarf galaxies in our local group will eventually merge together over the next few billion to ten billion years, everything else will recede farther and farther away. The cosmic microwave background will redshift to radio wavelengths, becoming undetectable to all but the most outlandishly large detectors. After a few hundred billion years, only the stars within our new, giant elliptical galaxy — Milkdromeda — will remain as observable phenomena within our cold, empty Universe (Fig. 11.2).

* * *

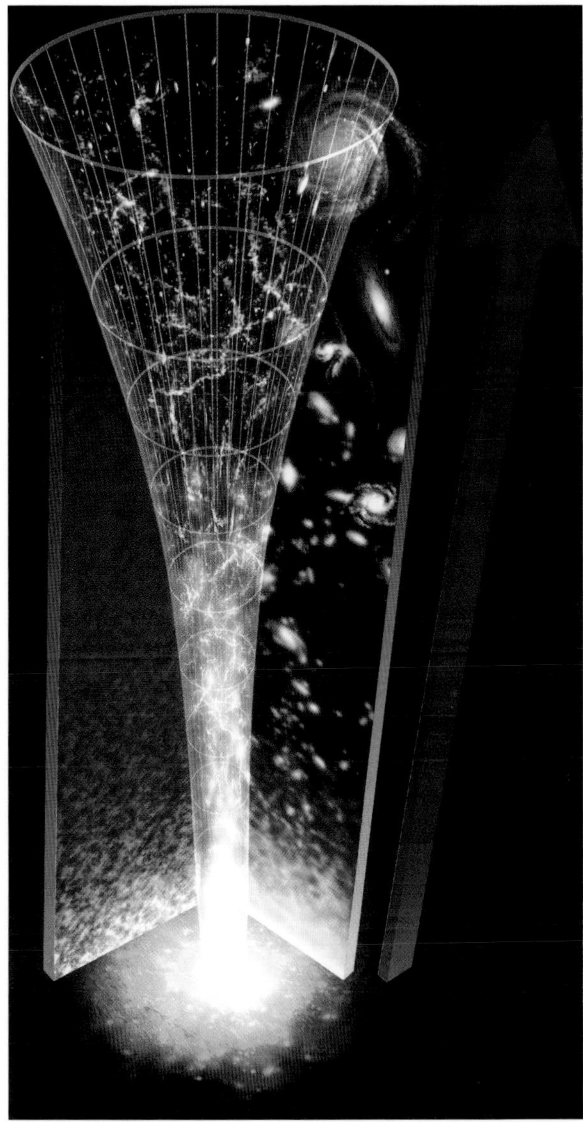

Figure 11.2 The cosmic story of our Universe can trace its origins from the end of inflation to the hot Big Bang and a continual expansion and cooling, giving rise to nuclei, atoms, gravitational clustering, stars, galaxies, heavy elements and eventually rocky planets around stars with life on them. The story does not end here with us, but will continue into the far future, with galaxies and clusters receding from one another in a Universe dominated by dark energy. Image credit: NASA/Goddard Space Flight Center.

What we have learned about the Universe in the past century has been truly revolutionary and amazing, but it is important to remember *how* we learned all of this. We learned it by asking the Universe questions about itself. We learned it by looking at the full suite of data that we had, at the full realm of physical possibilities, and when there was an apparent conflict between expectation and observation, we explored new theoretical avenues to account for the observed mismatch. We made progress by constantly looking for opportunities to revise our best ideas about the Universe and make novel, testable predictions. What we *did not* do is equally as important: we did not assume the answer before the telltale observations came in; we did not prefer one theory over another without sufficient evidence; we did not remain wedded to a theory when new, conflicting observations came in; we did not ignore the robust pieces of evidence that did not fit our preferred models.

This is not just the story of the Universe as we presently understand it, but also the story of how the scientific process works in general. Science is an ongoing story, and when we apply it even to the entire Universe, we continue to gain information that will enable us to understand it in more detail and at a deeper level. Scientific investigation is a process — a self-correcting process — that thrives when it has more data, when it has independent verification, when experiments and observations are repeated over time and when it is subject to the most intense scrutiny we can muster. It succeeds when there are tensions between observations and predictions, as those are often the omens of scientific advance. They are not always; when mistakes are made or incorrect conclusions are drawn, the solution to righting the ship lies in doing *more* and *better* science and paying attention to the new results. Even though scientists themselves may be flawed, one of the most beautiful things about the scientific process itself is that generations of mistakes or wrongheaded thought can be wiped away by one successful, new idea. If we are doing science properly, then when we all have access to the full suite of available data, we may all wind up drawing the same conclusions, so long as we are following the evidence and thinking scientifically (Fig. 11.3).

If we are lucky, many of the greatest open scientific questions may be answered over the coming generations. Perhaps we will be able to detect the gravitational waves predicted by inflation and determine even more information about how our Universe got its start? Perhaps we will learn

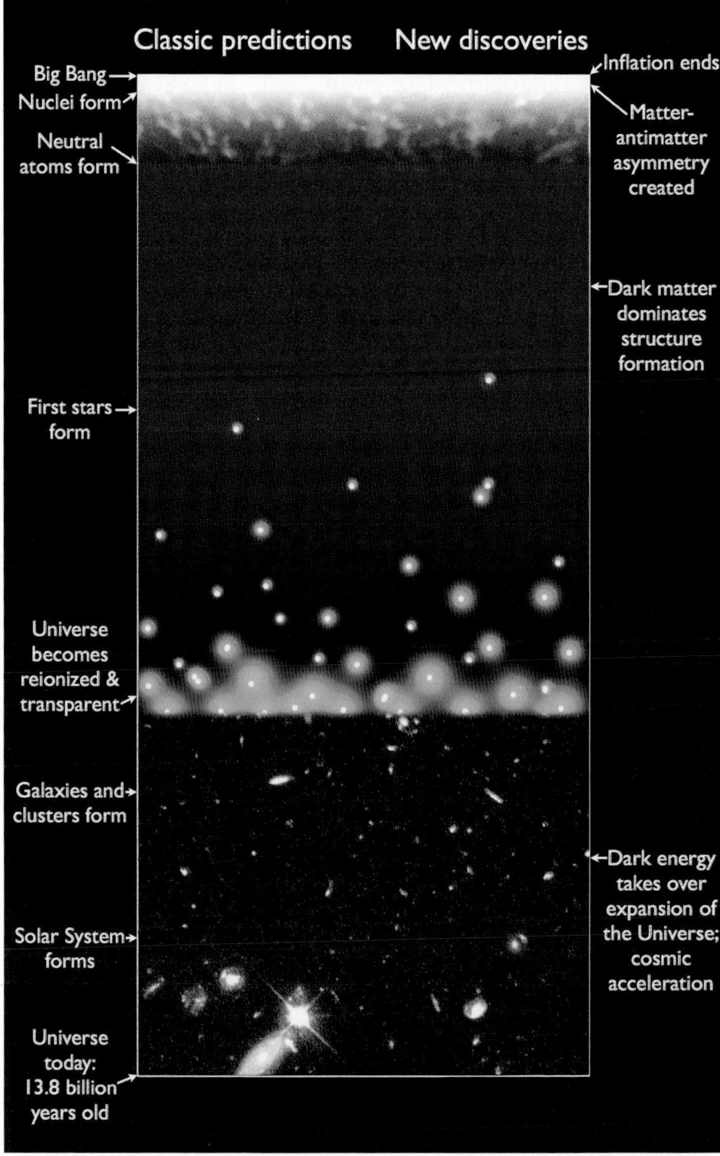

Figure 11.3 Even though the Big Bang made a number of classic predictions, the evidence we have uncovered about our Universe has led to a number of great refinements to Gamow's original model, including a period of cosmic inflation that started it all, a mechanism for baryogenesis to create a matter–antimatter asymmetry and the existence of dark matter and dark energy, among others. Image credit: E. Siegel, based on the original by S.G. Djorgovski, Digital Media Center, Caltech.

just how matter (and not antimatter) came to dominate the Universe today, and why we are made up of protons, neutrons and electrons instead of their antimatter counterparts? Perhaps we will uncover the secrets of dark matter and learn exactly what is responsible for the majority of the Universe's mass and gravity? Perhaps we will discover other planets in the Universe where life — or even *intelligent* life — took hold? Perhaps we will find a way to determine the true nature of dark energy, and better understand why our Universe is accelerating? Or maybe, just maybe, we will uncover something that truly surprises us, and be compelled to change our cosmic story in ways we have not even adequately imagined.

At present, we understand the cosmic story that the Universe tells us about itself in greater detail than at any moment in history. With each day that goes by, with each new measurement, new piece of data, new peer-reviewed paper and every new piece of information, we are attempting to push those frontiers back even farther. We may not all be scientists actively working to uncover the secrets of the Universe, but the story is there all the same, woven into the laws of nature and the cosmic tapestry itself, for us all to discover and share in. The story we have shared here is only what the best of our inquiries have revealed to us so far.

This is a story without end, limited only by what we have already discovered in the time we've been searching. The version told here is merely a progress report as we travel along the unending road towards scientific truth. There is a whole Universe left to discover.

Index